MECHANICAL D

FOR

SECONDARY SCHOOLS

BY

FRED D. CRAWSHAW, B.S., M.E.

PROFESSOR OF MANUAL ARTS, THE UNIVERSITY OF WISCONSIN

AND

JAMES D. PHILLIPS, B.S.

PROFESSOR OF DRAWING AND ASSISTANT DEAN COLLEGE OF ENGINEERING,
THE UNIVERSITY OF WISCONSIN

British Library Cataloguing-in-Publication Data
A catalogue record for this book is available from the
British Library

Technical Drawing and Drafting

Technical drawing, also known as 'drafting' or 'draughting', is the act and discipline of composing plans that visually communicate how something functions or is to be constructed.

It is essential for communicating ideas in industry, architecture and engineering. The need for precise communication in the preparation of a functional document distinguishes technical drawing from the expressive drawing of the visual arts. Whereas artistic drawings are subjectively interpreted, with multiply determined meanings, technical drawings generally have only one intended meaning. To make the drawings easier to understand, practitioners use familiar symbols, perspectives, units of measurement, notation systems, visual styles, and page layout. Together, such conventions constitute a visual language, and help to ensure that the drawing is unambiguous and relatively easy to understand.

There are many methods of constructing a technical drawing, and most simple among them is a sketch. A sketch is a quickly executed, freehand drawing that is not intended as a finished work. In general, sketching is a quick way to record an idea for later use, and architects sketches in particular (in a very similar manner to fine artists) serve as a way to try out different ideas and establish a composition before undertaking more finished work. Architects drawings can also be used to convince clients of the merits of a design, to enable a building constructer to use them, and as a record of completed work. In a similar manner to engineering (and all other technical drawings), there is a set of conventions (i.e particular views, measurements, scales, and cross-referencing) that are utilised.

As opposed to free-sketching, technical drawings usually utilise various manuals and instruments. The basic drafting procedure is to place a piece of paper (or other material) on a smooth surface with right-angle corners and straight sides – typically a drawing board. A sliding straightedge known as a 'T-square' is then placed on one of the sides, allowing it to be slid across the side of the table, and over the surface of the paper. Parallel lines can be drawn simply by moving the T-square and running a pencil along the edge, as well as holding devices such as set squares or triangles. Other tools can be used to draw curves and circles, and primary among these are the compasses, used for drawing simple arcs and circles. Drafting templates are also utilised in cases where the drafter has to create recurring objects in a drawing – a massive time-saving development.

This basic drafting system requires an accurate table and constant attention to the positioning of the tools. A common error is to allow the triangles to push the top of the T-square down slightly, thereby throwing off all the angles. Even tasks as simple as drawing two angled lines meeting at a point require a number of moves of the T-square and triangles, and in general drafting this can be a time consuming process. In addition to the mastery of the mechanics of drawing lines, arcs, circles (and text) onto a piece of paper – the drafting effort requires a thorough understanding of geometry, trigonometry and spatial comprehension. In all cases, it demands precision and accuracy, and attention to detail.

Conventionally, drawings were made in ink on paper or a similar material, and any copies required had to be laboriously made by hand. The twentieth century saw a shift to drawing on tracing paper, so that mechanical copies could be run off efficiently. This was a substantial development in the drafting process – only eclipsed in the twenty-first century with 'computer-aided-drawing' systems (CAD). Although classical draftsmen and women are still in high demand, the mechanics of the drafting task have largely been automated and accelerated through the use of such systems. The development of the computer had a major impact on the methods used to design and create technical drawings, making manual drawing almost obsolete, and opening up new possibilities of form using organic shapes and complex geometry.

Today, there are two types of computer-aided design systems used for the production of technical drawings; two dimensions ('2D') and three dimensions ('3D'). 2D CAD systems such as AutoCAD or MicroStation have largely replaced the paper drawing discipline. Lines, circles, arcs and curves are all created within the software. It is down to the technical drawing skill of the user to produce the drawing – though this method does allow for the making of numerous revisions, and modifications of original designs. 3D CAD systems such as Autodesk Inventor or SolidWorks first produce the geometry of the part, and the technical drawing comes from user defined views of the part. This means there is little scope for error once the parameters have been set.

Buildings, Aircraft, ships and cars are now all modelled, assembled and checked in 3D before technical drawings are released for manufacture.

Technical drawing is a skill that is essential for so many industries and endeavours, allowing complex ideas and designs to become reality. It is hoped that the current reader enjoys this book on the subject.

PREFACE

Mechanical Drawing is recognized today as an important part of a secondary education. For all classes of pupils it serves as an important means of developing visualization, strengthening the imagination, and forming habits of careful observation and perception. For those who will make use of it commercially, mechanical drawing is the accepted means of creating a conventional picture of objects.

This book analyzes mechanical drawing upon the basis of its elements, or natural divisions, such as Perspective Sketching, Orthographic Sketching, Pencil Mechanical Drawing, Inking, Tracing, and Reproducing. Each one of these divisions is treated separately in a chapter. Each chapter organizes the division of drawing which it represents. Hence in each chapter there is presented a progressive series of problems in one of the natural divisions of the subject.

The book contains six chapters and covers the first two years of mechanical drawing in Secondary Schools. The first four chapters are designed to occupy the time of a class for the first year of the two years' course. As there is a large element of flexibility in the selection of problems, no one individual is expected to solve all problems. The course may be easily extended over a period of more than two years, even to three or four years, depending upon the number of problems solved, whether a part or all of the chapters are included in the course, and the time devoted to the subject during each year.

The chapters are arranged in the order in which the divisions of drawing are dealt with in commercial drawing room practice. Problems, arranged in groups in each chapter, progress in the order of their difficulty. Each group of problems is chosen to emphasize the construction of a certain type of line, the use of particular instruments, and the application of commonly used conventions. It is believed that such a treatment both retains and extends all possible educational values attributed to mechanical drawing.

In those branches of vocational education which deal with industry, mechanical drawing is the means of showing the plan of construction or the method of assembling constructed parts. Therefore the authors of this book have taken the view that all problems presented must represent commercial industrial practice. They have selected problems which represent several common industrial materials, and the solutions required represent the best commercial drawing room practice. Consequently all abstract problems have been eliminated except in so far as they relate directly to practical problems. This feature, when coupled with the one of dwelling upon one division of drawing until a complete series of problems in it has been solved, makes the book unique in its presentation of *unit* courses. All of these units, when taken together, complete the field of mechanical drawing, and each one prepares the student for efficient service in a particular division of the whole field of drawing.

The course presented in this book has the following *subject matter* features:

1. Every problem represents typical industrial material, commercial construction, and the best drawing room practice.

2. Every chapter presents a complete course in one of the natural divisions of drawing.

3. All problems are arranged in groups depending upon the elements in drawing which are involved. The student may select or the instructor may assign any one or more of several problems in each group, depending upon student ability and community interest. This feature of flexibility makes it easy to adapt the course at any time to any student in a class and to any class in a community by any one or all of three means:

(a) A selection of problems within a group.

(b) The addition of problems to any group.

(c) The elimination of any section of subject matter or of any group of problems within a section.

4. All chapters, when completed in the order in which they are arranged, furnish a complete course, both in the subject of mechanical drawing and in the field of industry covered in secondary education, in which mechanical drawing plays a part. The course presented in this book has the following *method* features:

1. A type problem showing typical conventions and solutions is furnished for each group of problems.

2. Numerous data problems, given in the form of freehand sketches and finished mechanical drawings, present a standard in technique.

3. Each division in mechanical drawing is analyzed into its several elements which are presented in a series of well graded, practical problems, involving essential theory and its application.

4 Each division in drawing requires the concentration of the student upon one thing at a time until he has a fair mastery of both theory and practice. The next division reviews this theory and practice in related problems.

5. Each group of problems in each division in drawing is accompanied by explicit instruction and illuminating reading for the student, and suggestive demonstration material for the instructor.

6. Each chapter in the first year's course closes with a series of review problems and review questions.

In order to cover fully the field of mechanical drawing for secondary schools and to prepare students for commercial drawing room practice in the several divisions of the subject, the authors have given special attention in the second year of the two-year course to such subjects as Sheet Metal Drawing, Architectural Drawing, and Machine Drawing. The student who completes the work as outlined for the first year will therefore be able to devote his attention to any one of these subjects or to all of them depending upon his needs. This element of latitude of choice of subject matter makes the book particularly valuable in schools where drafting is taught for early vocational use.

A Teacher's Manual and an Outline of the Course of Study are furnished free to teachers using the text. The Manual gives brief but pertinent suggestions to assist the instructor. The Outline of the Course of Study shows clearly the plan of the text and indicates possibilities of modifying it to meet local conditions.

The authors wish to express their appreciation of the co-operation of H. D. Orth, Assistant Professor of Drawing and Descriptive Geometry, the University of Wisconsin. From the very beginning to the end of the book, he has been a co-author in its production.

THE AUTHORS.

TABLE OF CONTENTS

CHAPTER I

PERSPECTIVE SKETCHING

PROSPECTUS

It is the aim of this chapter to develop in a condensed but thorough manner the essential principles upon which perspective sketching is based. Furthermore, the presentation is intended to assist the student to develop a fair degree of skill in drawing perspectives of rectangular, angular, and cylindrical objects. Upon completion of the work of this chapter, he should be able to draw objects composed of a combination of these elementary forms. It is hoped that the student will have gained confidence in his ability to visualize and represent an object pictorially. If this has been accomplished he will find a use for perspective as an interpretation of orthographic drawing which will be treated in the succeeding chapters.

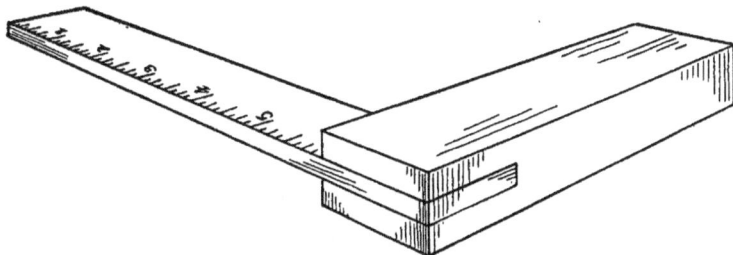

FIG. 1. SHADED PERSPECTIVE OF TRY-SQUARE

GENERAL PRINCIPLES

A perspective drawing of an object shows it as it appears when viewed from a given position. Fig. 1 is an example of a perspective drawing. This drawing gives the observer a correct idea of the form and proportion of the object.

The shading of the drawing, Fig. 1, while adding somewhat to its appearance, does not aid greatly in giving the correct impression of the form and proportion of the object. The shading may, therefore, be omitted, leaving the simplest kind of drawing—the outline drawing as shown in Fig. 2. Such drawings will be referred to in this course as *perspective sketches.*

Perspective sketches are valuable as a means of conveying information about the forms of objects to those who are not familiar with the more conventional means of representation used in mechanical drawing. The student will find the perspective sketch an aid in interpreting mechanical drawing.

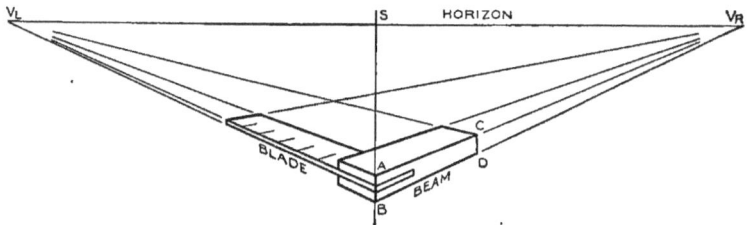

FIG. 2. PERSPECTIVE OF TRY-SQUARE

In this course all objects to be drawn in perspective will be represented as resting on a horizontal plane directly in front of the observer and below the level of the eye. The try-square is shown in Fig. 2 as an observer would see it when standing directly in front of A B with his eye on the same level as S.

In Fig. 2 the line marked *horizon* represents a line in space at an infinite distance in front of the observer. The eye of the observer is on a level with this line and is, consequently, above the level of the try-square, which rests on a horizontal plane.

The *horizon* or *horizon line* is therefore an imaginary horizontal line on a level with the eye of the observer and at an infinite distance in front of him. The apparent meeting of sky and water when one looks over a large body of water is an example of a horizon. Since the horizon is always on a level with the eye of the observer, it follows that, as the eye is raised or lowered to secure a different view of the object, the horizon will be raised or lowered the same distance.

Direction of Lines in Perspective. Referring again to Fig. 2, we note that:

1. In a perspective drawing all of the vertical edges of the object are represented by vertical lines in the perspective. Example: Lines A B and C D.

2. In a perspective drawing all of the horizontal edges of the object which are at *right angles* with the direction of sight of the observer are represented by horizontal lines in the perspective. Example: While not an edge of the object, the horizon line, Fig. 2, is an example of this case.

3. In a perspective drawing all of the horizontal edges which are parallel to each other, but not at right angles to the direction of sight of the observer, are represented by lines which converge to a point on the horizon. Example: Lines A C and B D.

4. Horizontal lines receding to right and left in a perspective, which make equal angles with the horizon line, meet the horizon line at points equally distant from the point on the line directly in front of the observer. Example: In Fig. 2, S is a point on the horizon line directly in front of the observer. The distance from S to V_R is equal to the distance from S to V_L.

45° Perspective. The angle between the beam and blade of the try-square is 90°. The try-square is so placed that the angles which the receding edges to the right and to the left make with the horizon are equal and must therefore be 45° angles. Because of this fact the try-square is said to be drawn in *45° perspective.*

All of the rectangular objects drawn in this course will be placed in a similar position to that of the try-square, i.e., in 45° perspective. This will insure comparative ease in the construction of perspective sketches, as will appear later.

The points V_L and V_R on the horizon toward which the horizontal receding edges of the try-square converge are called *vanishing points.*

A *vanishing point* is the common intersection of two or more converging lines which represent parallel receding edges of an object.

All parallel horizontal receding lines must converge to the

same point on the horizon. Example: The horizontal lines of the try-square converging to the right in its perspective meet in V_R. Likewise all the horizontal lines converging to the left meet in V_L.

Vertical Lengths in Perspective.

1. Equal distances on the same vertical edge of an object are represented by equal lengths in perspective. Example: In Fig. 2 the try-square blade is represented as entering the beam midway between the upper and lower surfaces. The distance from A to the blade is equal to the distance from B to the blade.

2. Equal distances on vertical edges of an object which are at unequal distances from the observer are represented by unequal lengths in perspective. Example: A B and C D represent equal lengths on the object but are unequal in the perspective.

3. Of two equal vertical distances on an object the one nearest the observer is represented by the greater length in perspective. Example: A B and C D which represent equal vertical lengths on the object are both included between two lines of the drawing which converge toward V_R. On account of the convergence of the two receding lines C D is shorter than A B.

Horizontal Lengths in Perspective.

1. Equal distances on a horizontal receding edge of an object are represented by unequal lengths in perspective. Example: The spaces between the lines representing the one-inch marks on the try-square blade, Fig. 2, are unequal.

2. Of the equal distances on a horizontal receding edge of an object, those farthest from the observer are represented by shorter lengths. Example: In Fig. 2 the spaces between the lines representing the one-inch marks grow shorter as they are farther away from the observer.

3 Equal distances on a horizontal receding edge of an object are represented by lengths which *appear* equal. Example: The spaces between the lines representing the one-inch marks on the try-square blade, Fig. 2, are made to appear equal.

The varying of the lengths of lines representing equal
distances on the object as described above is known as *foreshortening*.

Foreshortening is the process of shortening parts of a perspective of an object so as to give the impression of true form and
proportion.

The Cube in Perspective. Thus far only a general consideration of perspective has been given. The following is an application of the principles thus far developed to the representation of
a one-inch cube.

' In this course the cube will be regarded as the *basic form* for
all perspective drawing. The one-inch cube will be used as the
unit of measure and therefore it is essential that its proportions
and position with reference to the eye be well in mind. In Fig. 3
the eye of the observer is directly in front of the point S. The
vertical faces of the cube make 45° with the horizon and, also,
with the direction in which the observer is looking. This agrees
with the position of the try-square in Fig. 2 and is said to be in
45° perspective as defined on page 11. In 45° perspective the
distances from the point on the horizon directly above the nearest
point of the object to the vanishing points and to the eye must
be equal. In this course the vanishing points are taken 14″ to
the right and left of the point above the nearest corner of the
object.

The edges of the cube are one inch long. The front vertical
edge of the cube will be the longest line in the perspective of the
cube (See 3 under Vertical Lengths in Perspective). It will be
drawn in its true length, one inch.

The principles already developed are applied in the following analysis of the perspective of the cube.

Since the side faces are *equally inclined* to the direction in
which the observer is looking:

1. Angle D A E = angle D′A E′
2. Angle F B H = angle F′B H′.

Such angles will hereafter be referred to as the *angles of
inclination*.

The perspective of the corner G is directly above A.

Due to the *convergence* of A D with B F and A D′ with B F′:

Angles F B H and F′B H′ are greater than angles D A E and D′A E′.

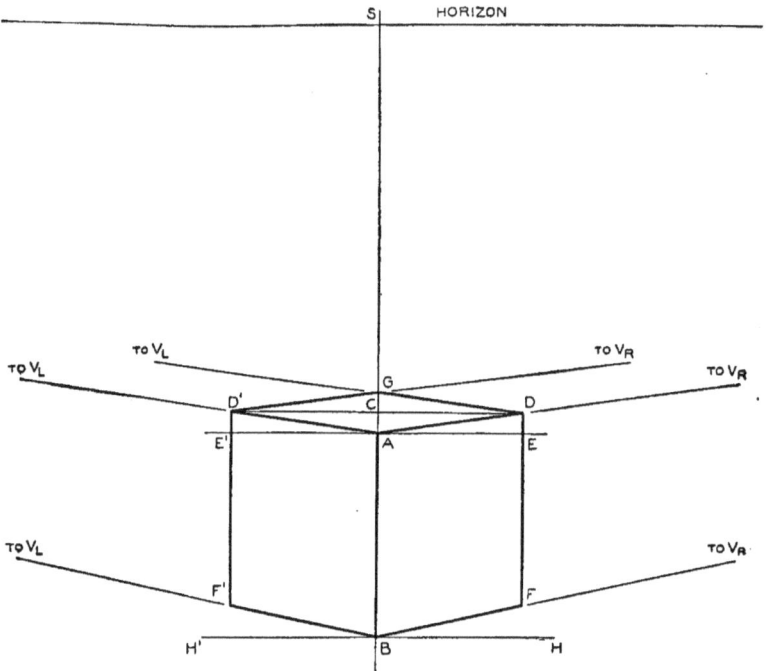

Fig. 3. Perspective of One-Inch Cube

Lines D F and D′F′ are shorter than A B. D F = D′F′. Due to the convergence of A D with D′G and A D′ with D G:

1. G D and G D′ are shorter than A D and A D′;
2. G C is shorter than C A.

Due to *foreshortening*:

A D and A D′ are shorter than A B.

RECTANGULAR OBJECTS

PREPARATORY INSTRUCTION FOR DRAWING PLATE 1

The following is a list of the materials needed to make the perspective sketches:

1. Drawing board.
2. High-grade drawing paper similar to Universal— 9″x12″ sheets.
3. High-grade pencils $\begin{cases} 3H \\ 5H \end{cases}$
4. Pencil pointer.
5. Erasers—Ruby and Flexible gray.
6. Thumb tacks.
7. A straightedge—ruler or triangle.

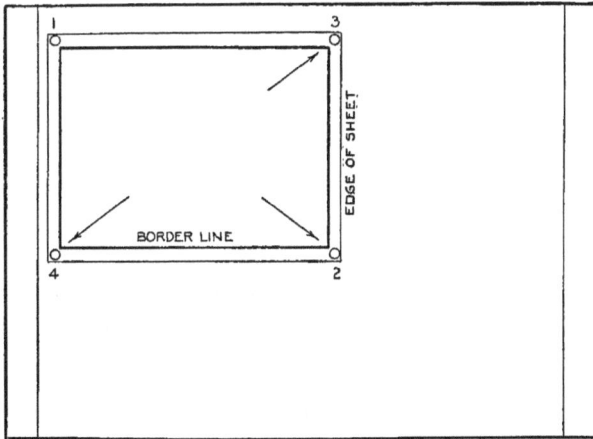

FIG. 4. POSITION OF SHEET ON DRAWING BOARD

The *drawing board* should be made of well-seasoned, straight-grained, soft wood, free from knots and cracks.

When in use the drawing board should be placed on the desk with the longer edges parallel to the front edge of the drawing table. It may be tilted to any convenient angle.

Drawing Paper. In selecting a drawing paper the draftsman should have in mind the purpose for which it is to be used. For *freehand* drawing, where it is desired to produce a porous, uniform line with a soft pencil, a slightly grained surface is satisfactory. It should stand erasing without injury.

In preparing to make a drawing, a sheet of paper should be tacked near the upper left hand corner of the board with the longer edges parallel to the longer edges of the board. Fig. 4. To fasten the sheet insert a tack in the upper left hand corner;

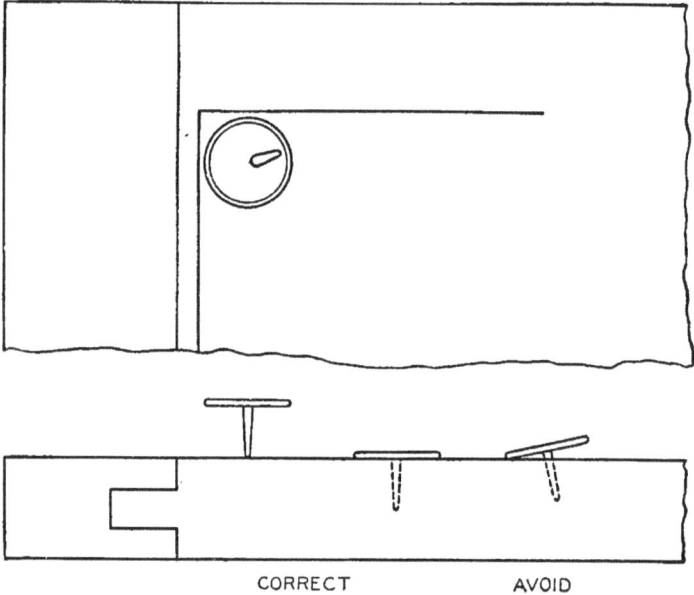

FIG. 5.   POSITION OF THUMB TACKS

square the paper with the board, and, stretching it diagonally, insert a tack in the lower right hand corner. Insert a tack in the upper right hand corner, stretch the sheet in the direction of the lower left hand corner, and insert a fourth tack. Press each tack down vertically until the head is firmly in contact with the paper. Fig. 5.

Pencils. The lead of the drawing pencil should be of firm, even grain. To secure the .desired effect in the drawing, the hardness of the pencil must be considered in connection with the surface of the paper. For freehand drawing a medium soft pencil should be used on a slightly grained surface. A soft pencil

is more easily controlled, and consequently there is more freedom
in drawing lines with it than can be secured with a hard pencil.

To sharpen the pencil, grasp it in the left hand as illustrated
in Fig. 6, and with the knife in the right hand, cut the shavings
by drawing the knife toward the body and through the *wood*

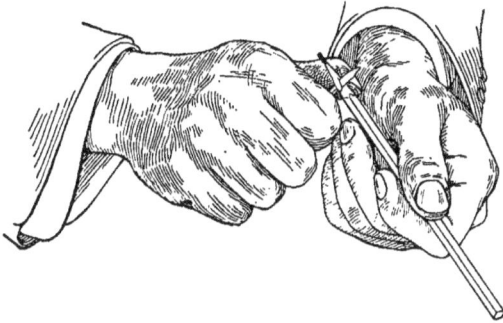

FIG 6. SHARPENING THE PENCIL. WHITTLING AWAY THE WOOD

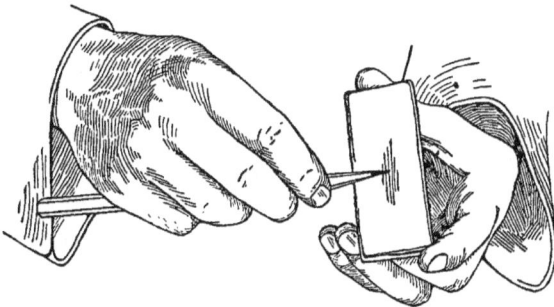

FIG 7. SHARPENING THE PENCIL. POINTING THE LEAD

only. About one-quarter inch of lead should be exposed, and the
wood tapered back about one inch from the lead. Sharpen the
lead on the surface of a sandpaper pad or file, rotating the pencil
so as to produce a conical point. Fig. 7. The sharpened lead
should be slightly rounded on the end in order that soft lines as
shown in Fig. 8 may be produced. This figure also shows the
sketching pencil properly sharpened.

The Constructive Stage. In making a freehand sketch all of the straight lines will first be drawn very lightly with the aid of a straightedge such as the edge of a triangle or ruler, using the 5H pencil. Fig. 12.

1. When two points on a line are known the edge of the triangle or ruler should be placed so that its edge passes through both points. The line may then be ruled lightly.

2. Sometimes only one point on the line and its general direction will be known. In this case the edge of the rule should be made to pass through the point with the edge adjusted to the proper direction.

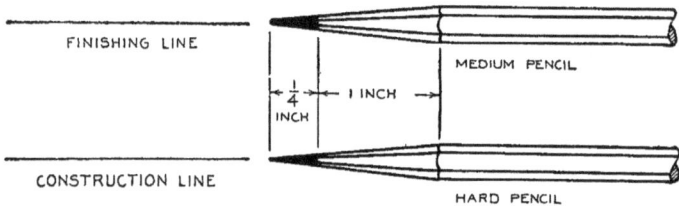

FINISHING LINE

MEDIUM PENCIL

$\frac{1}{4}$ INCH   1 INCH

CONSTRUCTION LINE

HARD PENCIL

FIG. 8. SKETCHING PENCIL PROPERLY SHARPENED

Ruling a Line. In ruling a line along a straightedge the pencil is held in the hand as indicated in Fig. 13. The line is drawn with a continuous motion from left to right with the tip of the fourth finger touching the ruler to steady the hand. *The forearm should always be at right angles with the line being drawn. The rule should preferably be between the draftsman and the line being drawn.*

The Finishing Stage. When the constructive stage has been completed all lines which will not appear in the finished drawing should be erased. The 3H pencil should then be properly sharpened and the lines of the drawing traced over *freehand*. They must be uniform in width and grayness of tone.

The Position of the Hand and Pencil in Sketching. In drawing a freehand line the pencil is held firmly, but not rigidly, between the first two fingers and the thumb as in writing.

In sketching a horizontal line the ends of the third and fourth fingers should rest upon the board to help support and steady the hand. Fig. 9. With the forearm resting on the drawing board, the hand should be moved from left to right, hinging at

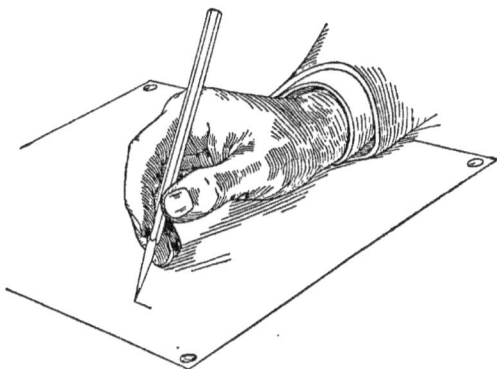

FIG. 9. SKETCHING A HORIZONTAL LINE

the wrist. This will permit only short strokes, about one inch long, to be taken. To sketch a long line, therefore, one must join together a series of one-inch lines. The position for each

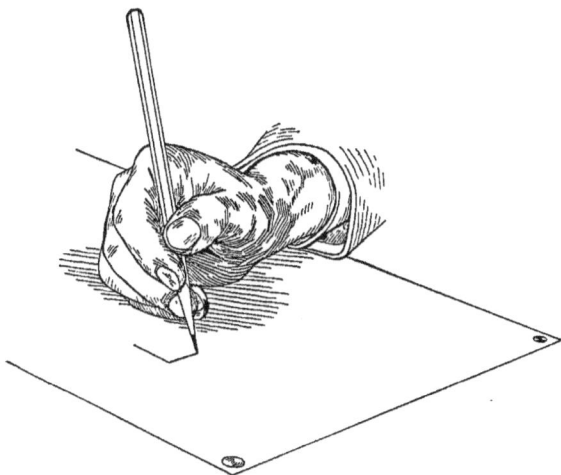

FIG. 10. SKETCHING A VERTICAL LINE

stroke should be obtained by moving the hand and forearm in the direction of the line. Each section should be joined to the preceding one, but not lapped upon it, as the lapping of sections produces an undesirable sketchy effect.

In sketching a vertical line the hand is placed in the position shown in Fig. 10. The hand rests upon its side instead of upon the ends of the third and fourth fingers. The pencil is moved downward. The strokes are made with a *finger movement* while the hand remains stationary. In sketching a vertical line the forearm should remain approximately in the position of a vertical line on the sheet.

The Border Rectangle. Before starting the drawing of the object on the sheet, draw a border line approximately one-half inch from each edge of the sheet. This may be done in the constructive stage by placing the straightedge parallel to each edge of the sheet at a distance estimated to be one-half inch in from the edge, and ruling a line lightly. This border rectangle should be traced over freehand in the finishing stage as are the lines of the sketch.

DATA FOR DRAWING PLATE 1

Given: The perspective of a cube, Fig. 11.

Required: To make a perspective sketch similar to that shown in Fig. 11, on a 9″ x 12″ sheet as explained below.

Instructions:

1. Draw lightly the border rectangle $\frac{1}{2}$″ from the edge of the sheet.

2. To locate the perspective of the cube use the 5H pencil with the ruler as a straightedge.

    a. Draw two very light horizontal lines $X V_R$ and $Y Z$ dividing the space between the upper and lower border lines into three equal parts.

    b. Draw a vertical line, V W, through the center of the sheet and S U midway between V W and the left border line.

    c. From B estimate one inch up on S U, thus locating A, the upper front corner of the cube.

3. V_R is about $\frac{1}{4}$″ from the right border line.

4. Draw lightly a horizontal line through A and connect A with V_R.

5. Get the direction of A D′ by drawing the angle of inclination D′ A E′ equal to angle D A E. This may be accomplished by placing the ruler so that the edge passes through point A and adjusting its direction until the angles appear equal. Fig. 12.

6. To obtain the width of a vertical face of the cube, draw D F so that the figure A D F B *appears* as a square. Fig. 13 shows the process of locating D F. A ruler or triangle is placed

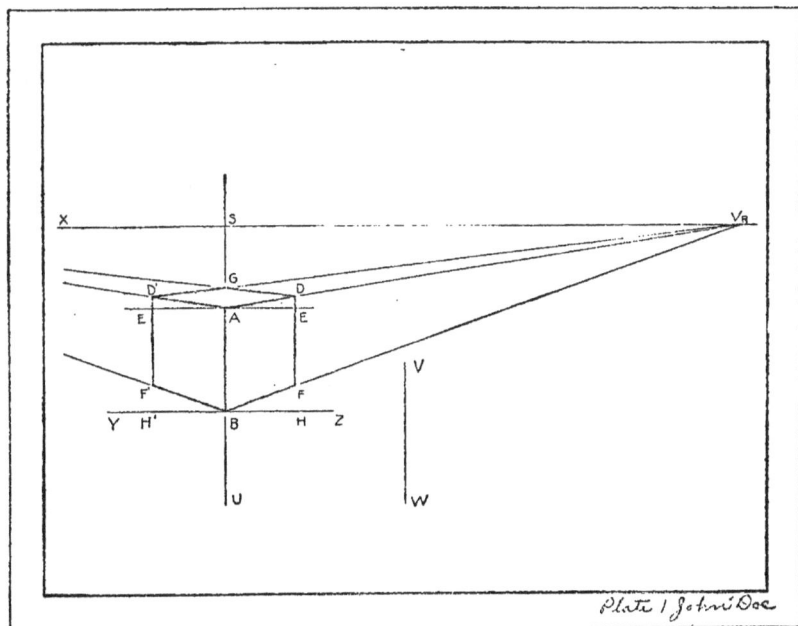

FIG 11. PERSPECTIVE OF CUBE

with an edge in a vertical position parallel to the line A B. It is then moved back and forth to right and left until, in the judgment of the draftsman, the figure A D F B appears as a square.

7. Draw D′F′ making A E′= A E. Complete the perspective of the cube by drawing D′V$_R$ and D G. Fig. 11.

This completes the constructive stage.

8. All lines not shown in Fig. 11 should now be erased. The cube and the border rectangle should be traced over *freehand* with a well sharpened 3H pencil to produce a line of even weight and uniform shade. The remaining lines of the drawing should be allowed to remain light, as drawn in the constructive stage. Omit all reference letters.

9. Write the plate number and name in the lower right hand corner of the sheet as in Fig. 11. Remove the sheet from the board, turn it over and, with a knife or other sharp instrument, press the paper back into the thumb tack holes.

The Method of Developing Lettering in the Course. One of the most difficult steps in making a drawing is the lettering of the notes, dimensions, and title. In this course lettering will be omitted from all drawings until the student has had considerable practice in forming and spacing the letters and figures. This practice will be had on small lettering plates. *Each drawing plate should be followed by the lettering plate of the same number.*

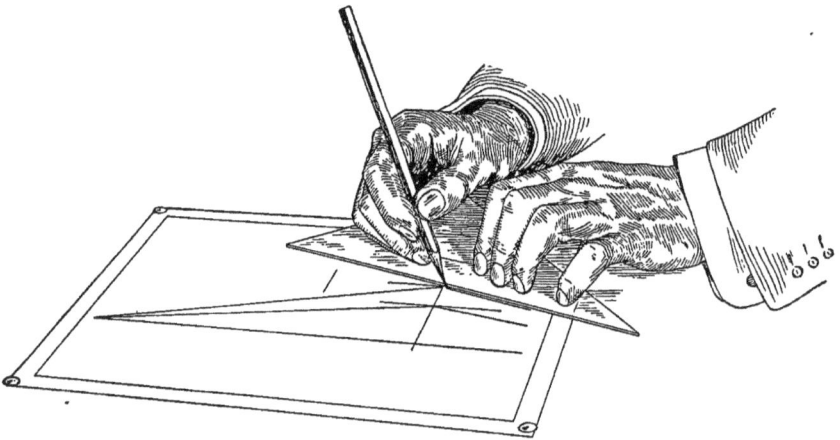

FIG. 12. SKETCHING THE ANGLES OF INCLINATION

LETTERING

Modern practice demands that the lettering done on working drawings be simple, legible, and capable of easy and rapid rendition. The simple Gothic style fulfils these requirements and is therefore quite generally used.

Form and Proportion. A careful study of the form and proportion of each letter must be made before the student can hope to make any considerable progress in lettering. Practice in drawing the letters will add something to his control of the media with which he works, but first of all he must have a distinct knowledge of what he is trying to accomplish.

Strokes. For convenience in forming letters they are divided into strokes. In most cases the strokes are natural divisions of the outline of the letter. Three things should be remembered about the strokes for each letter: (1) the number of strokes, (2) the order in which they are made, (3) the direction in which each stroke is drawn. The advantage in knowing and using a system of strokes lies in the fact that drawing the letters repeatedly in the same manner makes the forming of each letter more

FIG. 13. DETERMINING THE WIDTH OF THE FACE OF THE CUBE

nearly automatic. Hence it adds to the ease with which letters can be produced and aids in securing uniform results.

Spacing. Second only in importance to the forms of the letters is their relation to each other. The best effect is obtained when the areas included between the letters in a word *appear* equal. For the capital letters the area of these spaces should be equal to the area of a rectangle one-half the normal width of the H. The space between words should be about three times that between letters. Words set off by a comma should be spaced from one to one and one-half times the usual distance. The space between sentences should be about twice the space between words.

The final test of good spacing is legibility. The letters must be far enough apart to avoid a crowded effect and yet the spaces must not be so great that the letters appear scattered. In like manner words must be separated enough to stand out individually, but not enough to make reading difficult.

Lettering in Pencil. The pencil used for the freehand work on a drawing should be softer than the pencil used for the mechanical work. It should be of such grade that when properly sharpened a clear gray line can be produced with a single stroke. It should not be hard enough to cut into the surface of

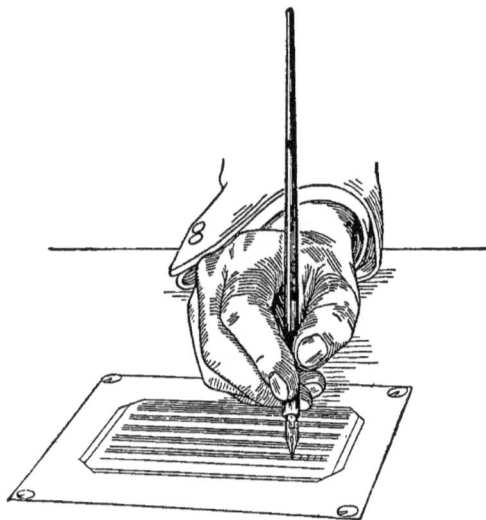

FIG. 14. CORRECT POSITION OF THE HAND AND PEN FOR LETTERING

the paper, as difficulty is then experienced in controlling the direction of the line.

The lead should be sharpened to a long taper, conical in form and rather blunt at the end. With one-quarter inch of lead exposed, and this tapered back to the wood, the section of the lead will be so nearly uniform near the end that it will stand considerable use without resharpening. The pencil should be held in the hand in the same position as the pen shown in Fig. 14, with the forearm nearly in the direction of the vertical stems of letters or, in the case of the inclined letters, nearly in the direction of the slant. The strokes should be drawn with a finger movement. The pencil should be turned about its axis frequently to keep the point round so as to produce a line of uniform weight. All strokes should be made with the hand held in the same position Shifting the arm to obtain advantageous positions for drawing strokes in different directions is a habit which will prevent the acquirement of commercial speed and at the same time will prevent the development of the professional type of lettering as distinct from the labored effect produced by the average novice.

Lettering in Ink. The beginner will find it more difficult to produce satisfactory results with pen and ink than with the pencil because of the complications which arise from the nature of the media. To secure a black line of uniform weight with a quick drying fluid such as India ink, and with an ordinary writing pen, presents a problem which usually requires a careful study of the methods of using these materials and considerable intelligent practice.

The pen should be held in the hand as shown in Fig. 14. In drawing a line the points of the pen should be side by side so that the width of the line can be controlled by the pressure applied to spread the nibs. The position of the pen in the hand should not be changed for strokes of different direction, but rather the weight of line should be kept uniform by varying the pressure on the pen. In lettering in ink as in lettering with the pencil, the hand should be held in the same position for all strokes. This will give a better general effect and will make it easier to develop commercial speed in forming the letters.

The pen should be filled by applying the quill attached to the stopper of the ink bottle to the under side of the pen. Enough ink should be put on the pen to last a reasonable length of time and to produce a wet line so that when it is dry, enough carbon will have been deposited to make it black. Overloading the pen, on the other hand, will cause the corners to fill at intersecting lines. The pen should be wiped frequently to remove the dry ink from the surfaces of the pen and between the nibs. Fresh ink and a clean pen are necessary to produce sharp clean-cut lines.

Titles. The title contains information by which the drawing can be identified, such as the name of the part or parts of the machine or structure, name of the complete machine or structure, manufacturer's firm name and address, drawing number, date, scale, and initials of draftsman, tracer, and checker.

The usual position of the title is in the lower right-hand corner of the sheet where it does not interfere with the drawing and at the same time may be read without taking the sheet from its place in a drawer or file. The relative importance of the items in the title is shown by varying heights and widths of the letters or the weight of their stems, or both.

The lines should be balanced, *i.e.*, the middle point of each line should fall on the same vertical line. To give the best effect

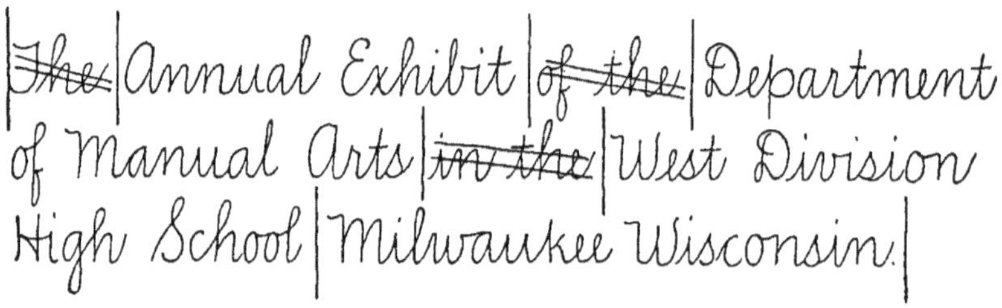

FIG. 15. TITLE MATERIAL DIVIDED INTO GROUPS OF WORDS

the lines should vary in length. The general contour of the title is very commonly oval or pyramidal in form.

The arrangement of the lines of the title and the determination of the height of each line present a problem in design for the solution of which the contour of the title should be kept in mind.

The space between the lines of letters for the single stroke capitals should be from three-fourths to one and three-fourths the height of the smallest adjacent letters.

The style of letter used for the title should be dignified. For this reason the capital letters are generally used.

The steps in designing a title should be taken in about the following order:

1. Assuming that the wording or at least the substance of the title is stated, write out the complete title and divide the words into logical groups for the different lines. Fig. 15.

2. Rewrite, tentatively arranging the lines as they will be in the printed title. Fig. 16.

3. Decide upon the relative importance of the lines and select heights of letters accordingly. It may now appear that a re-arrangement of the lines will give a better outline without affecting the meaning.

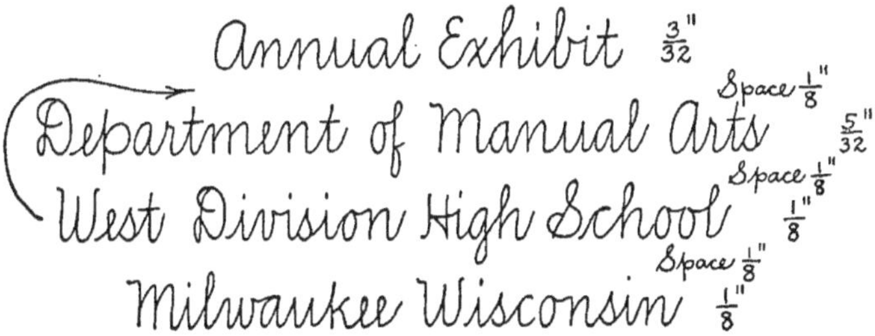

Annual Exhibit $\frac{3}{32}''$

Space $\frac{1}{8}''$

Department of Manual Arts $\frac{5}{32}''$

Space $\frac{1}{8}''$

West Division High School $\frac{1}{8}''$

Space $\frac{1}{8}''$

Milwaukee Wisconsin $\frac{1}{8}''$

FIG. 16. TENTATIVE ARRANGEMENT OF LINES OF THE TITLE

4. The title may be balanced by printing each line lightly in its proper space to obtain the spacing of the letters. Any adjustment necessary to make the middle point of each line fall on the center line of the title should be made. The letters should then be drawn in full weight. This method may be used with

ANNUAL EXHIBIT

WEST DIVISION HIGH SCHOOL

DEPARTMENT OF MANUAL ARTS

MILWAUKEE WISCONSIN

FIG. 17. FINISHED TITLE

success by those who have had considerable experience in lettering. The beginner will obtain better results with but little more work by lettering the lines first on a trial sheet to get the spacing and then by using these lines as a guide in balancing the lines and spacing the letters on the drawing, as described on page 140. Fig. 17 shows a balanced title.

In drafting offices or business firms where large numbers of drawings similar in general character are made, the items common to all titles are very often printed on the pencil drawing with a rubber stamp and on the tracing in type. Uniformity in treatment is thus secured and much time in lettering is saved. Fig. 144 illustrates commercial titles. These title forms are printed on the *under side* of the tracing cloth. Errors may thus be corrected and changes made in the lettering done by the draftsman without erasing the printed lines and letters.

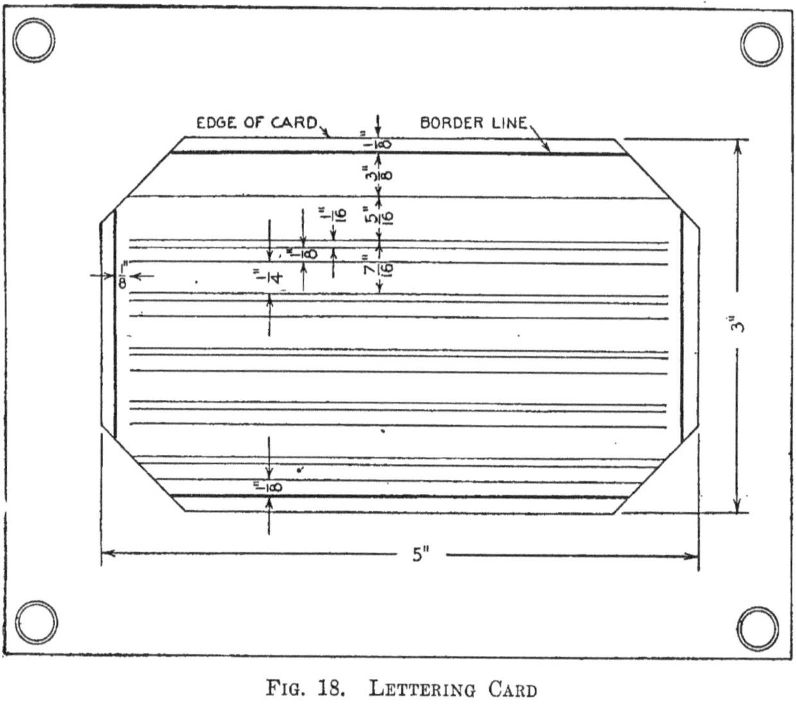

FIG. 18. LETTERING CARD

PREPARATORY INSTRUCTIONS FOR LETTERING PLATE 1

The Plate. The first ten lettering plates will be in pencil. Three by five cards of the regular drawing paper, ruled as shown in Fig. 18, will be used.

The Lettering Pencil. Use the 3H pencil for lettering, sharpened to a conical point as for freehand sketching. Fig. 8.

Number, Order, and Direction of Strokes. Each letter or numeral is made by one or more strokes. In general, vertical and inclined strokes are made downward and horizontal strokes to the right. Fig. 19 shows the number, order, and direction of strokes for the numerals 1, 4, 7, and the symbols used for the foot, inch, and dash. The relative width of numerals is shown in column 4.

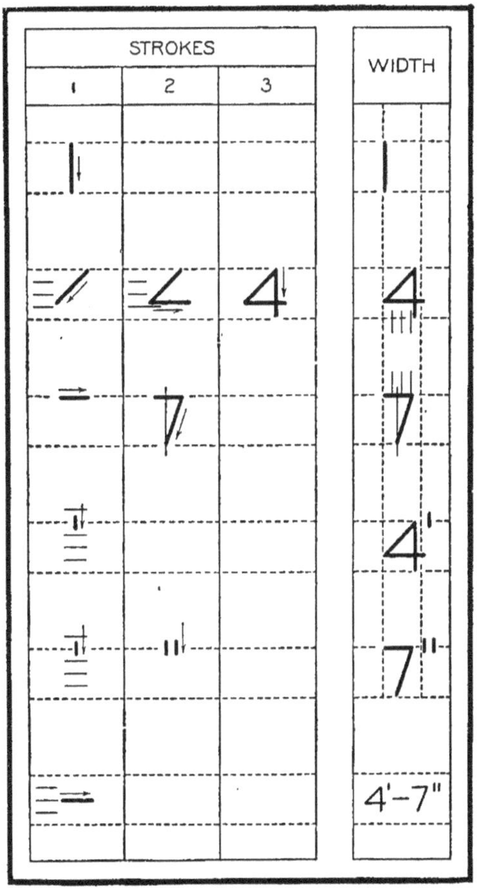

FIG. 19. ORDER, NUMBER, AND DIRECTION OF STROKES

The Scale of Heights. For convenience in estimating vertical distances the space between the guide lines is divided into four equal parts. Fig. 19.

A Scale of Widths. The width of the H is taken as the unit of width. The total letter distance is divided into four equal parts. Horizontal distances may be estimated by observing their relation to these divisions. Fig. 19.

Drawing the Strokes. Before starting a stroke, carefully *plan its position and direction.* Make each line with *one movement* of the pencil. A vertical stroke is made by drawing a line from one point to another *directly below it.* In case a stroke or letter is unsatisfactory it should be erased and redrawn.

Foot and Inch Marks. A short dash placed to the upper right of a numeral indicates feet. Two such dashes similarly placed indicate inches. A horizontal dash is placed between numerals representing feet and inches. See Fig. 19.

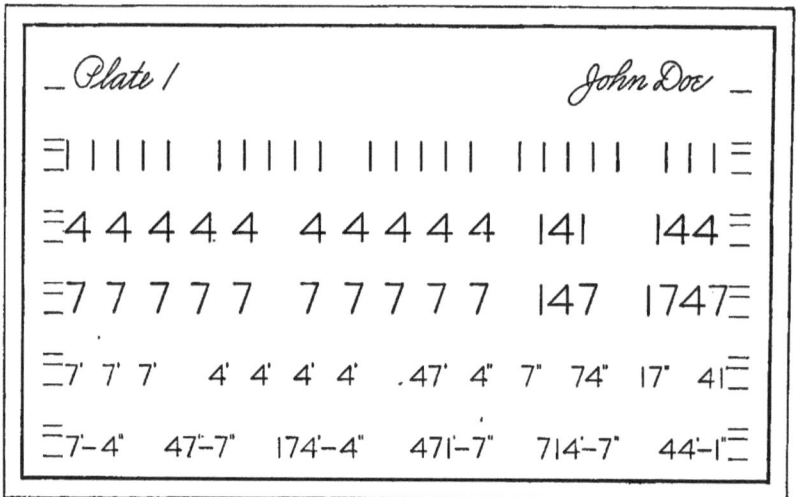

Fig. 20. Lettering Plate 1. 1, 4, 7

DATA FOR LETTERING PLATE 1

Given: Plate 1 to reduced size, Fig. 20.
Required: To make the plate to an enlarged scale.

Instructions:

1. Fasten the card to the board either with thumb tacks or by inserting its corners in diagonal slits cut in a larger piece of paper which is tacked to the board. Fig. 18.

2. Draw the numerals and symbols, using the number, order, and direction of strokes shown in Fig. 19.

3. Write in the plate number, followed by the name at the top of the sheet as indicated in Fig. 20.

PREPARATORY INSTRUCTIONS FOR DRAWING PLATE 2

Plate 1, page 21, gave practice in making a perspective of the unit of measure and basic form in perspective—the cube. The one-inch cube will be used in the following plates as a means of constructing and proportioning the perspective sketches of more complex objects.

A Scale of Levels. Fig. 21 shows a horizontal square at different levels in perspective. At the left in Fig. 22 the horizontal square is used as the top of a cube, represented at levels one-half inch apart. It will be noticed that in each of these figures the

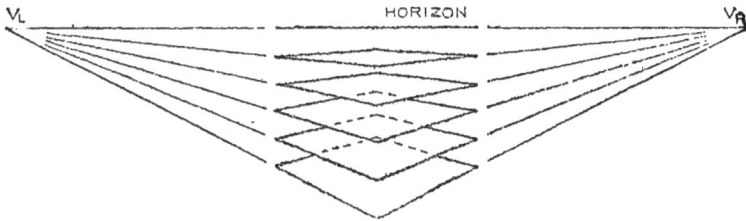

FIG. 21. VARIATION OF AREA WITH LEVEL

area of the figure representing the square and the angle of inclination increase with the distance below the level of the eye. The distance below the level of the eye of the front corner of each square, Fig. 22, is indicated by the numerals at the left of the figure. In this course the student will be aided in determining the level for the perspective of an object by referring to this scale. To the right of the scale is shown its application in representing a cube at different levels.

Vertical Measurements. Under, "Vertical Lengths in Perspective," page 12, the general facts regarding these measurements are given. In making vertical measurements in perspective the following rule must be observed.

All vertical distances on an object must be measured in perspective *on the line representing the front vertical edge of the object.* This is true for the following reasons:

1. The front vertical edge is drawn full length. In Fig. 23, A B is greater than the vertical line through I. To secure this length, one would determine A B and draw the vanishing lines.

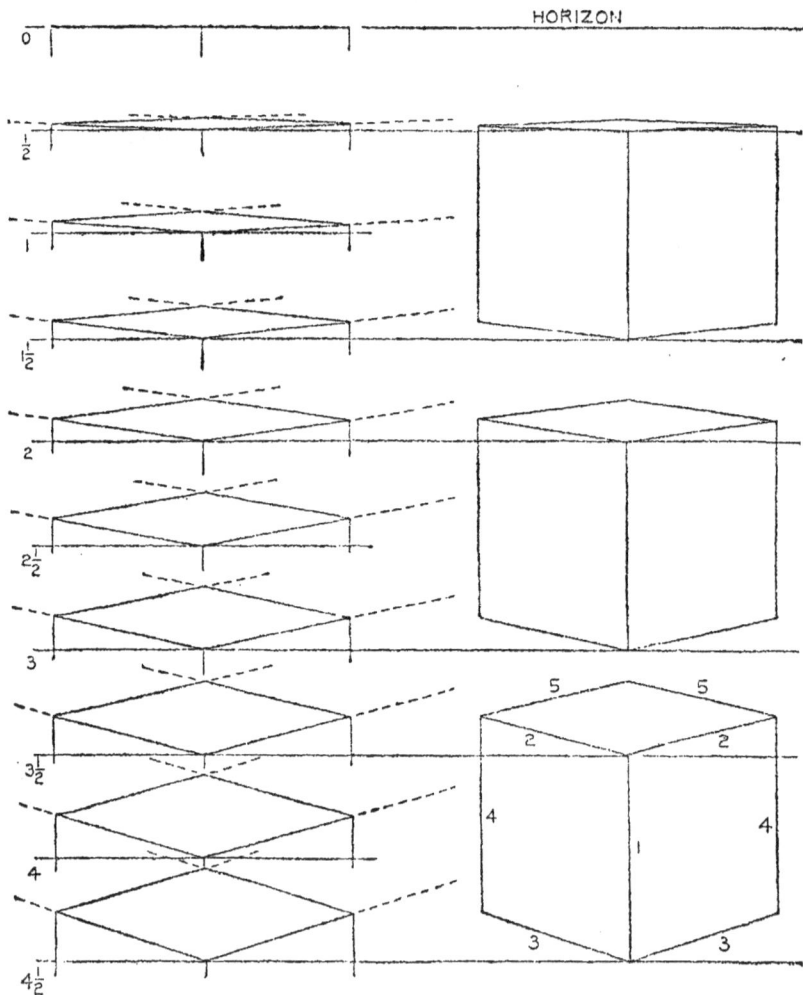

Fig. 22. Scale of Levels

2. In general, equal vertical distances are equal in perspective only when measured on the *same* vertical edge. Example: A B = B K = K L.

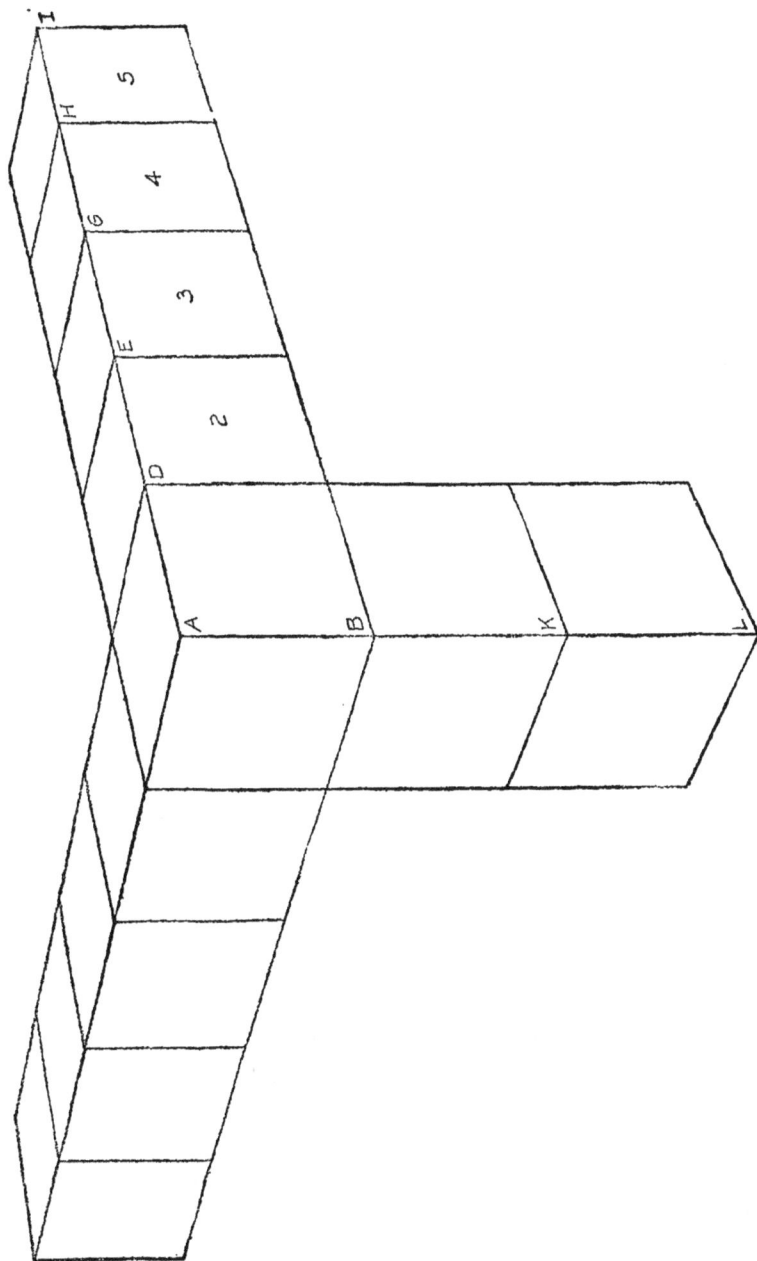

FIG. 23. FORESHORTENING SCALE

Horizontal Measurements. Under, "Horizontal Lengths in Perspective," page 12, the general facts regarding horizontal measurements are given. In making horizontal measurements in perspective the following method should be used:

Whenever possible, horizontal distances on an object should be measured in perspective by drawing the faces of a series of receding one-inch cubes so that they *appear* to be squares. In Fig. 23, lengths A D, D E, E G, G H, and H 1, representing equal horizontal distances, are measured by making faces 1, 2, 3, 4, and 5 appear as squares.

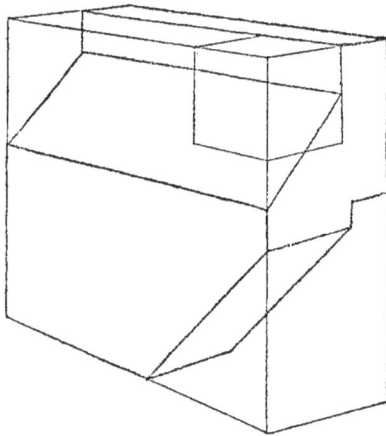

Fig. 24. Enclosing Solid

The Enclosing Solid. In using the methods of making horizontal and vertical measurements given above, one of the important steps in the construction of the perspective sketch will be the drawing of a rectilinear solid the edges and surfaces of which, so far as possible, are coincident with the edges and surfaces of the object. Fig. 24. This solid will be called the *enclosing solid.* This solid should be drawn completely before any attempt is made to construct the details of the object in perspective.

The Measure Cube. The first step in drawing the enclosing solid is to draw a one-inch cube with its upper front corner at the level required for the perspective of the object to be drawn. This one-inch cube will be at the upper front corner of the enclosing solid. The front vertical edge of the cube serves as the vertical unit of measure and the width of the side faces as the horizontal unit of measure. This cube is therefore called the *measure cube*.

The Table Line. When an object rests on a horizontal surface its position with reference to that surface is shown by a horizontal line called *the table line*. The position of this line as shown in Fig. 27 is taken arbitrarily. In its relation to the perspective it should represent the object as resting in a pleasing position on a horizontal plane. The table line should be drawn freehand.

DATA FOR DRAWING PLATE 2

Given: The dimensioned perspective of a sandpaper block, Fig. 27.

Required: To make a sketch of the sandpaper block, full size in perspective, omitting all dimensions and lettering, or any similar problem assigned by the instructor.

Instructions:

1. Draw the border rectangle as in Plate 1, page 21. Here and throughout the constructive stage use the 5H pencil.

2. To locate the center of the sheet proceed as follows: Place the ruler on the sheet with one edge in the position of one of the diagonals of the border rectangle. Rule a light, short line through the approximate center of the sheet. In like manner draw a part of the other diagonal. The intersecting lines will locate the center of the sheet.

3. With the aid of the ruler draw the measure cube with its upper front corner A at the center of the sheet and $3\frac{1}{2}''$ below the level of the eye. Fig. 25. Refer to the angle of inclination in Fig. 22 for the required level. Reproduce this angle as illustrated in Fig. 12.

4. Complete the enclosing solid by drawing the lines in the order indicated by the numerals. Fig. 25. Measure vertically and to the right and left as previously described under, "Vertical Measurements" and, "Horizontal Measurements" respectively, pages 31 and 34.

5. To sketch the open space through the sandpaper block which is to be occupied when the block is in use by a block of the

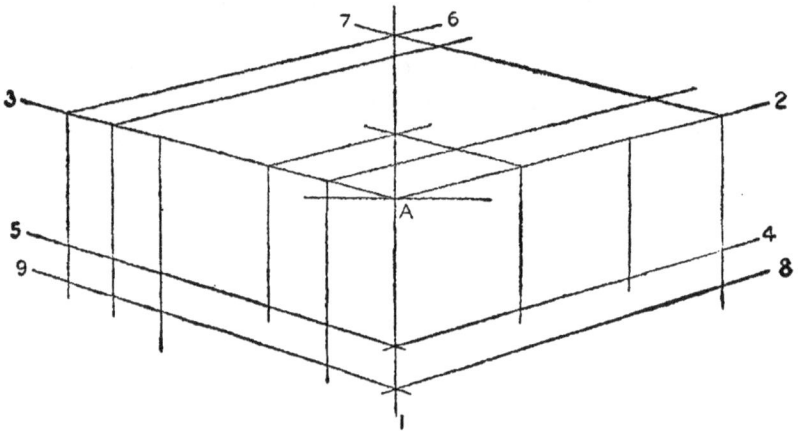

FIG. 25. CONSTRUCTIVE STAGE. ENCLOSING SOLID

same dimensions as the open space which holds the edges of the sandpaper, locate B, ½″ below A, and draw line 10 converging

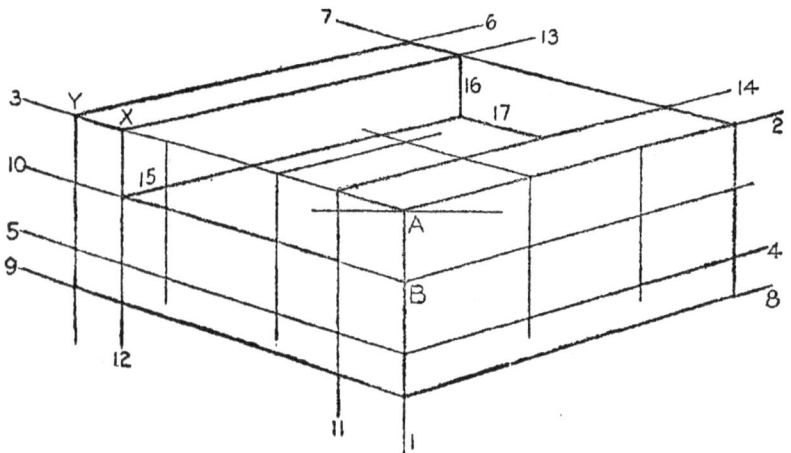

FIG. 26. CONSTRUCTIVE STAGE. COMPLETE

with line 3, Fig. 26. Lay off from A on line 3 a distance representing ½″. The principle of foreshortening applied here will make this distance slightly greater than one-half of the width of the face of the cube. Draw line 11. In the same manner locate and draw line 12. Draw lines 13, 14, and 15 converging with lines 2 and 6. Draw line 16 vertically from the intersection of lines 7 and 13. Draw line 17 converging with lines 3 and 7. This completes the constructive stage.

6. Erase all lines except the outline of the figure and trace over the sketch *freehand* with a carefully sharpened 3H pencil. Draw a table line as in Fig. 27.

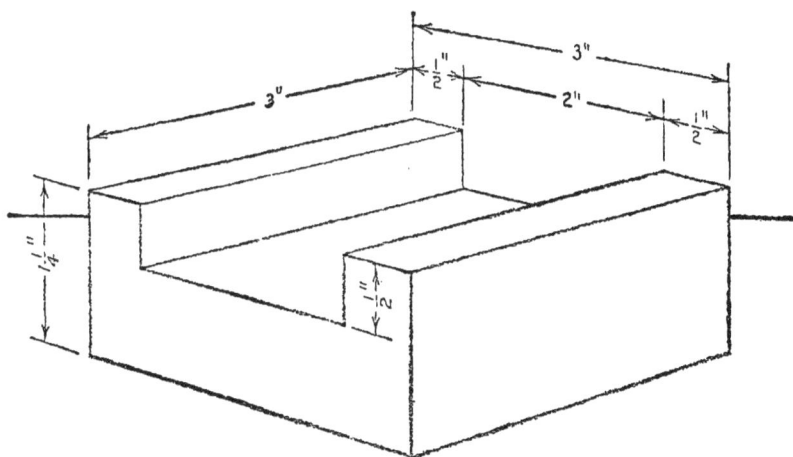

FIG. 27. SAND PAPER BLOCK

7. Write the plate number and name in the lower right hand corner of the sheet and press the paper back into the thumb tack holes as directed in Plate 1, page 21.

PREPARATORY INSTRUCTIONS FOR LETTERING PLATE 2

Curved Strokes. In making the curved strokes of the 5 and the 2 the student should have in mind the form of the *complete* oval.

The Dimension Form. The dimension form consists of the numerals designating feet and inches, the foot and inch marks, the dash, the dimension and extension lines, and the arrowheads as arranged in Fig. 77.

It will be seen that the arrowheads are placed on the dimension lines with their points touching the extension lines. They are composed of two slightly curved lines symmetrical with respect to the dimension line. The length of the arrowhead should be about $\frac{1}{8}''$ and the width $\frac{1}{16}''$. Fig. 77. Fig. 28 shows strokes for arrowheads pointing in different directions.

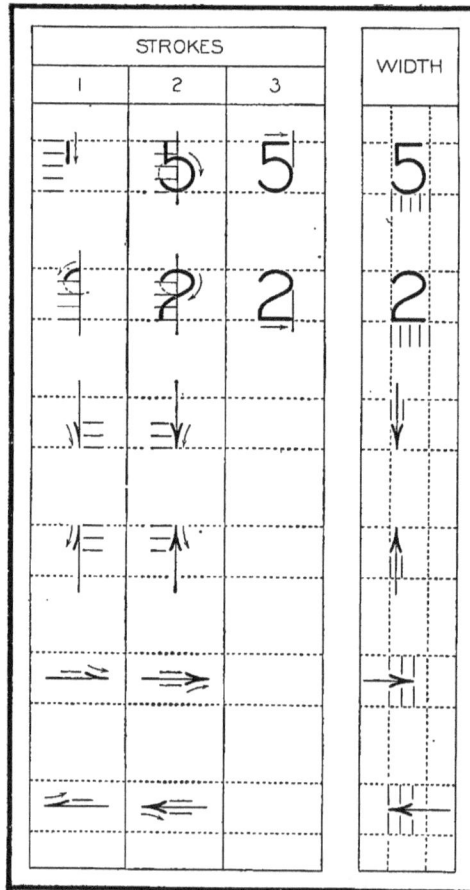

FIG. 28. LETTERING PLATE

DATA FOR LETTERING PLATE 2

Given: Plate 2 to reduced size, Fig. 29.

Required: To make the plate to an enlarged scale.

Instructions: Proceed as in Plate 1, page 30, following carefully the number, order, and direction of strokes.

DATA FOR EXTRA DRAWING PLATE

Given: A dimensioned perspective sketch of a clamping plate for lathe tail-stock.

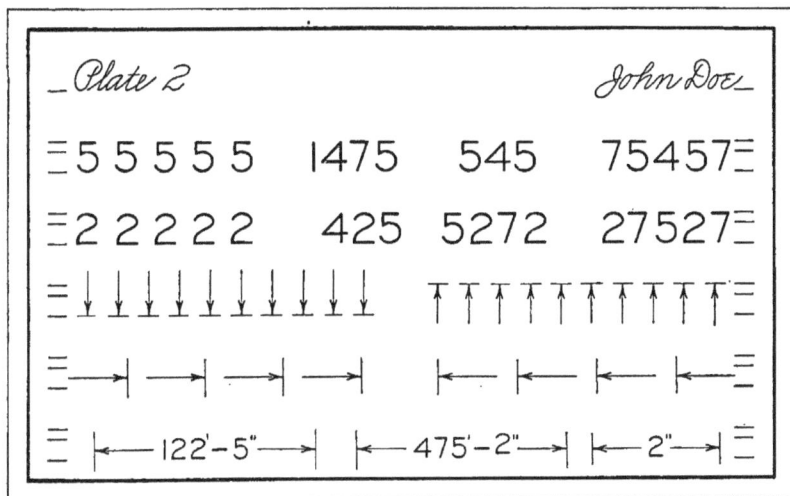

FIG. 29. LETTERING PLATE 2

Required: To make a sketch of the clamping plate full size, omitting all dimensions.

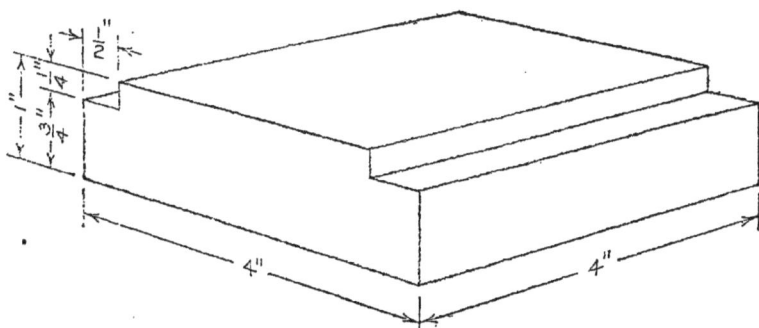

FIG. 30. CLAMP FOR TAIL STOCK

The upper front corner of the enclosing solid is in the center of the sheet and 3½″ below the level of the eye.

PREPARATORY INSTRUCTIONS FOR DRAWING PLATE 3

To Center a Perspective Sketch on a Sheet. For the preceding plates definite instructions have been given to center the sketch on the sheet. For the sake of appearance a sketch should be centrally located. The student should use considerable care, therefore, in locating the upper front corner of the enclosing solid. It cannot always be located at the center of the sheet. The following suggestions will be of value in locating the upper front corner of the enclosing solid.

A close approximation can be made to the correct position of the front vertical edge of the enclosing solid to right or left of the center of the sheet by referring to Fig. 23.

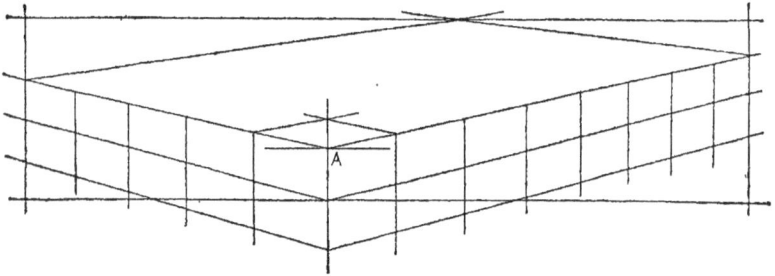

FIG. 31. ENCLOSING SOLID

1. On this figure the distance to be measured to the right and to the left of the front vertical edge of the enclosing solid may be marked off. If the horizontal distance between the extreme points is divided into two equal parts the division will come at the point in the perspective which should be at the center of the sheet.

2. The distance from A B, Fig. 32, to this middle point is the distance which the front vertical edge of the enclosing solid must be to the right or left of the center of the sheet.

In locating the upper front corner of the measure cube after the position of the front edge is determined, the length of the front edge of the enclosing solid and the distance of the back corner of the upper surface above the front corner of the enclos-

ing solid must be estimated. Half the sum of these two distances
should fall above and half below the center of the sheet.

DATA FOR DRAWING PLATE 3

Given: The dimensioned perspective sketch of a cord-wind
with the upper surface 3″ below the level of the eye. Fig. 33.
Required: One of the two solutions as stated below.
1. To draw the cord-wind, full size, with the upper surfaces
5″ below the level of the eye. Omit all dimensions and letters
from the finished sketch.

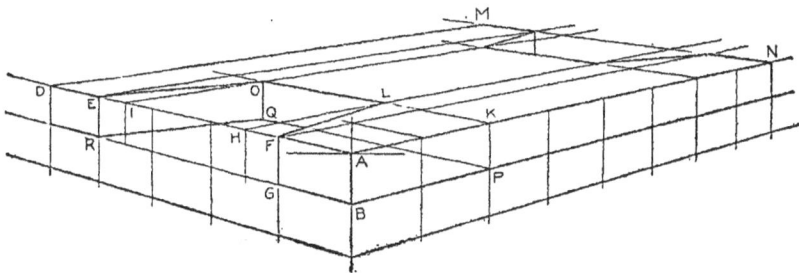

Fig. 32.   Constructive Stage Complete

2. To draw the cord-wind full size with the upper surface 5″
below the level of the eye, and turned so that the longer edges
vanish toward the left instead of toward the right.
3. To draw any similar object assigned by the instructor.

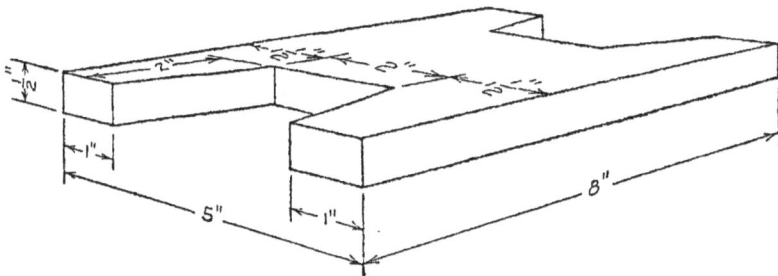

Fig. 33.   Cord-Wind

Instructions:
1. Locate the upper front corner of the measure cube **A** as
directed under, "To Center a Perspective Sketch on a Sheet,"
page 40.

2. Complete the enclosing solid as in Plate 2, page 35, taking care to secure the necessary convergence. As it is the aim that the student should learn to make perspective sketches entirely freehand he should now draw as many as possible of the lines of the constructive stage without the use of a straightedge.

3. To locate the lines representing the cut in the near end of the cord-wind lay off A F and D E, Fig. 32, to represent one

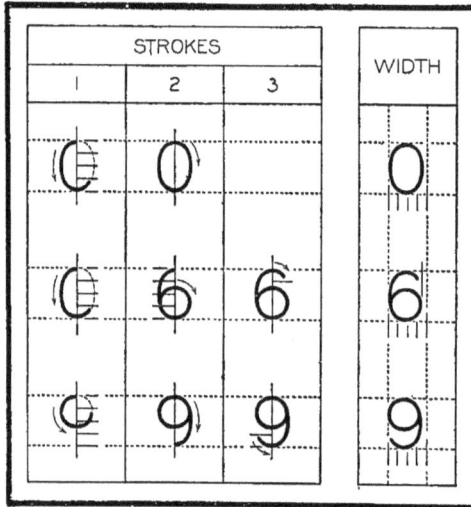

FIG. 34. LETTERING PLATE

inch and H F and E I to represent one-half inch. From H and I draw lines converging with A N and D M to the right. Lay off A K to represent two inches and draw a line from K converging with A D. This line meets the lines from H and I in L and O respectively. Connect E O and F L. From O draw a vertical line and from P a line converging with K O. These lines intersect in Q. Draw R Q.

By a similar method make the construction for the cut at the farther end of the cord wind.

4. Erase unnecessary lines and finish the sketch in the usual manner.

PREPARATORY INSTRUCTIONS FOR LETTERING PLATE 3

Curved Strokes. The oval of the numeral 0 is the basic form for the 6 and 9. In making the outline strokes of these numerals the student should have in mind the form of the *complete* oval. *Whole Numbers and Fractions.* The whole number in a dimension will be made $\frac{1}{8}''$ high.

The total height of the fraction should be twice the height of the whole number with a clear space between each numeral and the division line. Fig. 141. To check these heights mark off an eighth inch and a quarter inch space on the edge of a card and use it as a scale. Fig. 78.

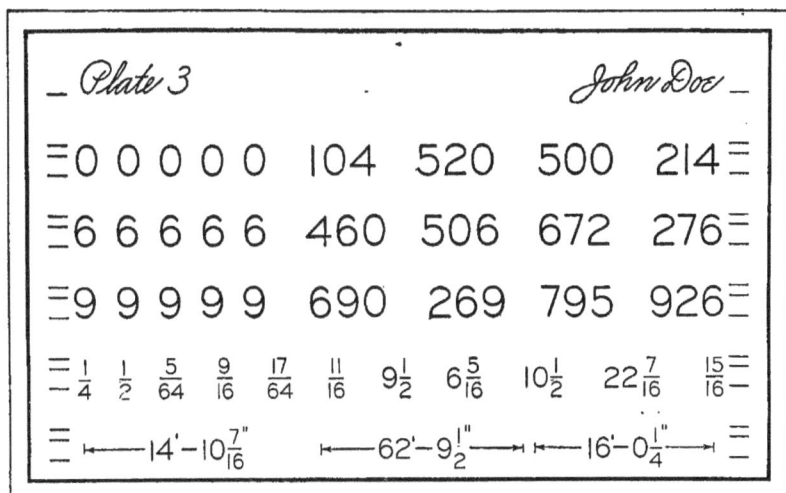

FIG. 35. LETTERING PLATE 3. 0, 6, 9

DATA FOR LETTERING PLATE 3

Given: Plate 3 to reduced size. Fig. 35.
Required: To make the plate to an enlarged scale.

DATA FOR EXTRA DRAWING PLATE

Given: The dimensioned sketch of a wall bracket, Fig. 36, with the upper surface of the enclosing solid $3\frac{1}{2}''$ below the level of the eye.

Required: To draw the wall bracket full size, with the surface as shown $3\frac{1}{2}''$ below the level of the eye, but with the longer horizontal edges receding toward the left instead of toward the right.

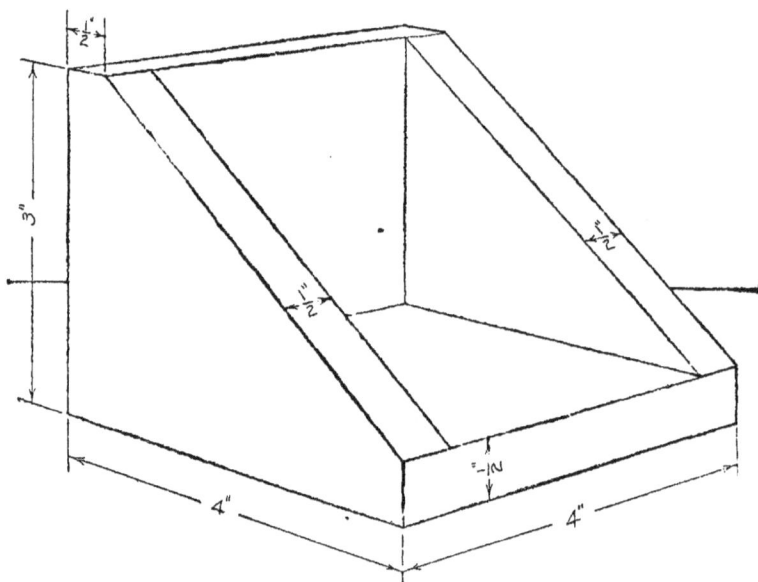

FIG. 36. WALL BRACKET

CYLINDRICAL OBJECTS

PREPARATORY INSTRUCTIONS FOR DRAWING PLATE 4

The Vertical Measure Cylinder. As stated before, the cube is the basic form for the perspective sketching in this course. To secure a measure unit for cylindrical objects a cylinder is inscribed in a measure cube as shown in Fig. 37. The cylinder is therefore one inch in diameter and one inch long. The principle

of foreshortening makes the axis of the cylinder and the major
axis of each of the ellipses representing its bases slightly less
than one inch. In sketching, these differences may be ignored.
In Fig. 38 these distances are one inch in length.

The following is an analysis of a cylinder which will be
referred to as the *vertical measure cylinder:*

1. The distance between the centers of the ellipses is equal
to their major axes or one inch. A B = C C' = D D'. Fig. 38.

2. The major axes C C' and D D' of the ellipses are at right
angles with the axis of the cylinder. These lines do not con-

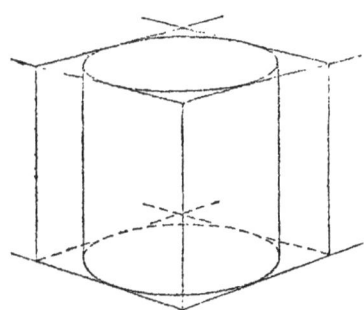

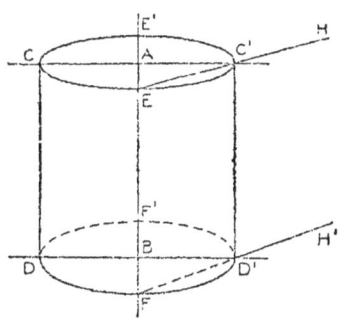

FIG. 37. VERTICAL CYLINDER IN-
SCRIBED IN A MEASURE CUBE

FIG. 38. VERTICAL MEASURE
CYLINDER

verge, since they represent lines at right angles to the direction
of sight of the observer. See, "Direction of Lines in Perspective," page 11.

3. The minor axes E E' and F F' are coincident with the line
representing the axis of the cylinder.

4. Due to the difference in level of the upper and lower bases
the minor axis F F' of the lower base is greater than the minor
axis E E' of the upper base. The minor axis of the upper base
may be determined for any level from the scale of levels discussed
in the following paragraph.

The half length of the minor axis of the lower ellipse may be
determined by drawing F H' through D', converging with E H.

A Scale of Levels. The left half of Fig. 39 is a *scale of levels*
showing the upper base of a measure cylinder at levels one-half

inch apart. It is evident that the area and minor axis of the ellipse increase with the distance of the ellipse below the level of the eye. The distance below the level of the eye of the center of

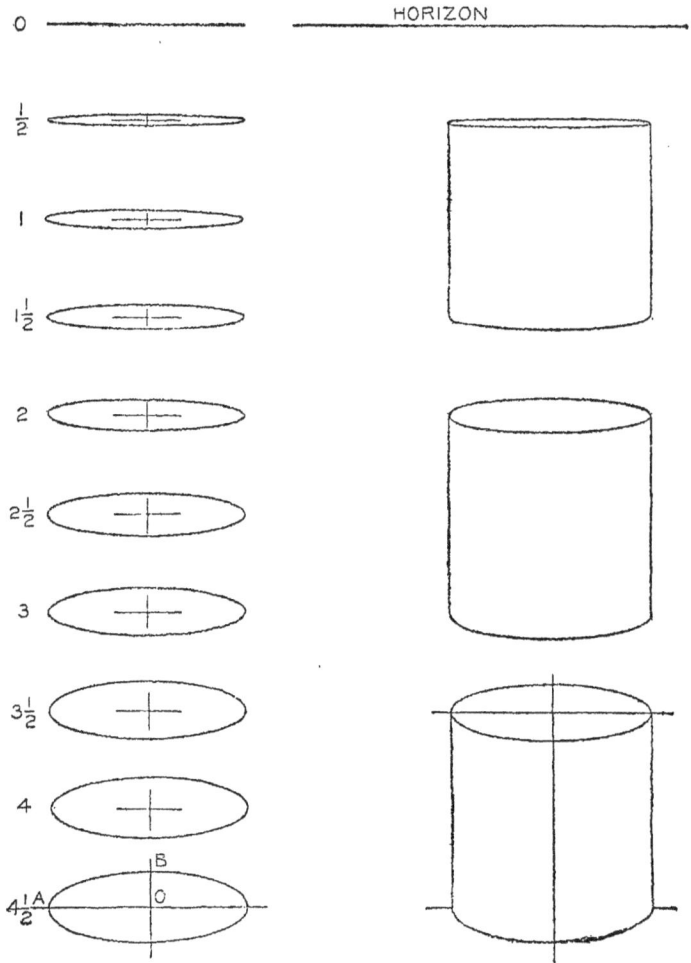

FIG. 39. SCALE OF LEVELS

each circle, Fig. 39, is indicated by the numerals at the left of the figure. In this course the student will be aided in determining the level for the perspective of a cylindrical object by referring to this scale. To the right of the scale is shown its application in representing a cylinder at different levels.

To Draw and Test an Ellipse Representing a One-Inch Circle.

1. Draw light indefinite lines at right angles to each other to represent the axes of the ellipse.

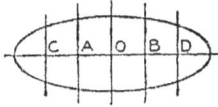

FIG. 40.   TESTING AN ELLIPSE

2. On the line representing the major axis lay off on either side of the intersection of the axes one-half the diameter of the circle.

3. Refer to the scale of levels; estimate and lay off the minor axis.

4. Sketch the ellipse lightly and freely, drawing corresponding parts in consecutive order, *i.e.*, draw the long sides of the

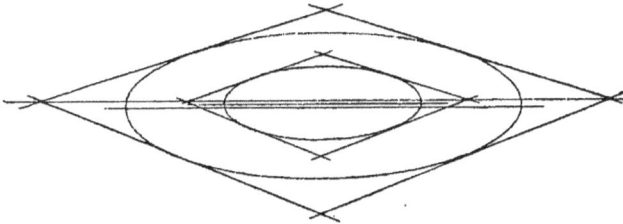

FIG. 41.   CONCENTRIC CIRCLES IN PERSPECTIVE

ellipse and then the ends. Compare the form thus secured with the corresponding ellipse in the scale of levels. Care should be taken to avoid sharp or blunt ends.

5. Ordinary defects in the form of the ellipse should be detected by examining it as follows:

a. Turn the sheet to the right, to the left, and upside down, and view the form carefully when the sheet is in each of these positions.

b. Locate two points as A and B, Fig. 40, on the axis equidistant from O. The vertical distances from these points to the ellipse should be equal. Compare these dis-

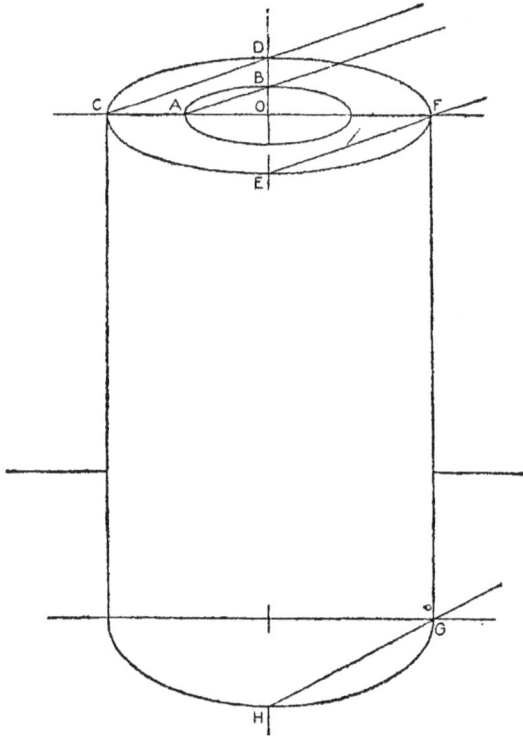

FIG. 42. SPLIT CORE BOX—AXIS VERTICAL

tances and make the necessary corrections. Likewise locate C and D equidistant from O and compare the vertical distances from these points to the ellipse. Make the necessary corrections as before.

CONCENTRIC CIRCLES IN PERSPECTIVE

The problem of drawing two concentric ellipses is more difficult than that of drawing a single ellipse.

Fig. 41 shows two concentric ellipses inscribed in concentric squares shown in perspective. The ellipses therefore represent circles. On account of foreshortening, the axes of the ellipses do not coincide with the line representing the diameter of the circles or with each other. In most cases the difference is so slight that it may be ignored. For very large ellipses, however, the construction shown in Fig. 41, where the major axis of the larger ellipse is

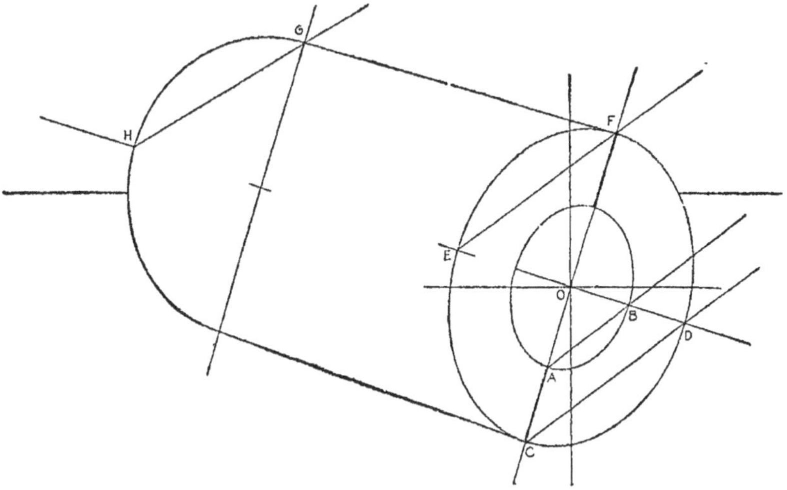

FIG. 43. SPLIT CORE BOX—AXIS HORIZONTAL

slightly in front of the major axis of the smaller ellipse, must be used.

In Figs. 42 and 43 the major axis C F is laid off equal to the diameter of the circle, as in the case of the ellipse representing a one-inch circle. Fig. 38. A one-inch ellipse should be drawn first and tested. In cases where the major axes of the ellipses are made to coincide, the half length of the minor axis of a larger or smaller ellipse may be determined as shown in Figs. 42 and 43. C D is drawn through C parallel to A B. In cases where the axes do not coincide the line corresponding to C D should be made to converge slightly with A B.

DATA FOR DRAWING PLATE 4

Given: The dimensions of a split core box for standard one-inch cores, which consists of a hollow cylinder split into halves; outside diameter 2″, inside diameter 1″, length 3″.

Required: A perspective sketch of the split core box with its axis vertical and the upper base 3″ below the level of the eye, Fig. 42, or any similar object assigned by the instructor.

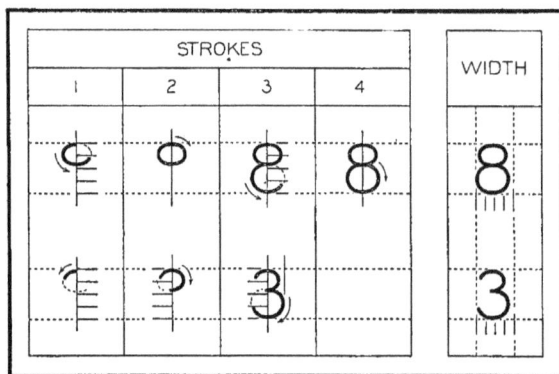

FIG. 44. LETTERING PLATE 4

Instructions:

1. Draw through the center of the sheet a vertical line to represent the axis of the cylinder. *All lines should now be drawn freehand.*

2. Through points 1½″ above and below the center of the sheet draw horizontal lines. as the major axes of the ellipses representing the ends of the bushing. The minor axes will coincide with the axis of the cylinder. Care should be taken to make the angle between these axes a right angle.

3. Draw the ellipse representing the smaller circle in the upper end of the core box at the required level. Refer to the *scale of levels* to estimate the major and minor axes of the ellipse. Draw the ellipse with these axes and test it as described under, "To Draw and Test an Ellipse," page 47.

4. Lay off the major axis and determine the length of the minor axis of the larger ellipse as described under, "Concentric Circles in Perspective," page 48.

5. Only one-half of the larger ellipse representing the lower end of the bushing will be seen. The length of the minor axis of this ellipse may be found by the method illustrated in Fig. 42. H G is drawn through G converging with E F. While only the lower half of the ellipse will be needed, the complete ellipse should be drawn as construction.

6. Complete the constructive stage of the sketch by drawing the vertical contour elements of the cylinder which join the ends of the major axes of the large ellipses.

7. Erase all construction lines and complete the sketch in the usual manner.

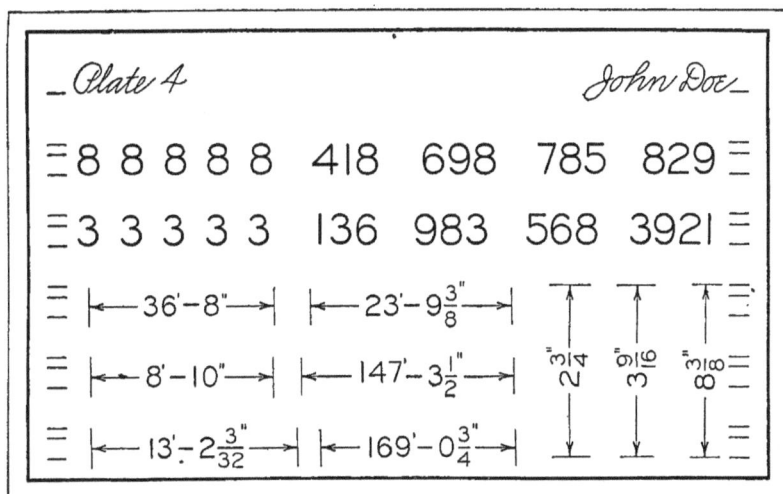

FIG. 45. LETTERING PLATE 4.   8, 3

PREPARATORY INSTRUCTIONS FOR LETTERING PLATE 4

The combination of ovals in the 8 serves as a basic form for the 3. In making the curved strokes of these numerals the student should have in mind the form of the complete oval.

DATA FOR LETTERING PLATE 4

Given: Plate 4 to reduced size. Fig. 45.

Required: To make the plate to an enlarged scale.

DATA FOR EXTRA DRAWING PLATE

Given: The dimensions of a picture frame as shown in Fig. 46.

Required: To draw the picture frame as though it were lying on a table, with its upper surface 4″ below the level of the eye. At this level a portion of the bottom of the hole will be visible.

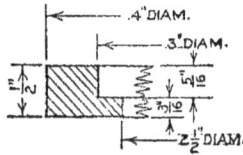

FIG. 46. PICTURE FRAME SECTION

PREPARATORY INSTRUCTIONS FOR DRAWING PLATE 5

The Horizontal Measure Cylinder. Fig. 47 shows a horizontal cylinder inscribed in a measure cube. This cylinder is therefore one inch in diameter and one inch long. Due to foreshortening,

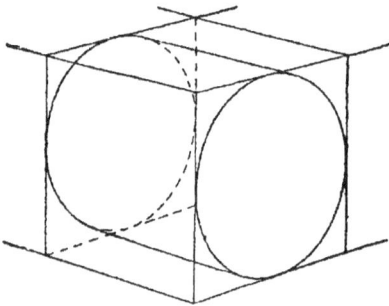

FIG. 47. HORIZONTAL CYLINDER IN-
SCRIBED IN A MEASURE CUBE

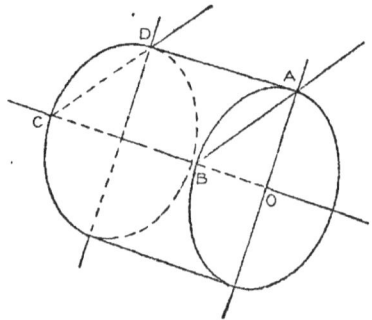

FIG. 48. HORIZONTAL MEASURE
CYLINDER

the major axis of the nearer base is slightly less than one inch, and the axis of the cylinder is shorter than the line representing the horizontal edge of the measure cube.

These differences are so slight that they will be disregarded in the following analysis of the cylinder, which will hereafter be referred to as the *horizontal measure cylinder.* Fig. 48.

Figs. 49 and 50 are similar to Figs. 47 and 48, respectively, but show a horizontal cylinder at a different level, with its axis receding to the right instead of to the left.

1. Since the axis of a horizontal cylinder in 45° perspective always extends toward a vanishing point, the inclination of the axis indicates the level at which the cylinder is drawn. Figs. 48 and 50.

2. The major axis of the bases are *perpendicular to* and the minor axis *coincident with*, the axis of the cylinder as in the vertical measure cylinder.

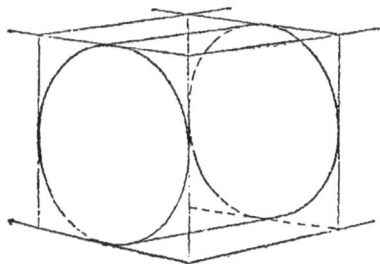

FIG. 49. HORIZONTAL CYLINDER INSCRIBED IN A MEASURE CUBE

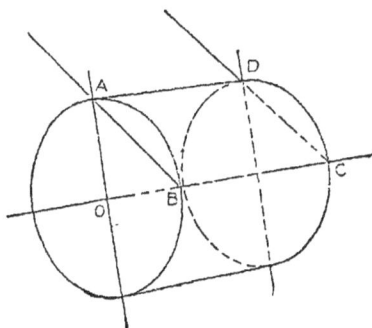

FIG. 50. HORIZONTAL MEASURE CYLINDER

3. The major axis of the nearer base is equal in length to the diameter of the cylinder, or one inch. The major axis of the farther base is shorter, on account of the convergence of the contour elements of the cylinder.

4. The distance between the centers of the bases is equal to the horizontal receding edge of the measure cube, which is approximately three-fourths of the front vertical edge of the cube. Since the major axis of the near base is drawn equal to the front vertical edge of the measure cube, or one inch, the distance between the centers of the bases may be taken as three-fourths the length of the major axis of the near base. This ratio remains constant for all ordinary levels.

5. It will be noticed in Figs. 48 and 50 that the minor axes of the nearer bases are practically equal to the distance between the centers of the bases or three-fourths of the major.axis of the

nearer base. When the nearer ellipse is drawn, the half length of the minor axis of the farther ellipse may be determined by drawing a line C D through D, converging with A B. Fig. 48. These lines do not converge toward a point on the horizon line.

DATA FOR DRAWING PLATE 5

Given: The Split Core Box represented in Plate 4. Fig. 43.

Required: To draw the Core Box in a horizontal position with its axis $4\frac{1}{2}''$ below the level of the eye, or any similar object assigned by the instructor.

Instructions:

1. Draw through the center of the sheet a line in the direction of one of the vanishing points, to represent the axis of the cylinder at the required level. The angle of inclination may be obtained from Fig. 22.

2. Refer to Fig. 23, estimate, and lay off three foreshortened inches on the axis. One-half of this length should fall on either side of the center of the sheet to locate the drawing centrally on the sheet.

3. Draw through the points thus determined the major axes of the bases at right angles to the axis of the cylinder.

4. Draw the ellipse representing the nearer end of the Core Box as shown in Fig. 43. C F is made equal in length to the outside diameter of the Core Box, $2''$. D is determined by drawing C D through C parallel to A B. Test each ellipse as described under, "To Draw and Test an Ellipse," page 47.

5. From the ends of the major axis of the larger ellipse draw contour elements converging with the axis of the cylinder to determine the ends of the major axis of the farther base.

6. The half length of the minor axis of the farther base may be determined as shown in Fig. 43. H G is drawn through G converging with E F. E F and H G do not converge toward a point on the horizon line. While only one-half of the farther ellipse will show in the finished drawing, a better result will be obtained by drawing the complete ellipse as construction.

7. Erase all construction lines, including the axes of the ellipses, and finish the sketch in the usual manner.

PREPARATORY INSTRUCTIONS FOR LETTERING PLATE 5

Horizontal and Vertical Strokes. As stated under Plate 1 vertical strokes are usually made downward and horizontal strokes to the right.

HIT

FIG. 51. SPACING OF ADJACENT VERTICAL STEMS

The direction of horizontal and vertical strokes must be exact. The relative width of the letters is shown in column 4. Fig. 52.

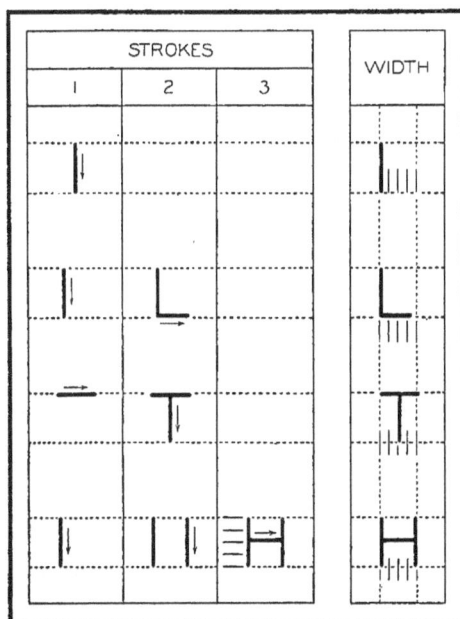

FIG. 52. LETTERING PLATE

Spacing. In this and the following plates practice in making individual letters will be followed by practice in making words. In order that the lettering may present a good appearance, it is as important that the letters be well spaced as that they be properly formed.

Correct spacing depends more on the judgment of the draftsman than on any rule which might be given. However, it may be said that as a rule the letters should *appear* to be equally spaced.

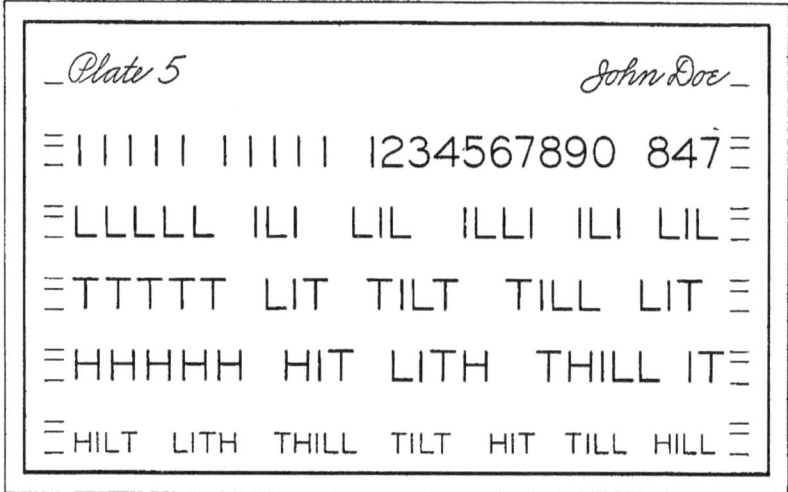

FIG. 53. LETTERING PLATE 5. I, L, T, H

For this style of letters adjacent vertical strokes should be a distance apart equal to one-half the width of the H. Example: H and I, Fig. 51. Letters of irregular form should be placed at such a distance that the space appears equal to that between the H and I. Example: I and T, Fig. 51.

When spacing a letter the beginning of the first stroke should be carefully located.

DATA FOR LETTERING PLATE 5

Given: Plate 5 to reduced size. Fig. 53.
Required: To make the plate to an enlarged scale.

DATA FOR EXTRA DRAWING PLATE

Given: The dimension of a picture frame as shown in Fig. 46.

Required: To draw the picture frame as though it were hanging flat against a vertical wall, with its center 3″ below the level of the eye.

EXTENSION OF PERSPECTIVE THEORY

PREPARATORY INSTRUCTIONS FOR EXTRA DRAWING PLATES

The Measure Cube in New Positions. In the preceding plates the measure cube was drawn at different levels, but always with its side faces at 45° with the horizon.

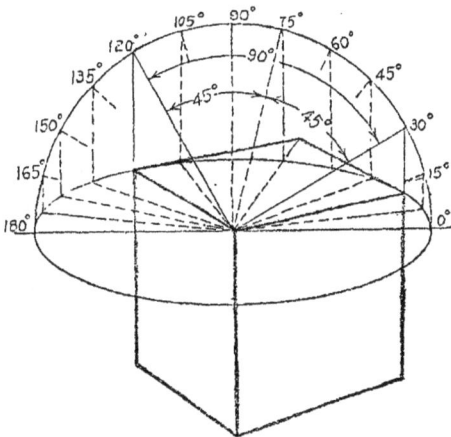

FIG. 54. DRAWING A CUBE AT ANY ANGLE

If the measure cube is turned with its side faces making other angles with the horizon, the number of positions in which an object may be drawn will be increased.

Fig. 54 shows a method of constructing a measure cube at any level and with its side faces at any desired angle. The steps in the construction are as follows:

1. Draw an ellipse representing a two-inch circle at the required level.

2. Draw a semi-circle of the same diameter as the circle represented by the ellipse, with its center at the center of the ellipse.

3. Mark off on the semi-circle the angles which the faces of the cube are to make with the horizon. These angles should be 90° apart.

4. Vertical lines through these points intercept the ellipse in the ends of the nearer edges of the upper face of the cube. These edges meet at the center of the ellipse which is the upper front corner of the cube.

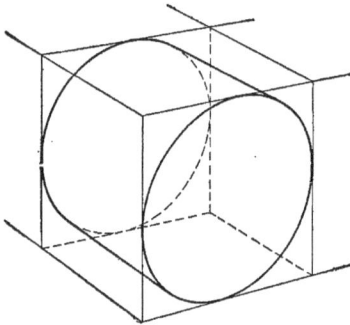

FIG. 55. CYLINDER INSCRIBED IN A   FIG. 56. MEASURE CYLINDER
MEASURE CUBE

5. Make the front vertical edge one inch long as in 45° perspective.

6. Complete the cube by drawing the remaining edges converging so as to give the faces of the cube the appearance of squares. It will be noted that the farther edges of the upper face intersect on the line making 45° with each of the side faces.

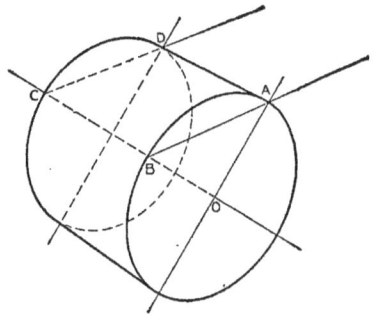

The Measure Cylinder in New Positions. Fig. 55 shows a horizontal cylinder inscribed in a measure cube with its side faces at other than 45° to the horizon. This cylinder is therefore one inch in diameter and one inch long. Due to foreshortening, the major axis of the nearer base is slightly less than one inch and the axis of the cylinder is shorter than the line representing the horizontal edge of the measure cube. These differences are so slight that they will be disregarded in the following analysis of the measure cylinder. Fig. 56.

1. The distance between the centers of the bases is equal to the horizontal receding edge of the measure cube. This distance will be shorter as the angle of the axis of the cylinder to the horizon increases.

2. The major axes of the bases are perpendicular to, and the minor axis coincident with, the axis of the cylinder.

3. The major axis of the nearer base is equal in length to the diameter of the cylinder, or one inch. The major axis of the farther base is shorter on account of the convergence of the contour elements of the cylinder.

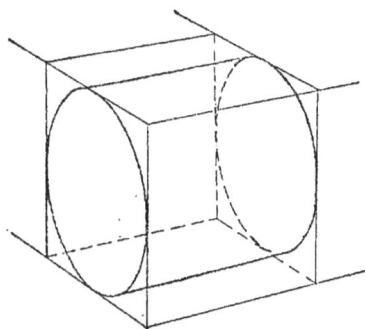

FIG. 57. CYLINDER INSCRIBED IN A   FIG. 58. MEASURE CYLINDER
MEASURE CUBE

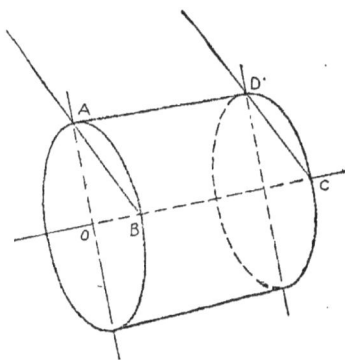

4. The length of the minor axis of the nearer base will depend upon the angle that the axis of the cylinder makes with the horizon. Fig. 56 illustrates the case in which the minor axis is lengthened, due to the axis of the cylinder making an angle greater than 45° with the horizon. Fig. 58 illustrates the case in which the minor axis is shortened, due to the axis of the cylinder making an angle less than 45° with the horizon. The length of the minor axis for other positions may be estimated by using Figs. 56 and 58 as guides. For any angle the axis of the cylinder makes with the horizon the length of the minor axis will remain the same for all levels. When the nearer ellipse is drawn, the half length of the minor axis of the farther base may be determined by drawing a line C D through D, converging with A B. Figs. 56 and 58.

DATA FOR EXTRA DRAWING PLATE

Given: The objects shown in Figs. 59, 60, 61, and 62.

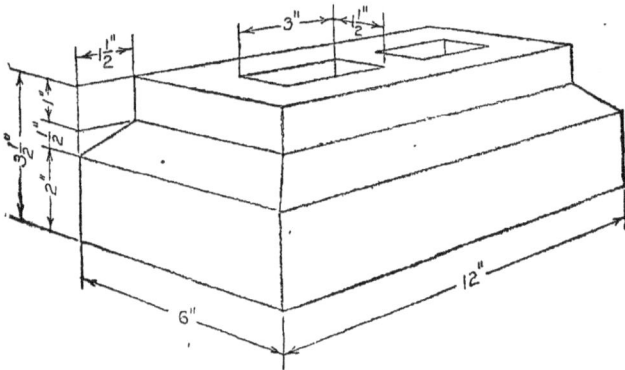

FIG. 59. CONCRETE BLOCK

Required: To draw one or more of the above objects in positions selected from the following table by the instructor.

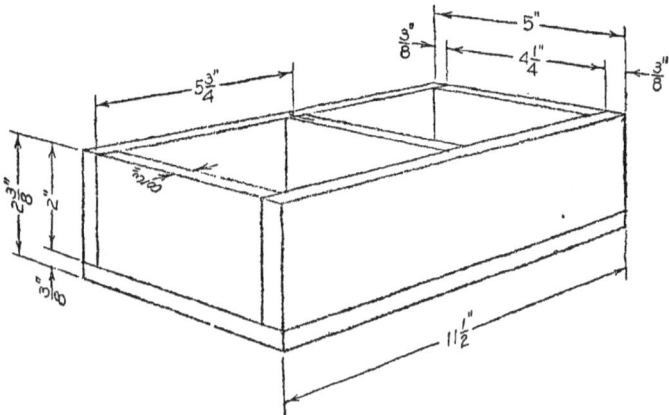

FIG. 60. NAIL BOX

The level at which the object is drawn may be assumed by the student.

The right vertical face of the enclosing solid makes one of the following angles with the horizon:

1. 15°.   2. 30°.   3. 60°.   4. 75°.

The objects should be centered on the sheet as in previous problems.

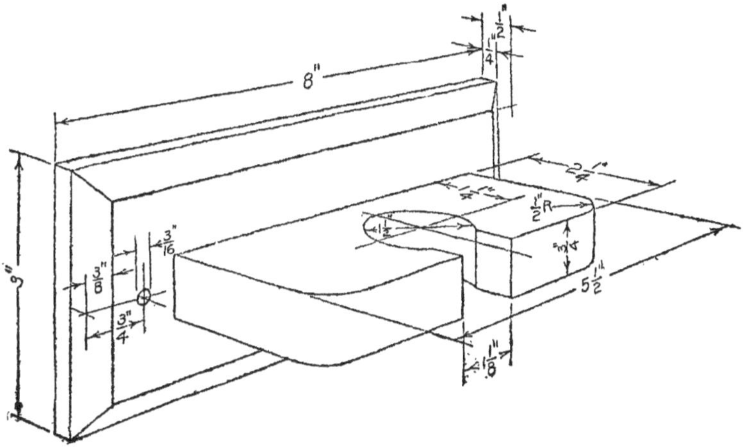

Fig. 61. Broom Holder

REVIEW QUESTIONS

1. (a) What is the horizon?   (b) How is it represented? (c) What is its relation to the eye?

2. (a) What is a vanishing point?   (b) Where is it located?

3. Where do parallel horizontal lines appear to meet in perspective?

4. Do vertical lines appear to converge in perspective?

5. (a) What is meant by foreshortening?   (b) Are the perspectives of equal lengths on the same vertical edge equal? (c) On the same horizontal edge?   (d) Are the perspectives of equal vertical lengths at different distances from the observer equal?

6. (a) What is the angle of inclination?   (b) How does it vary?

7. (a) In what position on the drawing board is the paper fastened? (b) How is it fastened?

8. Describe in detail how the pencil should be sharpened for sketching.

9. (a) What is the position of the hand and pencil in sketching horizontal lines? (b) Vertical lines? (c) What is the

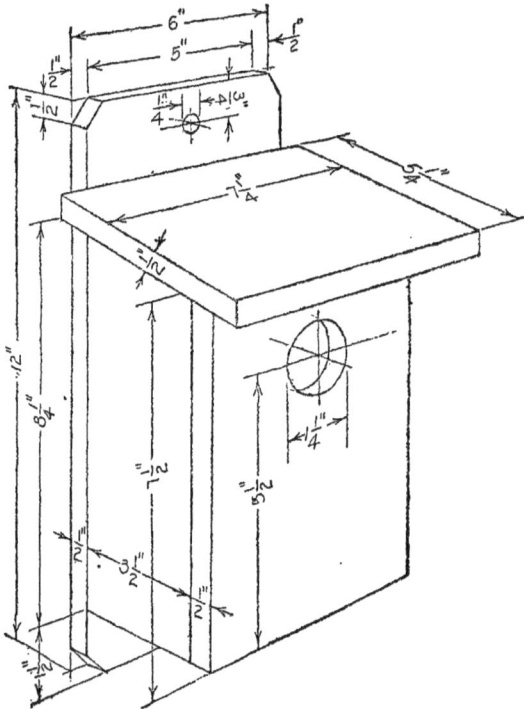

FIG. 62. BIRD HOUSE (DIMENSIONED PERSPECTIVE)

essential difference? (d) What movements are made to produce the line?

10. (a) What is meant by constructive stage? (b) Finishing stage?

11 In what way does a scale of levels assist in making a perspective of a rectangular object?

12. (a) Where are all vertical measurements laid off in perspective? (b) Why?

13. How are horizontal measurements made?

14. Explain what is meant by enclosing solid.

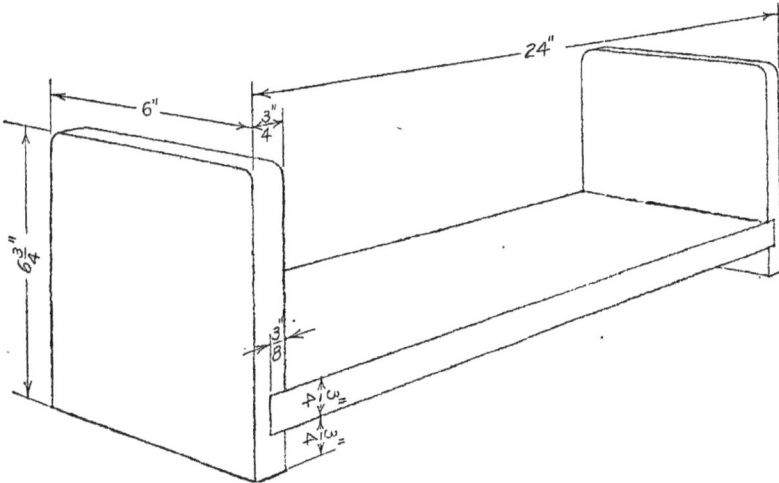

FIG. 63. BOOK RACK (DIMENSIONED PERSPECTIVE)

15. (a) What is a measure cube? (b) Why is it called a measure cube?

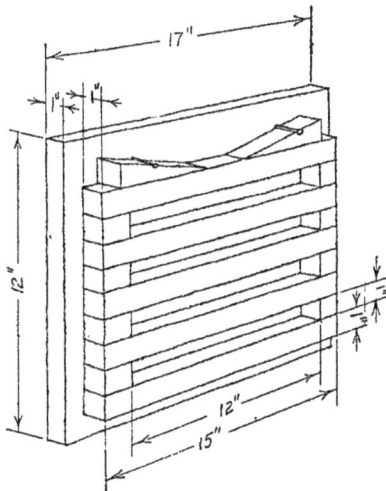

FIG. 64. FORM FOR TESTING CONCRETE PRISMS

16. Of what use is the table line?

17. (a) How do you proceed to locate the drawing centrally on the sheet?

18. How are the perspectives of the inclined lines located?

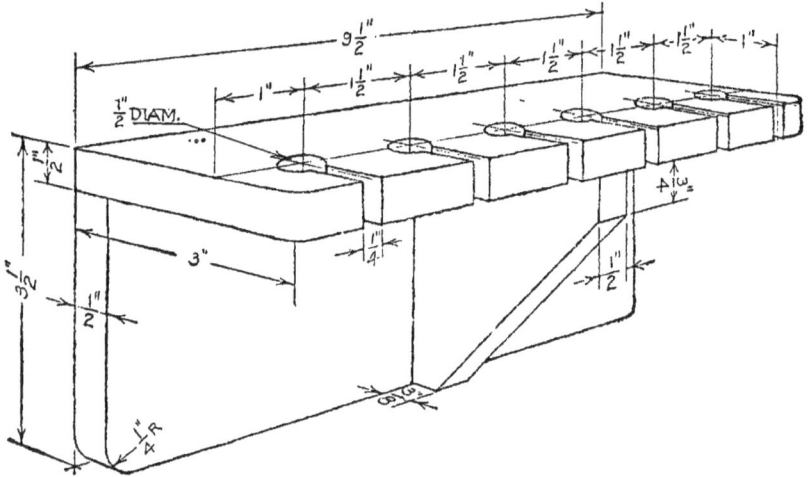

FIG. 65.   TOOTH BRUSH RACK

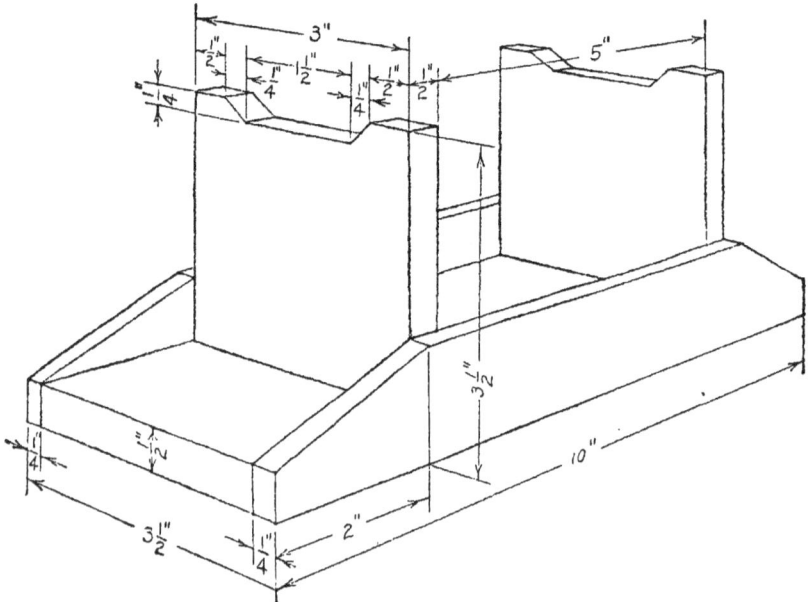

FIG. 66.   PEN RACK

19. (a) Give the proportions of the vertical measure cylinder. (b) The major axes of the bases are at what angle with the axis of the cylinder?

20. How does the difference in level affect the appearance of a horizontal circle in perspective?

21. Of what assistance is a scale of levels in drawing a vertical cylinder?

22. How is the ratio of the minor axes of two ellipses representing concentric circles determined?

23. (a) Give the proportions of a horizontal measure cylinder. (b) The major axes of the bases are at what angle with the axis of the cylinder? (c) What is the relative length of the major and minor axes of the nearer base?

DATA FOR REVIEW DRAWING PROBLEMS

Given: The objects shown in Figs. 63, 64, 65, 66.

Required: To draw one or more of the above objects in 45° perspective. The level at which the object is drawn may be assumed by the student.

CHAPTER II

ORTHOGRAPHIC SKETCHING

PROSPECTUS

In this chapter the work of the preceding chapter will be continued in order that the value of the perspective sketch as an aid in interpreting orthographic views may be apparent. At the same time more general application will be made of perspective principles and additional skill acquired in representing objects pictorially.

It is the chief aim of this chapter to familiarize the student with the method of representation generally used in working drawings. By the time the work of this chapter is finished the student should be able to read drawings of ordinary complexity as well as to make freehand orthographic sketches with a considerable degree of skill and confidence.

PREPARATORY INSTRUCTION FOR DRAWING PLATE 6

Views. In perspective sketching the object is viewed from one position, so chosen as to show its three general dimensions in one view. Such a means of representation does not show the principal surfaces of an object in their true form and proportion or the principal edges in their true lengths.

In order to represent the principal surfaces of an object in their true form and proportion and the principal edges in their true length, the object is usually viewed in two or more directions, viz.: from directly in front, directly above, or directly from the right or left. Each view thus secured will give the exact form and proportion of the surfaces and the true lengths of the edges toward which one is looking perpendicularly. Views thus secured are known as orthographic. In mechanical drawing orthographic views are generally used.

Fig. 69 shows two views of a bench stop. The view marked
TOP represents orthographically what is seen from directly
above the object and the view marked FRONT represents what is
seen from directly in front of the object. The top view shows
two general dimensions in horizontal directions, viz.: the dimension from left to right and the one from front to back. The front

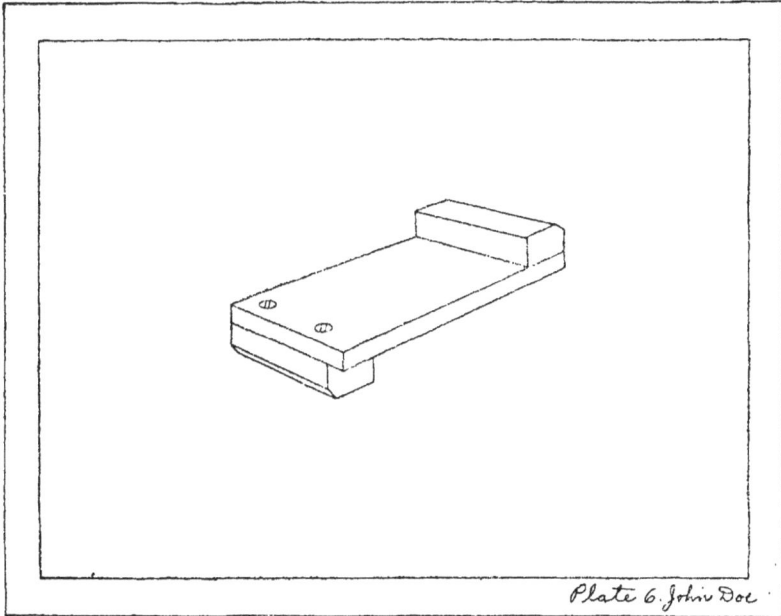

FIG. 67. TYPE PROBLEM. PERSPECTIVE OF BENCH STOP

view shows the horizontal dimension from left to right and the
vertical dimension. Thus the three general dimensions are given
in the two views and the proportions of the object are determined.
Relation of Top and Front Views. It should be clear from
the above statement that one of the general dimensions, viz.:
the horizontal dimension from left to right, is common to the
front and the top views. For this reason as a matter of convenience in making and interpreting the drawing it is essential
that the top view always be placed directly above the front view.

Under this condition all distances from left to right may be projected from one view to the other.

"Reading" the Drawing. To form a mental image of an object the relation of its surfaces, edges, and corners as represented must be studied. This process is called *reading the drawing* and is illustrated under the four following headings: (The present discussion is confined to rectangular solids.)

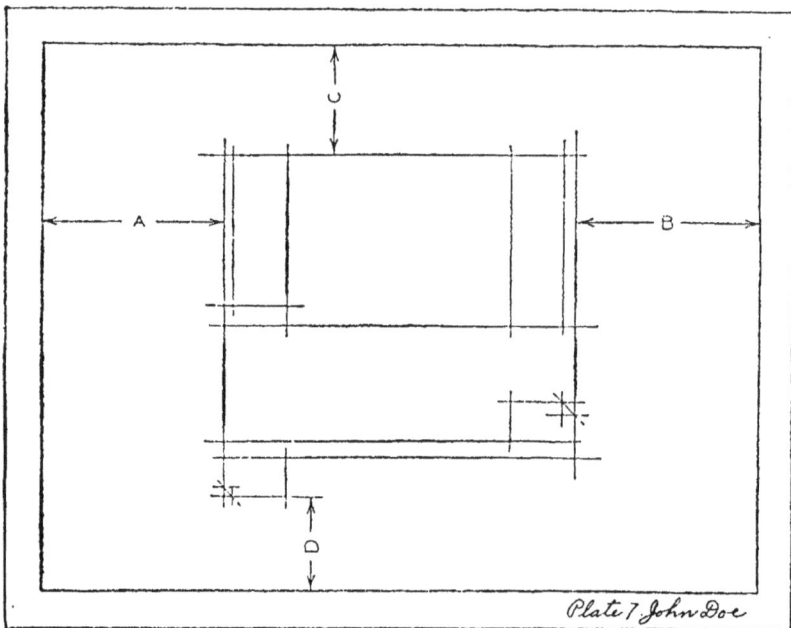

FIG. 68. TYPE PROBLEM. CONSTRUCTIVE STAGE OF THE ORTHOGRAPHIC
SKETCH

Plane Surfaces. Fig. 69 represents an object having plane surfaces.

1. When the observer is looking perpendicularly at a surface it appears in its true form and proportion. Example: The rectangular top surface A B C D of the bench stop, Fig. 69, is represented in its true form and proportion in the top view.

2. When the observer is looking edgewise at a plane surface it appears as a straight line. Example: Line E F is the front view of the top surface A B C D. Fig. 69.

Straight Edges.

1. A straight edge viewed at right angles to its length shows as a line in its true length. Example: The front edge of the top surface of the bench stop shows in its true length in line A B in the top view and in line E F in the front view.

2. A straight edge viewed endwise appears as a point. Example: Point F is the front view of the edge B C. Fig. 69.

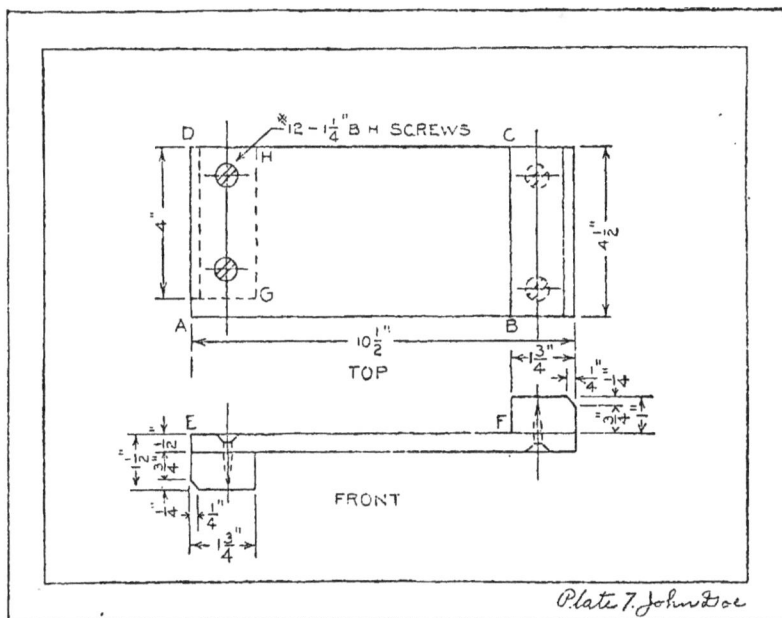

FIG. 69. TYPE PROBLEM. FINISHED SKETCH OF BENCH STOP

Corners. A corner appears as a point when viewed from any direction. Example: The upper front corners at the left of the bench stop are represented by A in the top view and E in the front view.

Invisible Edges. Hidden edges or hidden surfaces viewed edgewise are represented by dotted lines to distinguish them from visible edges or surfaces. Example: G H in the top view. Fig. 69.

The student should test his knowledge of the orthographic principles just stated by answering the following questions: See Fig. 70.

1. (a) Where is the front view of the horizontal surface 9, 10, 15, 16? (b) Of 10, 12, 13, 15? (c) Of 9, 11, 14, 16?

2. (a) Where is the top view of the horizontal surface 5, 4? (b) Of 8, 1?

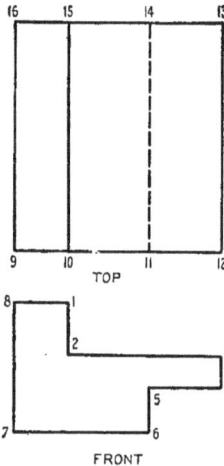

FIG. 70. REVIEW PROBLEM

3. (a) Where is the top view of the front vertical surface 1, 2, 3, 4, 5, 6, 7, 8? (b) Of the rear vertical surface 1, 2, 3, 4, 5, 6, 7, 8?

4. (a) Where is the top view of the vertical surface 7, 8? (b) Of 3, 4? (c) Of 5, 6?

5. (a) Where is the top view of the front horizontal edge 2, 3? (b) Of 7, 6?

6. (a) Where is the front view of the rear horizontal edge 15, 13? (b) Of 16, 14?

7. (a) Where is the front view of the upper horizontal edge 10, 15? (b) Of 12, 13?

8. (a) Where is the top view of the edge 5? (b) Of 6?

9. (a) Where is the front view of the upper front corner 12? (b) Of 9? (c) Of the upper back corner 15?

10. (a) Where is the top view of the front corner 2? (b) Of 5?

The Type Problem. In each of the following problems presented for solution the methods to be employed and the results to be obtained will be illustrated by a *type problem.* This type problem will consist of two parts:

1. A drawing of an object similar to, and represented in the same manner as, the one given for solution.

2. A solution of the problem corresponding to that required of the student.

Example: Fig. 69 is the type problem for the first orthographic sketch. Fig. 67 is the perspective of the bench hook shown in Fig. 69 and corresponds to the kind of a drawing the student will make from Fig. 71, 72, or 73.

Materials. The materials used for the plates in this chapter are the same as those used in perspective sketching (see page 15) except that in this case the 5H pencil will be used for both the constructive and finishing stages.

Perspective Sketches. In this chapter perspective sketches will be drawn preceding the orthographic sketches as a means of interpreting the orthographic views and at the same time to continue the practice necessary to develop skill in representing objects in perspective.

DATA FOR DRAWING PLATE 6

Given: An orthographic sketch, Fig. 71, 72, or 73.

Required: To draw a 45° perspective sketch of the object shown in Fig. 71, 72, or 73 as assigned by the instructor, with the upper front corner of the enclosing solid 3½″ below the level of the eye or any similar object assigned by the instructor.

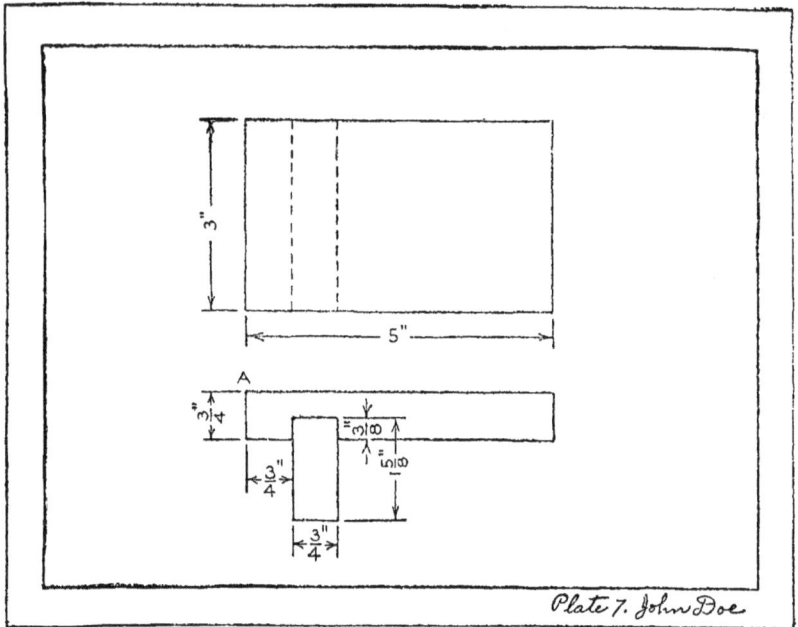

Plate 7. John Doe

FIG. 71.   GAIN JOINT

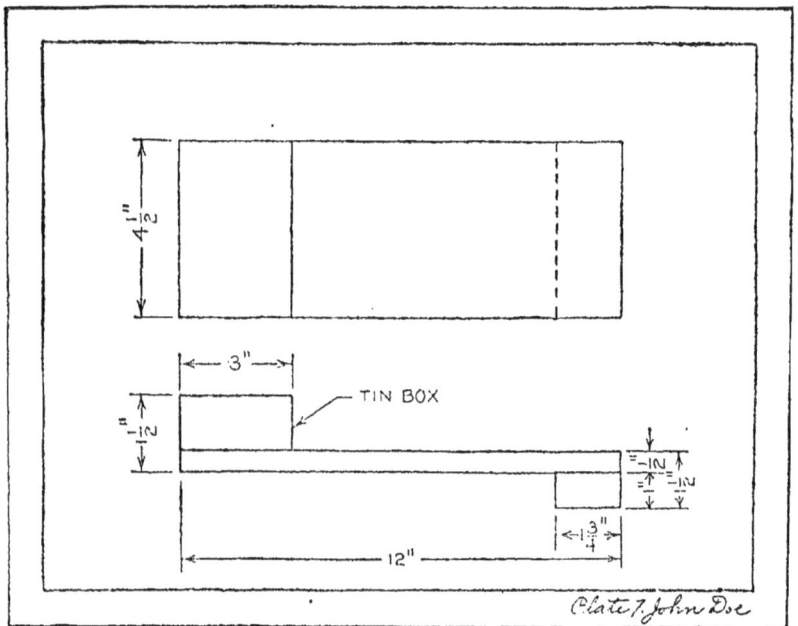

TIN BOX

Plate 7. John Doe

(72)

FIG. 72.   SCOURING BOARD

Instructions:

Use the corner marked A as the upper front corner of the object. All lines of this drawing are to be made *freehand*, including the light lines in the constructive stage. Omit all dimensions.

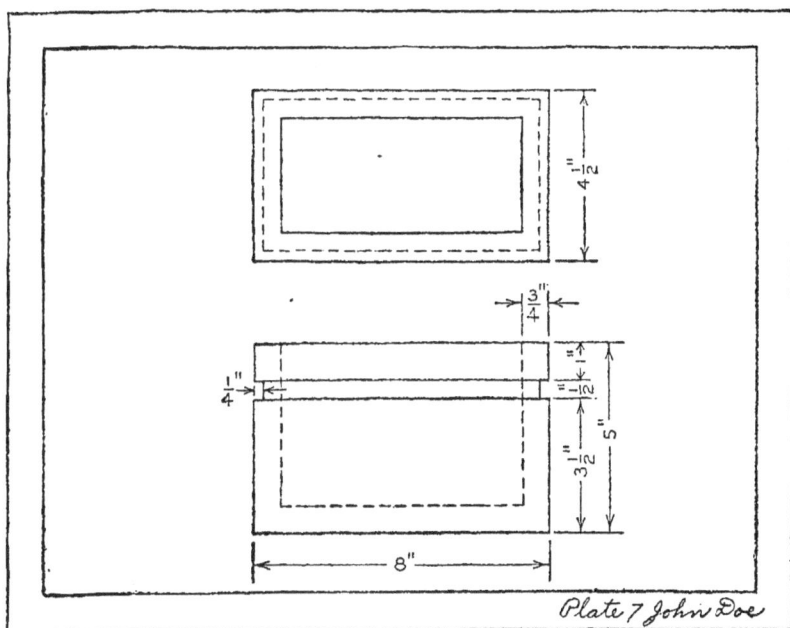

FIG. 73. CEMENT FERN JAR

PREPARATORY INSTRUCTIONS FOR LETTERING PLATE 6

Inclined Strokes. Before starting an inclined stroke, the student should sense its direction, moving the pencil between its two ends without touching the paper.

DATA FOR LETTERING PLATE 6

Given: Plate 6 to reduced size. Fig. 75.
Required: To make the plate to an enlarged scale.

PREPARATORY INSTRUCTIONS FOR DRAWING PLATE 7

The Constructive Stage. This stage in orthographic sketch-ing is similar to the constructive stage in Perspective Sketching described on page 18. It consists of drawing all lines of the sketch lightly and full. No attempt should be made to make the lines exactly the right length in this stage. When drawing

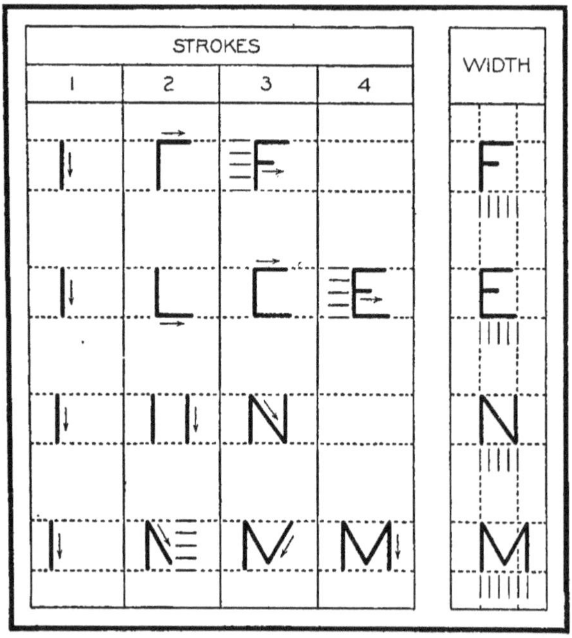

FIG. 74. LETTERING PLATE

each line it should be made long enough to give all necessary intersections with other lines.

By this time the student should have gained such facility in drawing freehand lines that he will not need the rule to draw straight lines.

If a straightedge is used at all in this chapter it should be necessary only in ruling long lines in the constructive stage and in locating one view directly opposite another.

In laying out the views of an orthographic sketch on the sheet, proceed in the following manner:

1. Referring to Fig. 68, mark off tentatively the position of the extreme right and left of each view. Shift both marks to the right or left, if necessary, to make A equal B.

2. In like manner mark off the vertical dimensions of each view, leaving a space between the two views proportional to that which is shown in the figure. This distance should be from $\frac{3}{4}''$

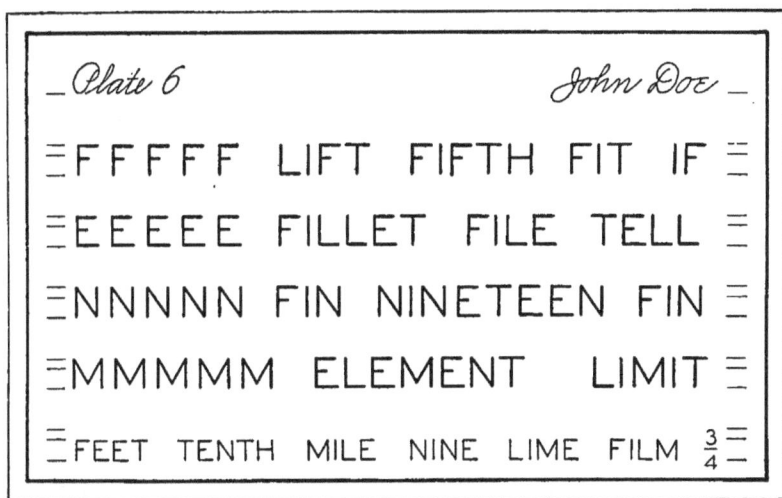

FIG. 75. LETTERING PLATE 6

to 1". Shift all marks up or down if necessary to make C equal D.

3. Make any necessary adjustments in the general proportions of the views.

4. In proportioning the details of the views, a comparison of the dimensions of each detail with the dimensions of the views in which it appears will aid in securing good results. Example: In the front view, Fig. 69, the width of the cleats is about one-sixth of the total length of the bench hook and their thickness is twice that of the board to which they are fastened.

Finishing Stage. As in Perspective Sketching, the finishing stage consists in erasing unnecessary lines made in the constructive stage, tracing over the outline of the drawing, and otherwise giving it a finished appearance. The student should proceed as follows:

1. Erase all construction lines and retrace the lines of the drawing, using a 5H pencil.

2. Represent all invisible edges by dotted lines which are composed of $\frac{1}{8}''$ dashes with $\frac{1}{32}''$ spaces between them. The ends of the dashes should be made definite by placing the pencil on the

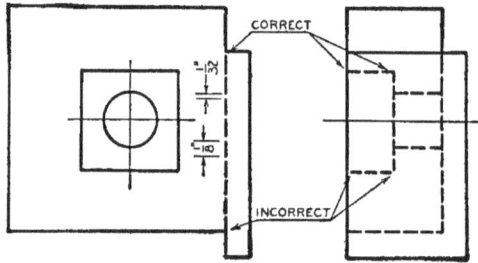

Fig. 76. Dotted Lines

paper, moving it the required length, and then removing it as nearly as possible vertically from the paper. Fig. 76 shows the correct method of joining dotted lines to full lines.

3. Place the dimensions on the sketch as described below under, "Arrangement of Dimensions."

Arrangement of Dimensions. Dimensions are placed on a drawing as shown in Figs. 71 and 72 to show the size of the object represented Only those dimensions are given which are necessary to determine completely the size of the object.

An over-all dimension is one which shows the distance from one extreme point to another. Example: The five inch dimension in Fig. 71. A detail dimension is one which shows the distance between two points on some part or detail of the object. Example: The three and one-half or one inch dimension. Fig. 73.

When detail dimensions and a dimension representing their
sum are given, they should be grouped in parallel lines.   The
shorter dimension should be near the outline of the object to avoid
the confusion arising from the crossing of lines.   Example: Those
below the front view, Fig. 69, are properly arranged.
The Dimension Form.   Fig. 77 shows what is known as the
dimension form.   It includes all of the elements of the convention
used in indicating linear dimensions on a drawing.   The following
points should be noted:

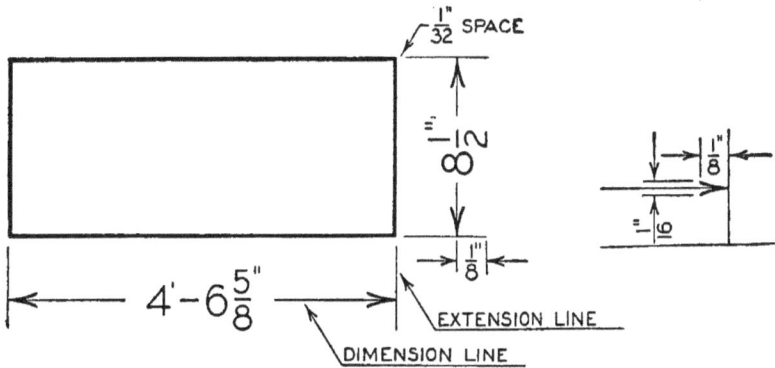

FIG. 77. DIMENSION FORM

1. Horizontal dimensions read from left to right.
2. Vertical dimensions read from the bottom toward the top
of the sheet.
3. Extension lines begin about $\frac{1}{32}''$ from the outline of the
object and continue $\frac{1}{8}''$ beyond the arrowhead.
4. The space between the outline of the object and the nearest
dimension line or between two parallel consecutive dimension
lines is about $\frac{1}{4}''$.
5. Arrowheads are placed on the dimension lines at their
extreme ends.
6. Arrowheads are composed of two slightly curved lines symmetrical with respect to the dimension line.   The length of the
arrowhead should be about $\frac{1}{8}''$ and the width $\frac{1}{16}''$.   Fig. 77.   The

strokes for arrowheads pointing in different directions are shown in Fig. 28.

7. The whole number in the dimension figure will be made $\frac{1}{8}''$ high.

8. The total height of the fraction in the dimension figure is twice that of the whole number with a clear space between each numeral and the division line.

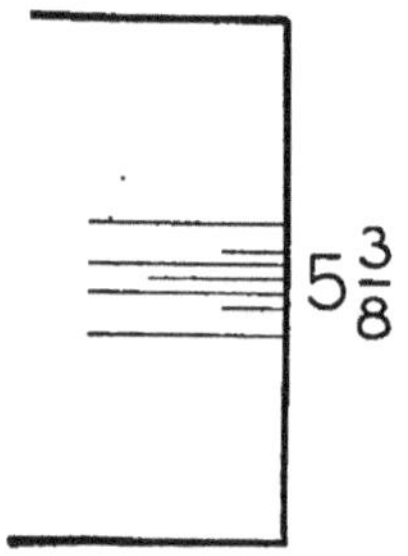

FIG. 78.　Showing Actual Heights of Whole Number and Fraction

To check these heights of numerals in a dimension figure, mark off an eighth-inch and a quarter-inch space on the edge of a card and use it as a scale. Fig. 78.

9. The dimension figure is generally located centrally in the dimension line, which is broken sufficiently to admit it.

DATA FOR DRAWING PLATE 7

Given: Orthographic sketches, Figs. 71, 72, and 73.

Required: To make an orthographic sketch of the object shown in Fig. 71, 72, or 73; or any similar object as assigned by the instructor, on a 9''x 12'' sheet.

Instructions:

1. Draw a border line as in perspective sketching.

2. The rectangles shown about the drawing in Figs. 71, 72, and 73 are proportional to the size of the 9''x 12'' sheet and the border rectangle. The over-all lengths of the view of the sketch

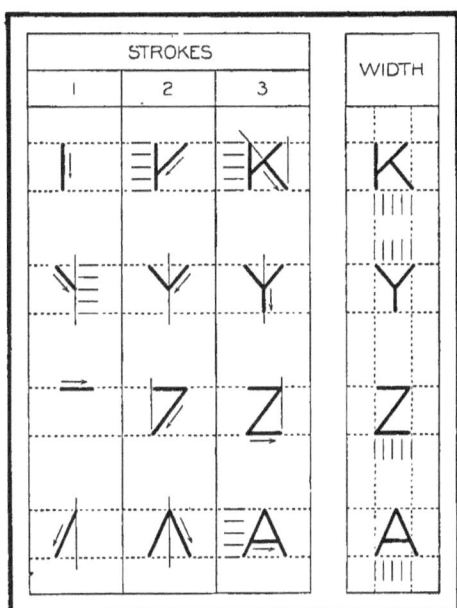

FIG. 79. LETTERING PLATE

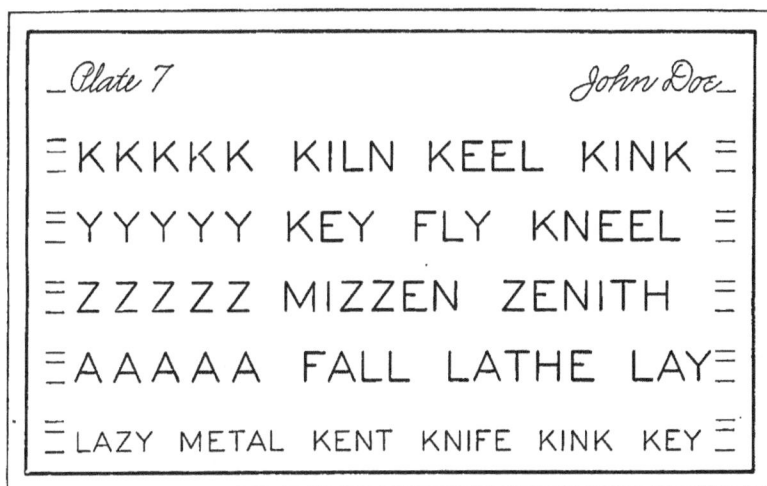

Plate 7   John Doe

≡KKKKK KILN KEEL KINK ≡
≡YYYYY KEY FLY KNEEL ≡
≡ZZZZZ MIZZEN ZENITH ≡
≡AAAAA FALL LATHE LAY≡
≡LAZY METAL KENT KNIFE KINK KEY≡

FIG. 80. LETTERING PLATE 7. K, Y, Z, A

should bear the same ratio to the dimensions of the sheet as the corresponding dimensions in the figure bear to the size of the rectangle representing the sheet. With this in mind proportion the views and locate them centrally on the sheet as previously explained.

3. Draw in the details and finish the drawing of the views as usual.

4. Draw in the extension lines, dimension lines, arrowheads, and numerals following the directions given under, "Arrangement of Dimensions," page 76.

PREPARATORY INSTRUCTIONS FOR LETTERING PLATE 7

As stated under Plate 6, the student should sense the direction of an inclined stroke before drawing it.

The spacing between irregular letters should appear equal to the area of one-half the H. Fig. 81.

KEY

FIG. 81. SPACING OF IRREGULAR FORMS

DATA FOR LETTERING PLATE 7

Given: Plate 7 to reduced size. Fig. 80.
Required: To make the plate to an enlarged scale.

PREPARATORY INSTRUCTIONS FOR DRAWING PLATE 8

It is customary to draw the *top* and *front* views of an object when these views will show the form and proportions satisfactorily and when only two views are needed. Some objects are of such a form, however, that a front view and a view from one side are needed to determine completely the form of the object. The particular side view is selected which will represent the object by the use of the least number of dotted lines. Fig. 82 represents an object which would be well defined by the use of a front view and one side view. The right side view would be the one chosen in this case. Since the two side views contain the same information, if one is given the other may be drawn. As previously explained, an observer sees all vertical dimensions and the horizontal dimension from left to right in the front view. All vertical dimensions and the horizontal dimensions from front to back are seen in the side view.

A right side view is always placed *directly to the right* and a left side view *directly to the left* of the front view. This is done both for the sake of convenience in making and reading a draw-

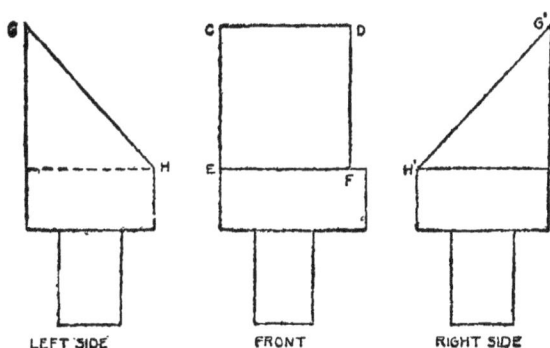

FIG. 82. BEVELER

ing and because an observer, when viewing an object, would, after obtaining the front view, naturally step to the right for a right side view or to the left for a left side view. To secure the front view after the side view is drawn, or to secure the side view after the front is drawn, all vertical distances may be projected from the first of the two views drawn. Fig. 82 shows the front view and the two side views of an object in their proper relative positions.

Inclined Surfaces. Any surface which is at right angles to the line of sight, when an object is being viewed, will show in its true form and size. A surface which makes other than a right angle with the line of sight is called an *inclined surface.* Such surfaces do not show in their true form and size. Fig. 82 represents an object having inclined surfaces. If one surface is rectangular and two of its edges are at right angles to the direction in which it is inclined, as in the case of the surface C D E F, Fig. 82, the vertical dimension of the rectangle representing the surface in the front view is less than the actual width of the surface.

The inclined surface C D E F is represented by the inclined line G H in the left side view. G H is equal to the true width of the surface. G'H', representing the same surface in the right side view, is also equal to the true width of the surface. It must be evident from a study of Fig. 82 that in representing any rectangular inclined surface which has two of its edges at right angles to the line of sight, the dimension represented by these

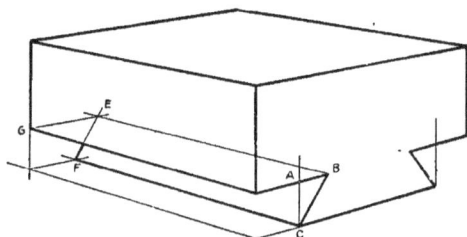

Fig. 83. Construction for Angles and Hidden Corners (Perspective)

edges will show in its true length. C D and E F, perpendicular to the direction of sight in the front view, show the true length of the rectangle in this view.

The end edges, G H and G'H' of the surface C D E F, are perpendicular to the direction of sight in the side views and therefore show the true width of the surface in these views.

Inclined Edges. A straight edge which is not at right angles to the direction in which it is viewed is represented by a line shorter than the actual length of the edge. Example: The end edges of the surface C D E F are represented in the front view, Fig. 82, by lines C E and D F. These lines are shorter than the actual lengths of the edges, as shown by lines G H and G'H' in the side views.

In sketching an angle where the direction of the edge is given by dimensions locating two points on the edge, the line representing the edge should be determined by laying off the dimensions given to locate the points on the line. Where the dimension is given in degrees the ends of the inclined lines should be located by estimating the lengths of the legs of the right triangle of which the inclined line is the hypotenuse. Example: The length of the lines A B and A C, Fig. 83, are laid off to determine the direction of B C. For a 45° angle A B and A C represent equal distances. For a 60° angle A B is roughly $\frac{9}{16}$ of A C.

In determining the position of a line passing through an invisible corner, such as E F, Fig. 83, make a construction for the invisible corner by drawing lines B E and E G.

The student should test his knowledge of the orthographic principles just stated by answering the following questions:

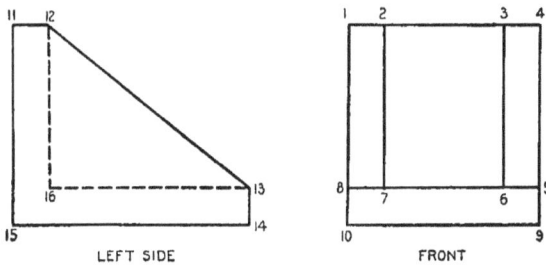

FIG. 84. REVIEW PROBLEM

PROBLEMS AND QUESTIONS IN ORTHOGRAPHIC PRINCIPLES
Refer to Fig. 84.

1. Where is the side view of the inclined surface 1, 2, 7, 8?
2. (a) Is line 1, 2 equal to the true width of the inclined surface? (b) Where is its true length shown? (c) Why?
3. Where is the inclined edge 1, 8 shown in its true length? Why?
4. (a) Is the vertical surface 11, 15 on the front or back of the object? (b) Why? (c) Where is it shown in the front view?
5. Where is the vertical surface 14, 13 shown in the front view?

6. Where is the horizontal surface 13, 16 shown in the front view?

7. Where is the horizontal surface 6, 7 shown in the side view?

8. Where is the vertical surface 4, 9 shown in the side view?

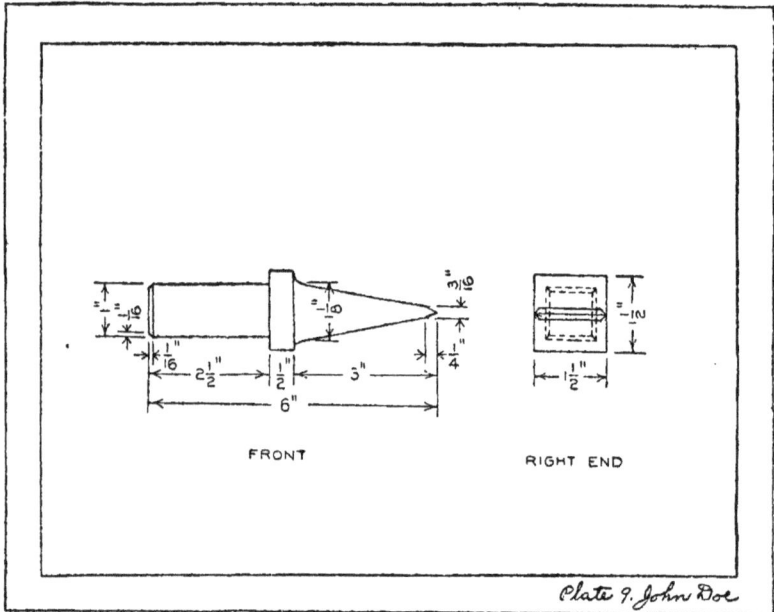

FIG. 85. TYPE PROBLEM. HARDIE. GIVEN VIEWS

DATA FOR DRAWING PLATE 8

Given: Orthographic sketches, Figs. 88 and 89, 90 and 91.

Required: To draw a 45° perspective sketch of the object shown in Fig. 88, 89, 90, or 91, or any similar object as assigned by the instructor.

The upper front corner of the enclosing solid is 2½″ below the level of the eye. Use the point marked A as the upper front corner of the measure cube. All lines of this sketch including the constructive stage should be made entirely freehand. Omit all dimensions.

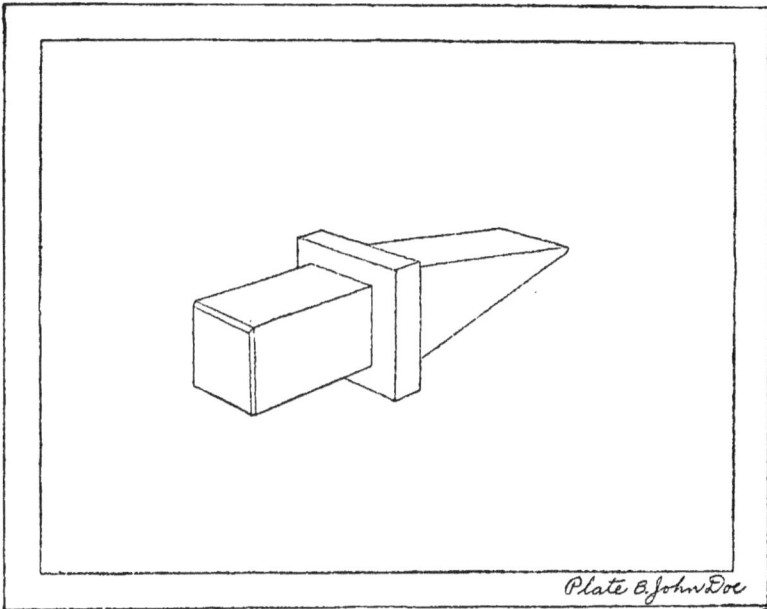

FIG. 86. TYPE PROBLEM. HARDIE. PERSPECTIVE SKETCH

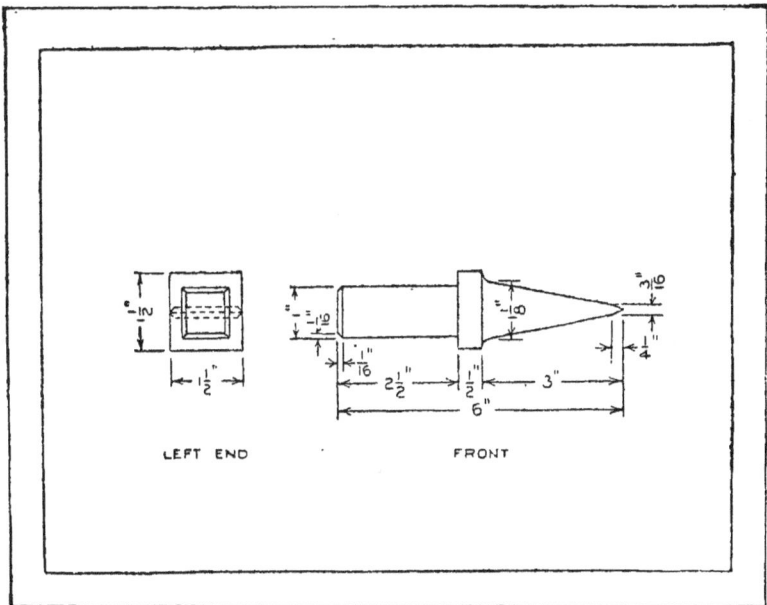

LEFT END   FRONT

FIG. 87. TYPE PROBLEM. HARDIE. REQUIRED VIEWS   (85)

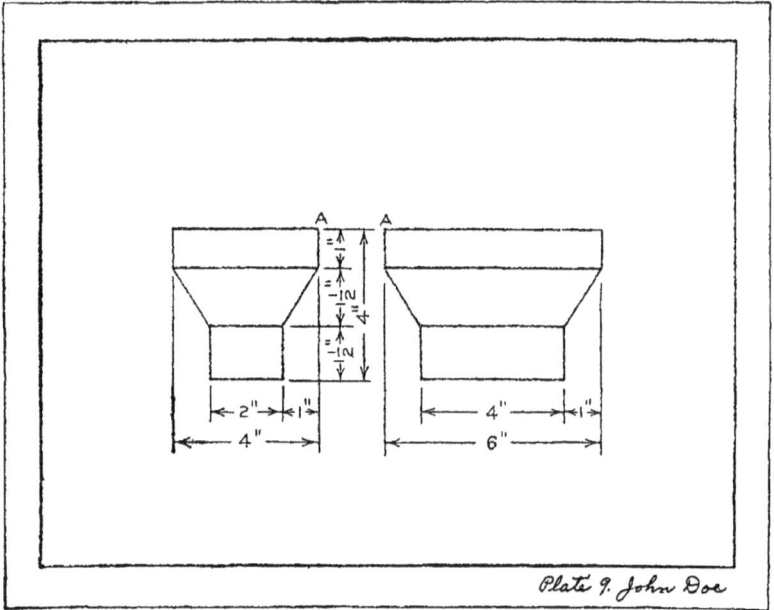

FIG. 88. SHEET METAL HOPPER

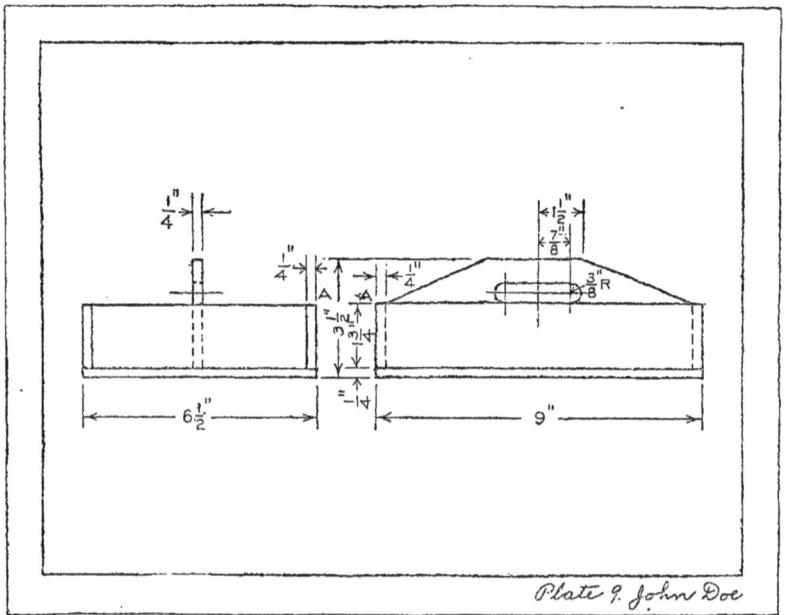

(86)

FIG. 89. KNIFE AND FORK BOX

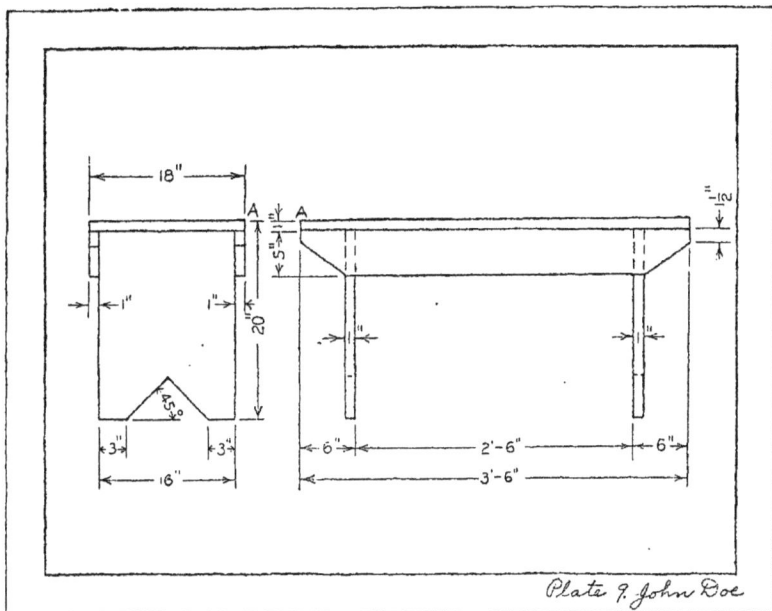

FIG. 90. BENCH

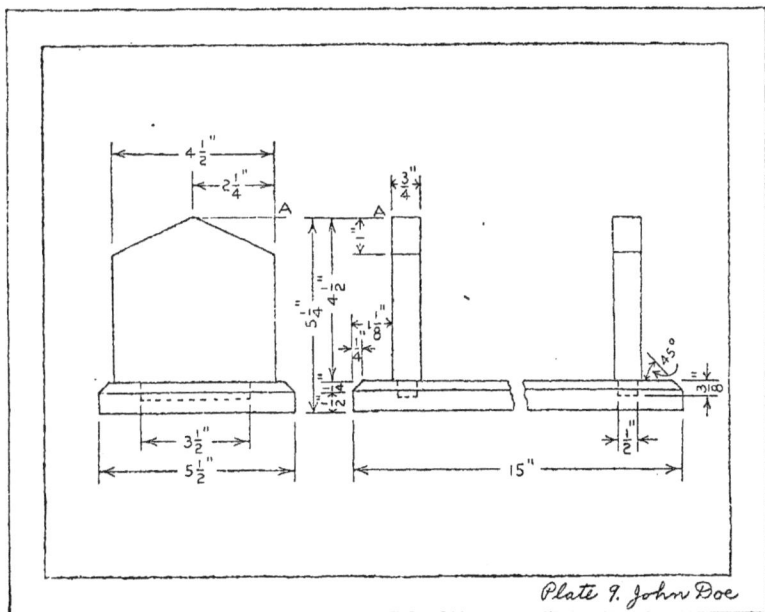

FIG. 91. BOOK RACK

(87)

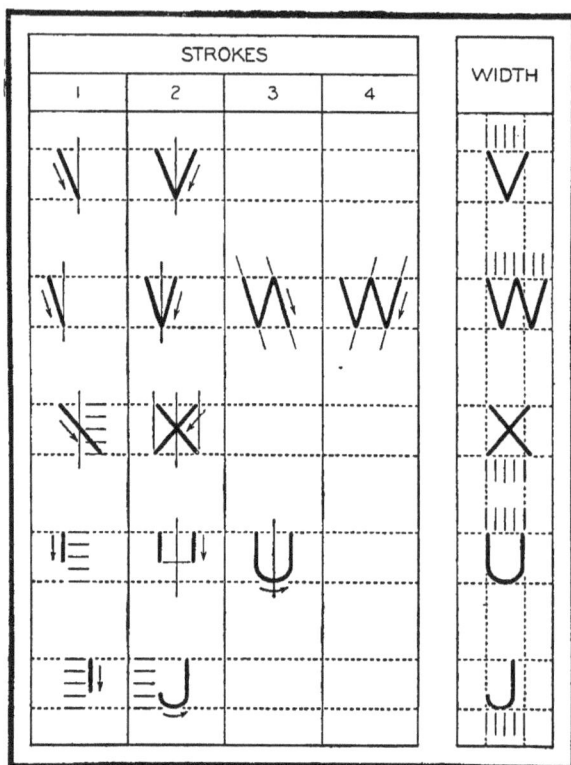

FIG. 92. LETTERING PLATE

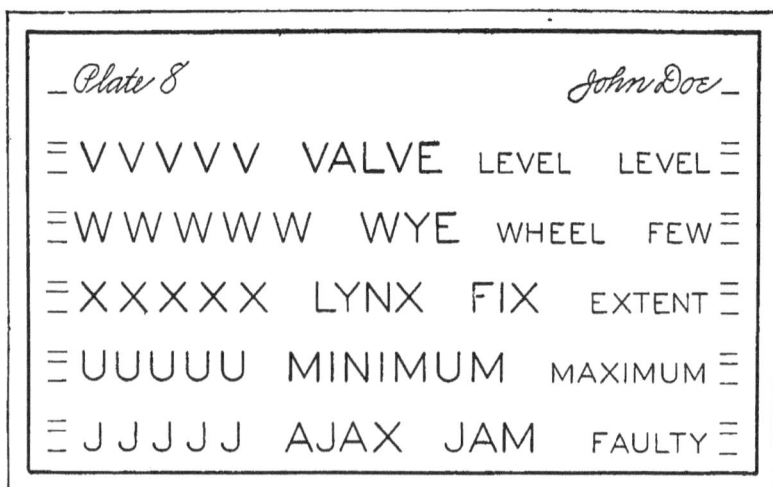

Plate 8   _John Doe_

≡VVVVV VALVE LEVEL LEVEL≡

≡WWWWW WYE WHEEL FEW≡

≡XXXXX LYNX FIX EXTENT≡

≡UUUUU MINIMUM MAXIMUM≡

≡JJJJJ AJAX JAM FAULTY≡

FIG. 93. LETTERING PLATE 8. V, W, X, U, J

DATA FOR LETTERING PLATE 8

Given: Plate 8 to a reduced size. Fig. 93.

Required: To make the plate to an enlarged scale.

PREPARATORY INSTRUCTIONS FOR DRAWING PLATE 9

Dimensioning Angles. The inclination of an edge or surface is commonly determined by giving dimensions which fix two points on the line, usually its ends. Example: The wedge end of the hardie. Fig. 85. In some cases it is desirable to give the inclination of an edge or surface in degrees. In this case the dimension line is an arc with its center at the intersection of the two lines forming the angle. Example: The 45° angle in the end of the bench. Fig. 90.

Solution of the Problem. Attention is called to the fact that in the following problems the student is required to draw different views from those given. Read the statement of the problem carefully before starting to draw.

DATA FOR DRAWING PLATE 9

Given: Orthographic sketches, Figs. 88, 89, 90, and 91, showing the *front* and *left side* of each of the objects.

Required: To draw the *front* and *right side* views of the object shown in Fig. 88, 89, 90, or 91, or any similar object as assigned by the instructor.

Instructions:

1. Block in the views of the object as described on page 75 and as carried out in Plate 7, so that they are in the center of the sheet.

2. Complete the details of the views in light lines.

3. Trace over the lines as explained on page 76, making them the proper weight.

4. Draw in the dimension lines and put in the arrowheads and figures in the order given on page 77.

5. Write in the plate number and name as usual. Press the paper back into the tack holes.

FIG. 94. SPACING OF CURVED FORMS

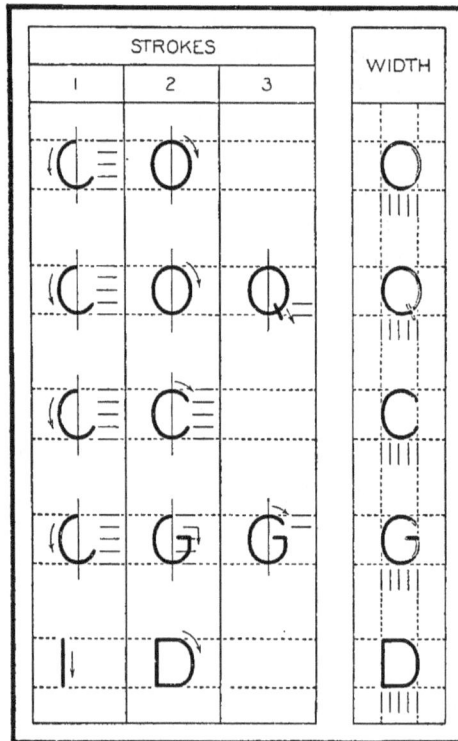

FIG. 95. LETTERING PLATE

PREPARATORY INSTRUCTIONS FOR LETTERING PLATE 9

The letter O is wider than the numeral 0.
The forms of the Q, C, and G are based on the oval of the O.

Spacing Curved Stroke Letters. As stated in the instructions for Plate 5, the area included between the contour of two adjacent letters should appear equal to the area of one-half of the H. When a vertical stroke and a curved stroke are properly spaced the clear distance between them is slightly less than one-half the width of the H. Example: The I and O. Fig. 94.

The clear distance between two curved strokes will be less than that between vertical and curved strokes. Example: The O and O in Fig. 94.

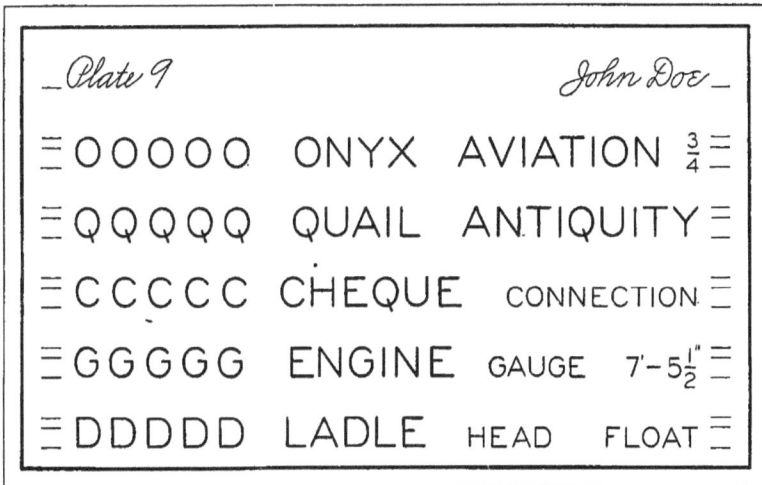

FIG. 96. LETTERING PLATE 9. O, Q, C, G, D

When spacing a letter having a curved outline the beginning of the first stroke should be carefully located. In planning the letter, the clear space between it and the previous letter should be held in mind.

DATA FOR LETTERING PLATE 9

Given: Plate 9 to reduced size. Fig. 96.
Required: To make the plate to an enlarged scale.

PREPARATORY INSTRUCTIONS FOR DRAWING PLATE 10

Objects thus far sketched for which orthographic views were drawn have had plane surfaces. In Plate 10 an object having cylindrical surfaces is to be represented orthographically.

Cylindrical Surfaces. In Fig. 97 an object having cylindrical surfaces is represented by orthographic views.

1. The outline of the front view represents cylindrical surfaces when viewed at right angles to their axes.

2. A simple cylinder, when viewed in this direction, appears as a rectangle.

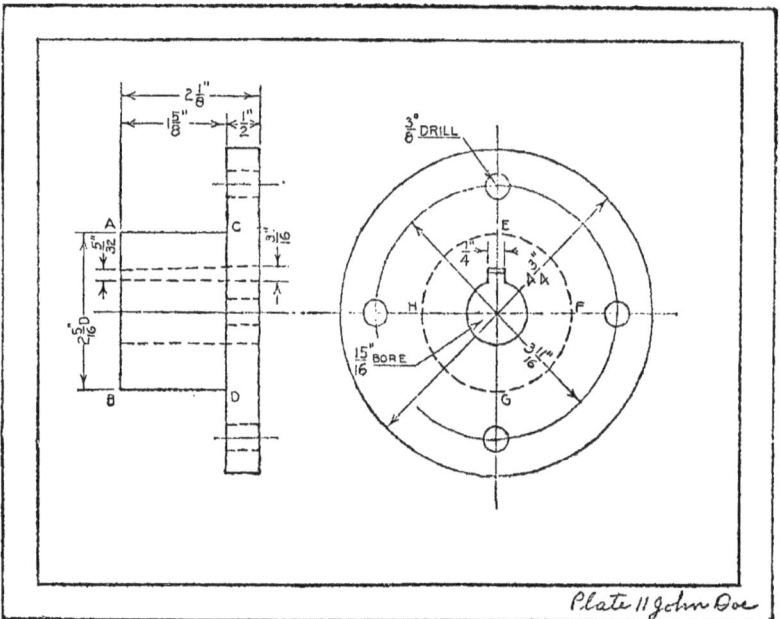

FIG. 97. TYPE PROBLEM. GIVEN VIEWS OF A SHAFT COUPLING

3. The straight lines representing the bases of the cylinder, A B and C D, Fig. 97, are equal in length to the diameter of the cylinder and represent the bases of the cylinder viewed edgewise.

4. The straight lines representing the contour elements of the cylinder A C and B D are the elements of the cylinder which divide the visible part of the surface from that which is invisible. They are viewed at right angles to their direction and are therefore shown in their true length. See page 81 under "Inclined Surfaces."

As stated in discussing the representation of plane surfaces, a surface viewed edgewise is represented by a line. In the case of a plane surface the line representing the surface is a straight line.

When a cylindrical surface is viewed in the direction of its axis the observer is looking at the surface edgewise and it therefore appears as a line, which in this case is a circle. Example: E F G H, Fig. 97.

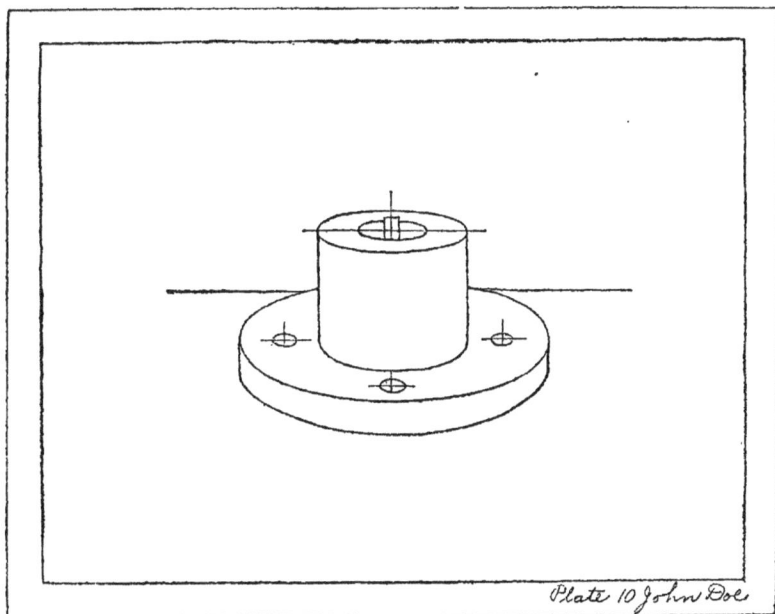

FIG. 98. TYPE PROBLEM. PERSPECTIVE OF A SHAFT COUPLING

Circular Edges.

1. A circular edge viewed at right angles to its plane shows as a true circle. Example: E F G H in Fig. 97.

2. A circular edge viewed in the direction of its plane shows as a straight line equal in length to the diameter of the circle. Example: A B and C D in Fig. 97.

The student should test his knowledge of the orthographic principles just mentioned by answering the following questions: See Fig. 102.

1. (a) Where is the left end view of the cylindrical surface 3, 4, 11, 12? (b) Of 1, 2, 13, 14?

2. Where is the front view of the cylindrical surface 25, 26, 27, 28?

3. (a) Where is the circular surface 1, 16, 15, 14, shown in the end view? (b) Where is the surface 2, 3, 12, 13 shown in the end view?

4. (a) Where is the circular edge 6, 9 shown in the end view? (b) Where is the circular edge 4, 11 shown in the end view?

5. What surface would be crosshatched if a quarter section were made cutting on the lines 0, 17 and 0, 18?

6. What surface would be crosshatched if a half section were made cutting on the line 17, 19?

7. (a) Where is the left end view of the extreme element 3, 4? (b) 11, 12?

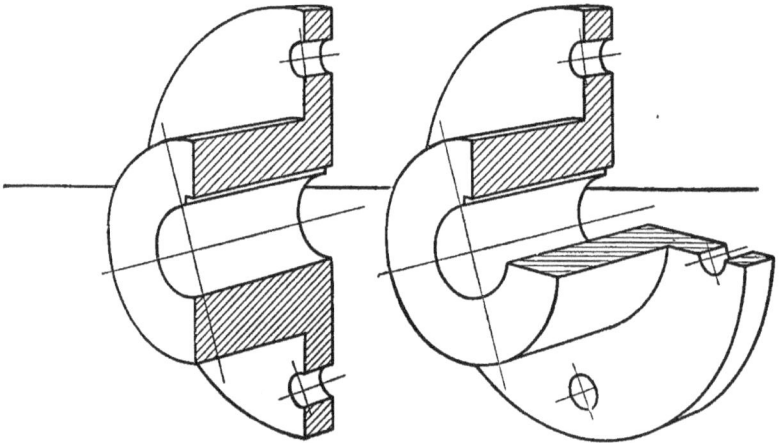

Fig 99. Half Section. Quarter Section. Illustrated in Perspective

DATA FOR DRAWING PLATE 10

Given: Orthographic sketches, Figs. 103 and 104.

Required: To draw a perspective sketch of the object shown in Fig. 103, 104, or any similar object as assigned by the instructor, with its axis vertical. The upper end of the object is $2\frac{1}{2}''$ below the level of the eye.

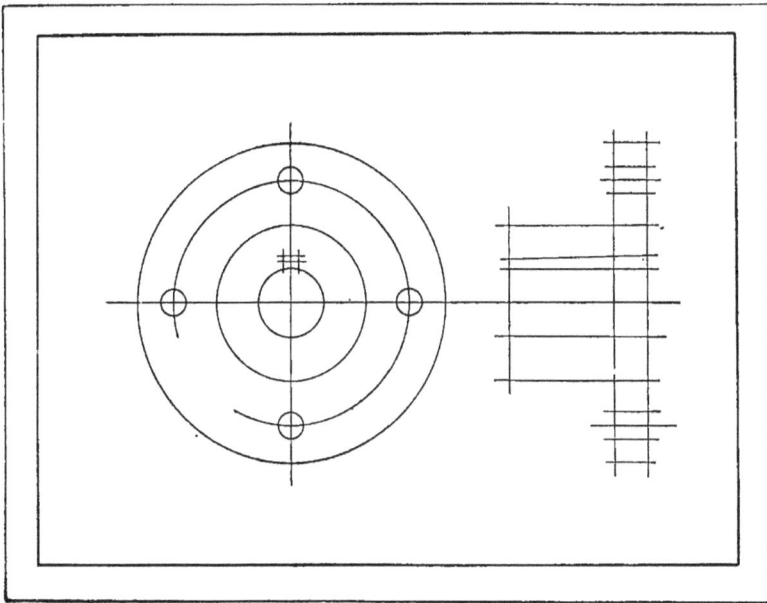

FIG. 100. TYPE PROBLEM. CONSTRUCTIVE STAGE OF THE ORTHOGRAPHIC
SKETCH

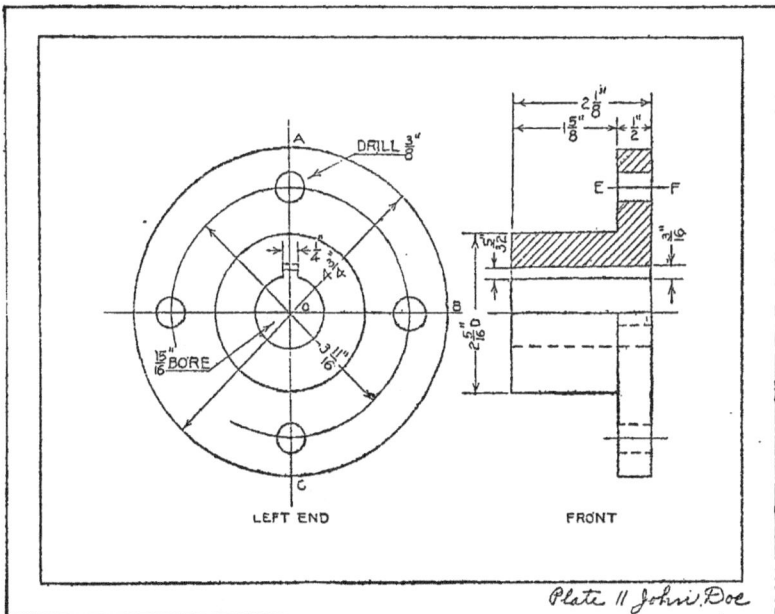

FIG. 101. TYPE PROBLEM. FINISHED SKETCH OF A SHAFT COUPLING

(95)

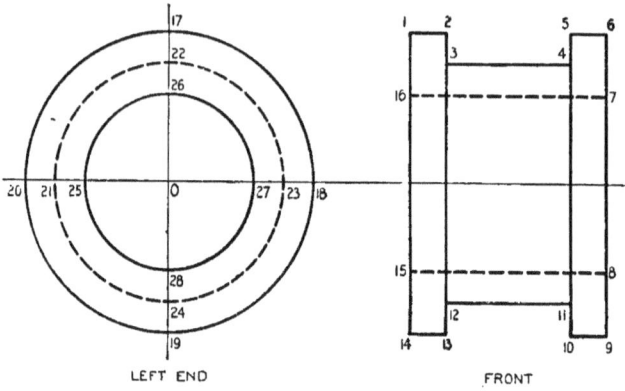

FIG. 102. REVIEW PROBLEM

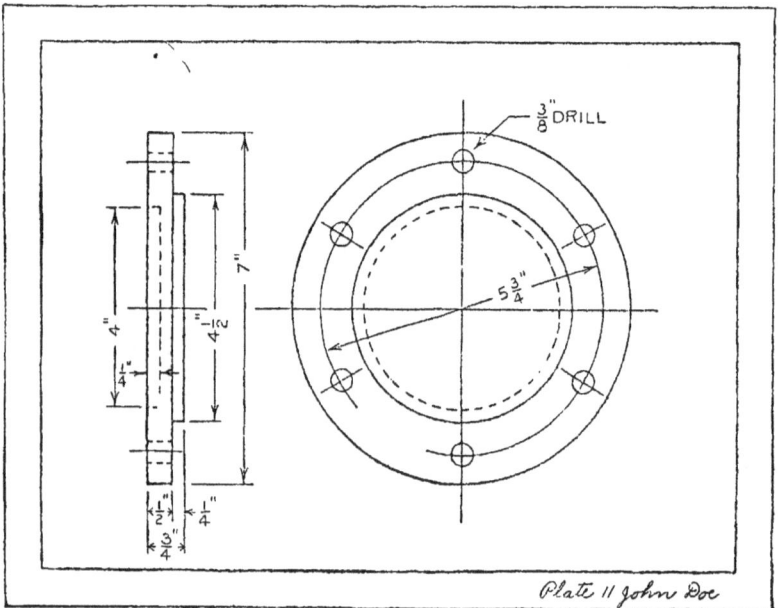

FIG. 103. CYLINDER HEAD

PREPARATORY INSTRUCTIONS FOR LETTERING PLATE 10

The first two strokes of the P, R, and B are exactly alike. The basic form of the S is a combination of two ovals. When drawing the strokes of the S these ovals should be held in mind.

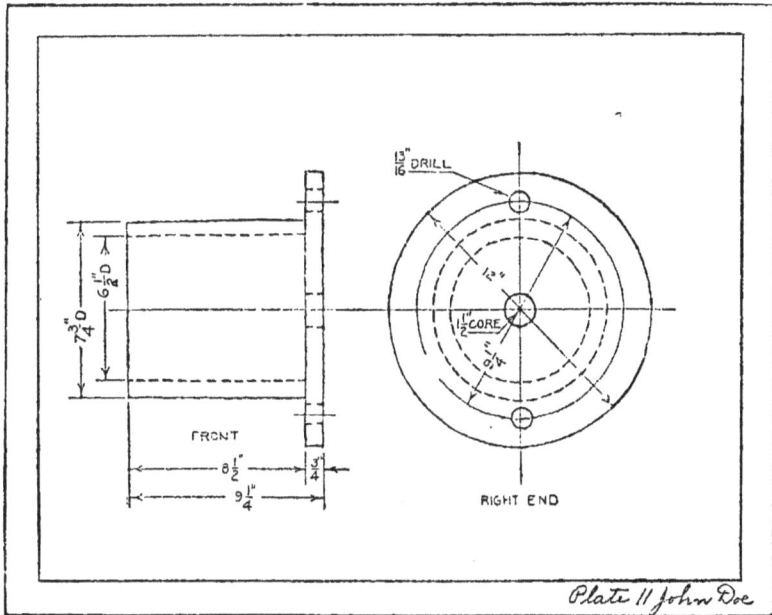

FIG. 104.   SPRING CASING

DATA FOR LETTERING PLATE 10

Given: Plate 10 to a reduced size.   Fig. 106.
Required: To make the plate to an enlarged scale.

PREPARATORY INSTRUCTIONS FOR DRAWING PLATE 11

Objects thus far drawn have been shown in their complete form. By referring to Fig. 97 it will be noticed that all hidden contour elements are represented by dotted lines. When there are too many of such lines they tend to confuse the reader of the drawing.

Half Section. If an object is represented as though a portion of it has been removed, the drawing can often be made much clearer because of the reduction in the number of dotted lines.

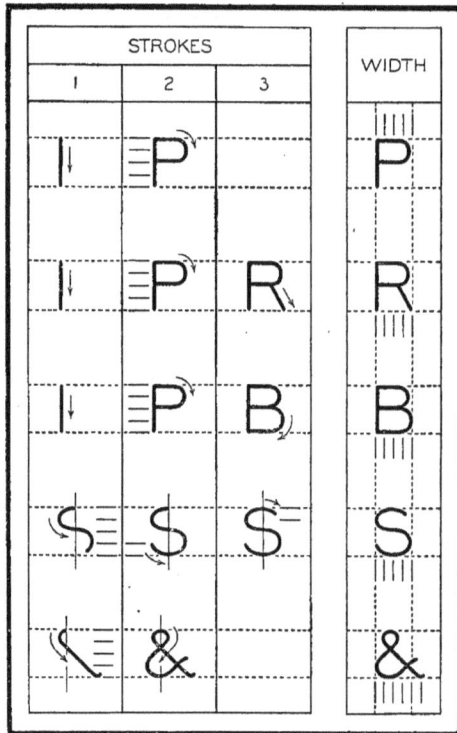

FIG 105. LETTERING PLATE

A common method of showing a part removed is to imagine the object cut into two similar parts through an axis of symmetry. Fig. 99.

The cut is made in a plane at right angles to the line of sight and the near half of the object is imagined removed. The observer then sees the cut surfaces of the remaining half in their true form and proportions. The view thus obtained is called a *half section* view. Example: See the front view. Fig. 101.

It should be noticed that a part of the object is imagined removed only in the drawing of one view. The end view, Fig. 101,
represents the complete object.

Quarter Section. Another common method of representing
an object is to imagine it cut on two planes at right angles to
each other in to an axis of symmetry. Fig. 99. One of the cut
surfaces is at right angles to the line of sight and the other is
parallel to it. The quarter of the object included between the

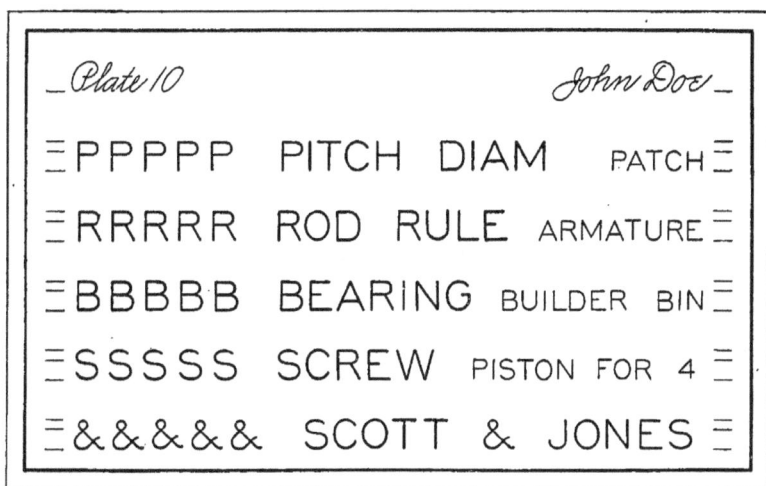

Fig. 106. Lettering Plate 10

two cut surfaces is considered removed. The observer then sees
the cut surfaces in one plane in their true form and proportions,
and those in the other plane appear as a line. Fig. 101. The
view thus obtained is called a *quarter section.* It should be
noted here again that a part of the object is imagined removed
only in the drawing of one view.

Crosshatching. The cut surfaces in the section view are represented conventionally by crosshatching, which consists of
drawing very fine parallel lines, equally spaced, over a surface
represented as cut. In the student problems the lines should be
drawn about $\frac{1}{16}''$ apart and at an angle of 45° with the horizontal. Both the distance between the lines and the angle at

which they are drawn should be estimated—not measured. It is suggested, however, that after drawing the first few lines the student check the spacing with a scale. The angle of the crosshatch lines should be carefully checked in a corner where horizontal and vertical boundary lines meet at right angles. When the distances from the corner to each end of the same crosshatch line are equal, the angle is 45°.

Center Lines. A line which represents an axis of symmetry is called a *center line.* Center lines may be straight or curved. Of the straight lines there are two classes, principal and secondary. A principal center line is one about which the entire view is symmetrical. Example: A C in Fig. 101. The principal center lines should extend about $\frac{1}{2}''$ beyond the outline of the view. A secondary center line is one about which only part of the object is symmetrical. Example: E F, Fig. 101, is the center line for the hole only. Secondary center lines should extend about $\frac{1}{4}''$ beyond the outline of the part of which they represent the axes.

A circular center line usually passes through the centers of a number of holes grouped at a certain distance from a central point. It is not quite a complete circle. Example: See left end view, Fig. 101.

In general, every circle in a drawing must have two center lines at right angles with each other. Example: Lines 17, 19 and 18, 20. Fig. 102. When one of the center lines is circular, as in the case of the center line for the drilled holes in the end view, Fig. 103, the other center line is at right angles to the tangent of the circular center line, at the center of the hole. This line is therefore a radial line from the center of the circular center line.

Dimensioning Cylindrical Surfaces and Circles. The diameter of a cylinder may be given by placing the dimension on a diameter of the circle representing the cylindrical surface. Example: See Fig. 101. It may be given between extension lines drawn from the rectangular view. Example: See Fig. 101. In this case the dimension figures should be followed by a *D* or *Diam.* to indicate that the dimension is a diameter. A hole to be drilled, cored, or bored may be indicated by printing the word showing how it is to be obtained or finished, together with the dimension and arrow pointing to the hole. The word and the dimension should be placed in an open area near the hole represented. The line drawn under the word should be about $\frac{1}{32}''$ below the letters. Example: See Fig. 101.

To Sketch a Circle.

1. Through the center of the circle draw two light lines at right angles, and on each lay off points at a distance from the center equal to the radius of the circle.

2. Repeat this process by drawing another pair of lines which make 45° with the first pair. In case the circle is large other similar lines may be drawn which bisect the angles made by the first lines.

3. Sketch in the circle through the points located on the several lines. This should be done with great care. The first lines drawn should be very light.

4. When the circle is complete observe its form carefully and true it up by erasing and redrawing any portion which is untrue.

5. In the finishing stage the corrected light lines should be traced over to produce a line of the proper weight.

DATA FOR DRAWING PLATE 11

Given: Orthographic sketches, Figs. 103 and 104, or any similar problem selected by the instructor, showing the front and right end views.

Required: To draw the front quarter section and left end views of the objects shown in Fig. 103, 104, or any similar object as assigned by the instructor.

Instructions:

1. Proportion the views on the plate.

2. Draw the principal center lines for the circular view. Draw the circles in the following order: (1) The larger circles.

(2) Circular center line. (3) The circles representing the small holes.

3. To draw the rectangular view, first determine its vertical dimensions by projecting from the end view. Complete the view.

4. Retrace the lines as in the finishing stage, giving particular attention to those affected by the section.

5. Draw in extension and dimension lines and put in the dimensions. Finish the sketch by crosshatching the cut surfaces. Care should be taken to make the section lines parallel to each other and at 45° to the horizontal.

LETTERING IN INK

PREPARATORY INSTRUCTIONS FOR LETTERING PLATE 11

The following is a list of the materials used in making lettering plates in ink.

1. Tracing Cloth, 4″ × 6″ sheets.

2. One of the following or any similar pen which will give satisfactory results may be used:

> 303 Gillott's
> 404 Gillott's } Three of each.
> Spencerian No. 1
> Lady Falcon

3. Penholder.

4. Black waterproof drawing ink.

Square one of the three by five inch cards on the board and stretch the tracing cloth over it with the dull side up. The surface of the cloth should be prepared for inking by being rubbed with chalk dust. All superfluous chalk must be removed to prevent its clogging the pen. The guide lines for the letters should be drawn on the cloth in pencil. When the plate is finished a border rectangle should be drawn and the sheet trimmed to 3″ × 5″. Fig. 18. The space outside the cutting lines may be used to try the pen on during the process of lettering the plate. A pen should be selected which will give a width of line suited to the height of letters to be made. The proper width of line should be secured with but little spreading of the nibs of the pen. Fig. 14 illustrates the position of the pen

ROUGH ROUND RODS
OHIO CORLISS ENGINE

FIG. 107. EXAMPLES OF WORD SPACING

in the hand while lettering. Note that the forearm is nearly parallel to the vertical strokes. Vertical strokes should be made with a finger movement. In making the horizontal and curved strokes this movement is combined with a turn of the wrist.

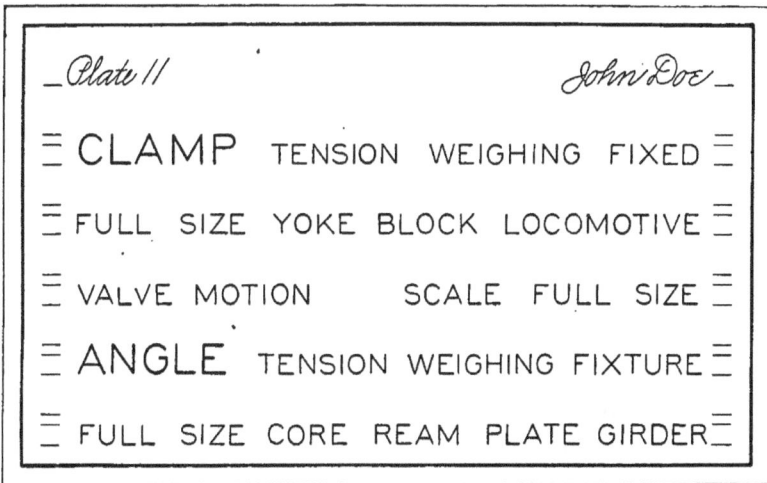

Plate 11   _John Doe_

$\equiv$ CLAMP TENSION WEIGHING FIXED $\equiv$

$\equiv$ FULL SIZE YOKE BLOCK LOCOMOTIVE $\equiv$

$\equiv$ VALVE MOTION   SCALE FULL SIZE $\equiv$

$\equiv$ ANGLE TENSION WEIGHING FIXTURE $\equiv$

$\equiv$ FULL SIZE CORE REAM PLATE GIRDER$\equiv$

FIG. 108. LETTERING PLATE 11

To fill the pen, place the ink on the under side by means of the quill attached to the stopper of the bottle. The stopper should be returned to the bottle since the ink dries rapidly.

Composition. In this and the following plates in lettering, words will be combined into phrases and sentences. The spacing of words plays an important part in securing a good general effect in a line of letters. The space between words should appear equal to three times that between letters or one and one-half times the width of the H. Adjacent vertical strokes will therefore be separated by a space one and one-half times the width of the H. The clear distance between two words having vertical strokes adjacent to a curved stroke will be less than one and one-half times the width of the H. The clear distance between two words having adjacent curved strokes will be still less. Example: See Fig. 107.

DATA FOR LETTERING PLATE 11

Given: Plate 11 to reduced size. Fig. 108.

Required: To make the plate to an enlarged scale. In this plate the wording of the titles for the first pencil mechanical drawing plates is used. The letters are approximately the height used in the title.

REVIEW QUESTIONS.

1. (a) In what direction does one look at an object in making its orthographic views? (b) How does this differ from the way it is viewed in making a perspective of it?

2. (a) How many general dimensions does each orthographic view show? (b) How many orthographic views are necessary to show three general dimensions?

3. (a) What is the position of the top with reference to the front view? (b) Why? (c) Which general dimension is common in the top and front views?

4. What is meant by "Reading" a drawing?

5. (a) Under what condition does a surface appear as a line in a view? (b) When a hidden surface is viewed edgewise how is it represented?

6. (a) When is a plane surface shown in its true form in one view and as a straight line in the other? (b) When is a plane surface shown in less than its true size in one view and as a straight line in the other?

7. (a) When is a cylindrical surface represented as a rectangle? As a circle? (b) When is a circular surface represented as a straight line in one view and as a circle in the other?

8. (a) When is a straight edge of an object shown in its true length in two views? (b) When in its true length in one view and as a point in the other? (c) When in its true length in one view and in less than its true length in the other?

9. How are the corners of an object represented?

10. Describe the process of proportioning the views of an object and locating them centrally on the sheet.

11. What are the lengths of dashes and spaces in dotted lines?

12. (a) Illustrate by a sketch how detail and over-all dimensions are grouped. (b) What space is allowed between the outline of the object and the nearest dimension line? (c) Between dimension lines?

13. (a) Where is the right side view placed with respect to the front view? (b) Where is the left side view placed with respect to the front view?

14. (a) What general dimensions of an object are shown in the right side view? (b) In the left side view?

15. What determines the choice between a right and left side view?

16. (a) Why is an object sometimes shown with a part removed? (b) Define quarter-section. Define half-section. (c) Is the part cut by the section planes shown as removed in both views?

17. (a) What is the purpose of crosshatching? (b) At what angles are the crosshatching lines drawn? (c) What is the usual distance between crosshatching lines?

18. (a) What is a principal center line? (b) A secondary center line?

19. (a) When is a straight center line used? (b) Circular center line?

20. (a) How many center lines must be drawn for each circle? (b) At what angle to each other?

21. Illustrate how the two views of a cylinder may be dimensioned.

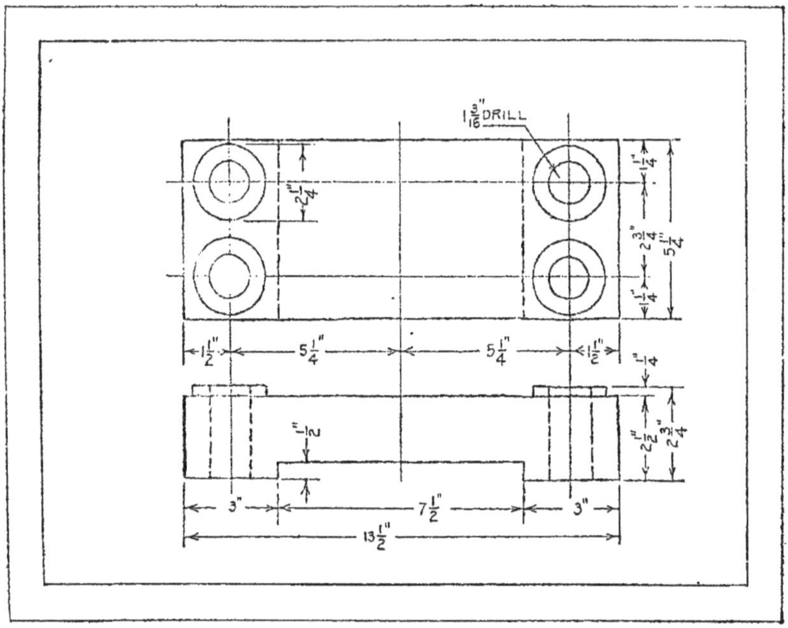

FIG. 109. BEARING CAP

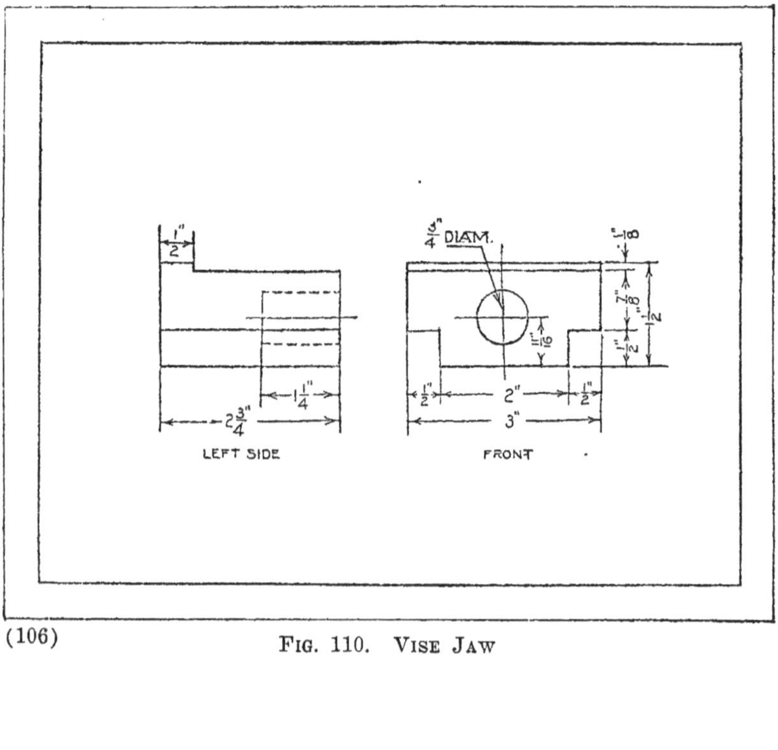

LEFT SIDE FRONT

FIG. 110. VISE JAW

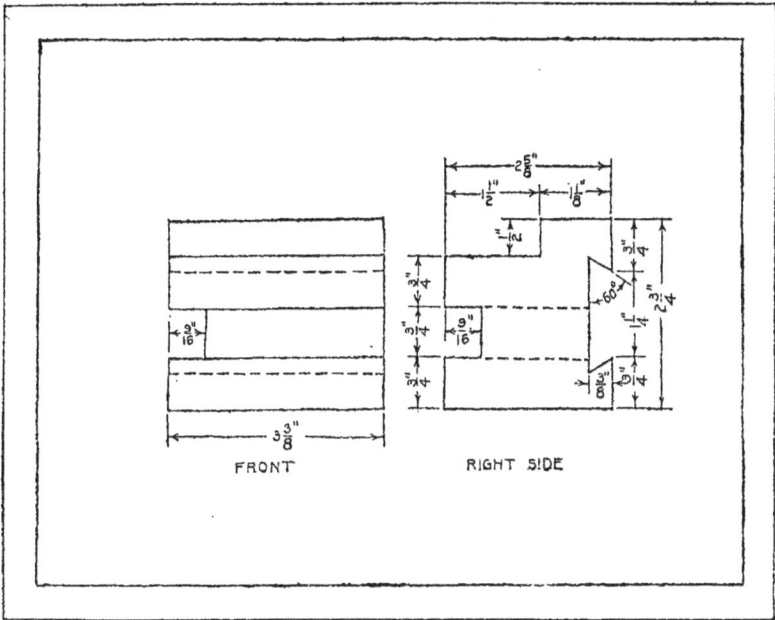

FRONT   RIGHT SIDE

FIG. 111. TURNING TOOL HOLDER

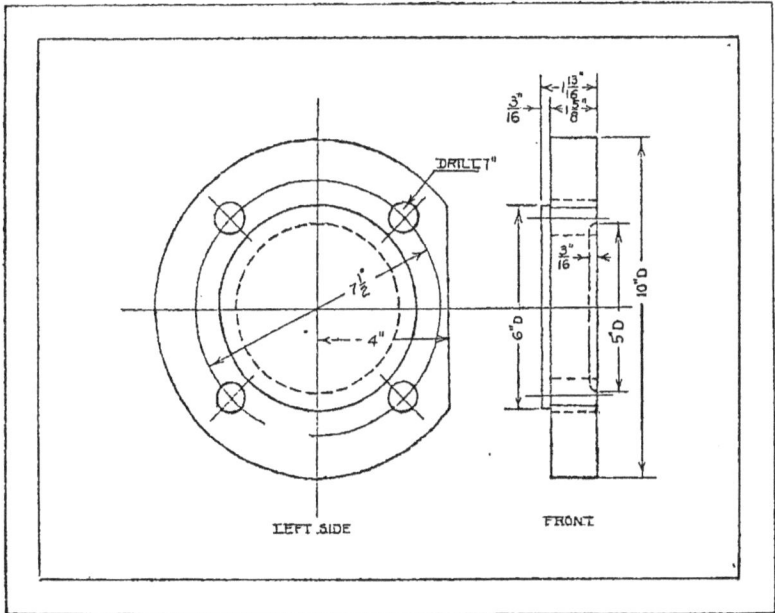

DRILL 7"

LEFT SIDE   FRONT

FIG. 112. VALVE BONNET   (107)

DATA FOR REVIEW PROBLEMS

Given: An orthographic sketch, Fig. 109.

Required: To make the orthographic sketch shown in Fig. 109 to an enlarged scale.

Given: An orthographic sketch, Fig. 110, showing the front and left side views of the object.

Required: To draw the front and right side views of the object shown in Fig. 110.

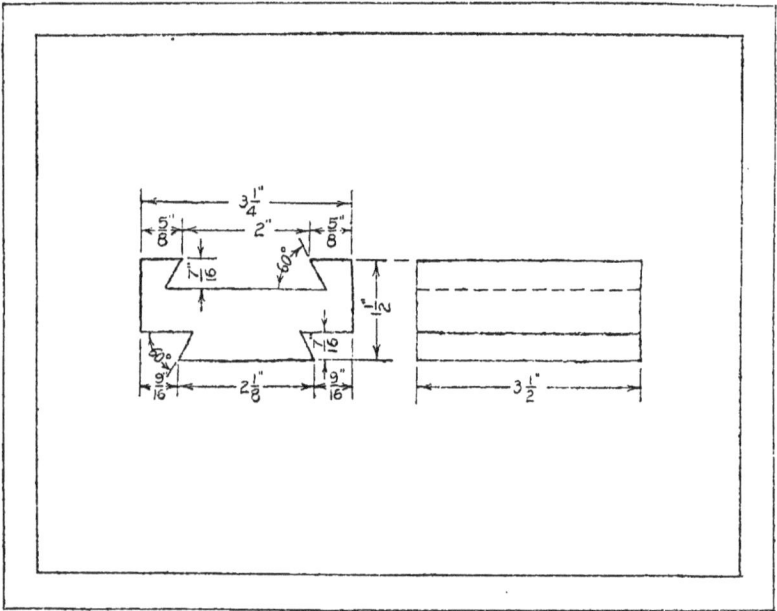

FIG. 112A.   DOVETAIL CROSS SLIDE

Given: An orthographic sketch, Fig. 111, showing the front and right side views of the object.

Required: To draw the front and left side views of the object shown in Fig. 111.

Given: An orthographic sketch, Fig. 112, showing the front and left side views of the object.

Required: To draw the front quarter section and right side views of the object shown in Fig. 112.

Given: Orthographic sketches, Fig. 112A, showing the front and left side of each of the objects.

Required: To draw the front and right side views of the object shown in Fig. 112A as assigned by the instructor.

CHAPTER III

PENCIL MECHANICAL DRAWING

PROSPECTUS

In this chapter orthographic sketching is continued. A more general application of the principles of orthographic drawing is made. This is done principally by introducing problems requiring three views from perspective sketches. It is the chief aim of this chapter to give considerable practice with some of the common instruments and materials used in making mechanical drawings and to fix a standard of technique. When the work of this chapter is completed, the student should be able to make neat, accurate mechanical drawings of simple objects. The technique of the lettering, arrowheads, and figures should be of a standard comparable with that secured in the mechanical line work.

PREPARATORY INSTRUCTIONS FOR DRAWING PLATE 12

Three-view Problems. From the principles developed in Chapter II it is evident that each view shows two of the general dimensions of an object and therefore only two views are necessary to obtain all three of the general dimensions of any object. However, in some cases all of the general dimensions, length, height, and thickness may be given and still the form of the object will not be clearly defined. When this is true a third view is necessary. A top, front, and side view are drawn with the top *above* the front view and the side view *to the right or left* of the front view as in the problems of Chapter II.

Example: The front and side views of the part of a Sash Joint in Fig. 113A do not show the form of the tenon. Hence a top view is necessary. Also, the front view is necessary to show the notch in the tenon, and the left side view to show the bead.

As stated under Plate 8, the right and left side views convey the same information and therefore either may be drawn. The one is usually selected which requires the fewer dotted lines. Example: Comparing Fig. 113, A and B, the right side view is preferable for this reason. Ordinarily the right or left side view is drawn opposite the front view In some cases, however, a better arrangement will be secured by placing the side view oppo-

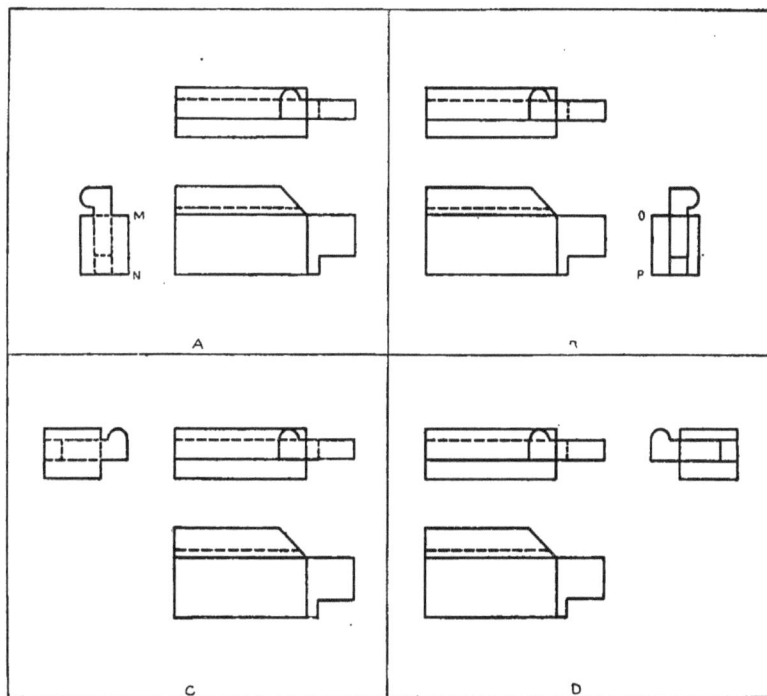

FIG. 113. RELATION OF FRONT, TOP, AND SIDE VIEWS

site the top view instead of opposite the front view. Fig. 113, C and D. In this case the views are so related that horizontal distances from front to back, which are common to the top and side views, may be projected from one view to the other.

To relate properly the side view to the front view of an object, attention should be given to the following conditions:

1. In all cases the side views of the front surfaces are adjacent to the front view of the object. Example: M N and O P in Fig. 113 represent the side views of the front surface.

2. In securing the views of an object, *one should never move the object but should himself move* from the position taken in securing a front view, viz., *to the left,* to secure the left side view or *to the right* to secure a right side view.

The student should test his knowledge of the orthographic principles just stated by answering the following questions: See Fig. 114.

1. (a) Why is the top view of the object necessary? (b) The front view? (c) The right side view?

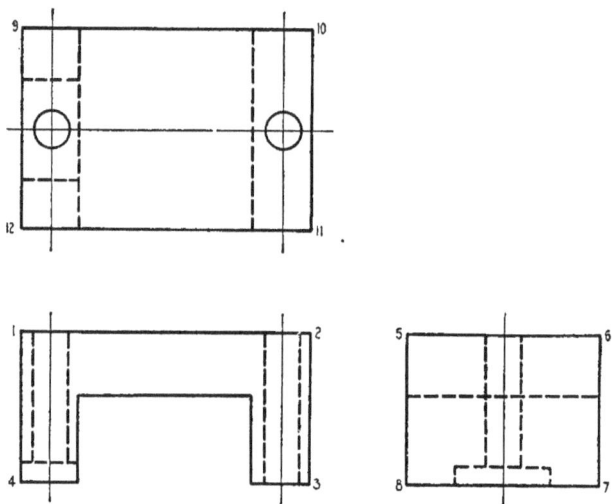

FIG. 114. REVIEW PROBLEM

2. (a) Is the right side view preferable to the left side view? (b) Why?

3. (a) Where is the near horizontal edge 5, 6, shown in the top view? (b) In the front view?

4. (a) Where is the vertical surface 5, 8, shown in the front view? (b) In the top view?

5. (a) Where is the back vertical surface 1, 2, 3, 4, shown in the side view? (b) In the top view?

6. Draw the right side view opposite the top view.

7. Draw the left side view opposite the front view.

8. Draw the left side view opposite the top view.

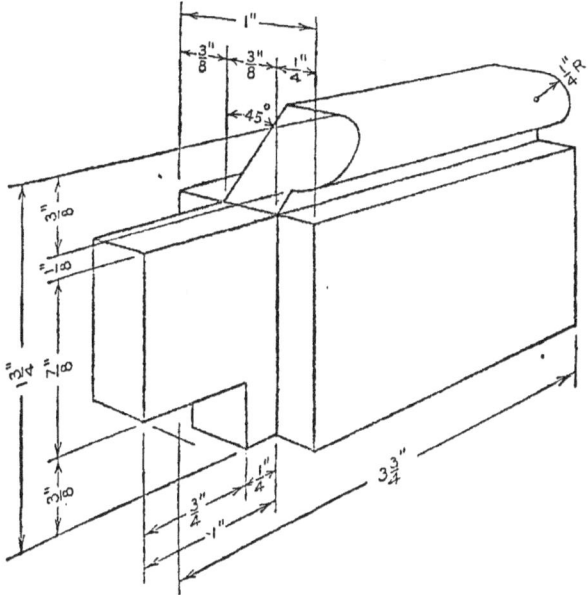

FIG. 115. TYPE PROBLEM. PERSPECTIVE OF SASH JOINT

In this chapter an orthographic sketch will be required preceding each mechanical drawing. This freehand practice will

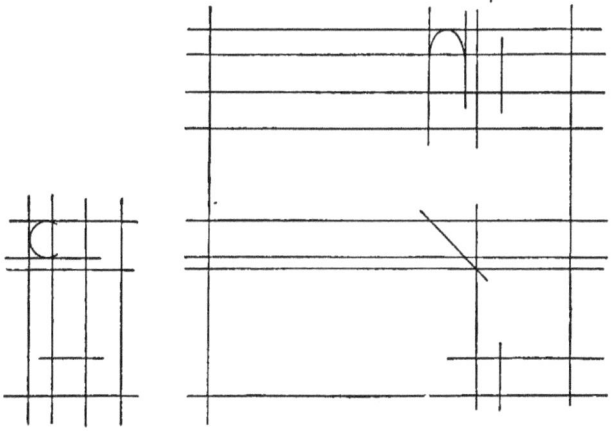

FIG. 116. TYPE PROBLEM. CONSTRUCTIVE STAGE OF PENCIL MECH. DRAWING

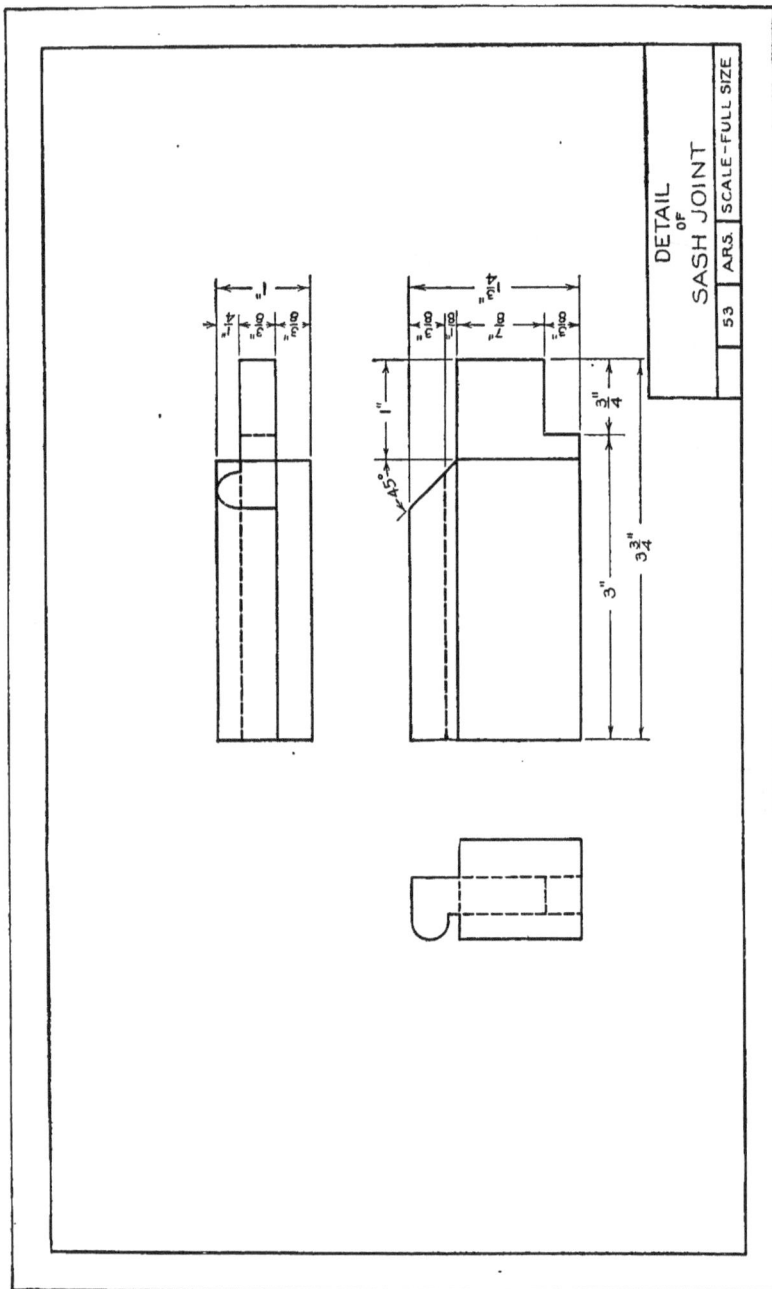

DETAIL
OF
SASH JOINT

| 53 | ARS. | SCALE–FULL SIZE |

FIG. 117. TYPE PROBLEM. FINISHED DRAWING OF SASH JOINT

(114)

develop further skill in orthographic sketching and will make the student familiar with the problem. As a result time will be saved in making the mechanical drawing.

DATA FOR DRAWING PLATE 12

Given: Perspective sketches, Figs. 118, 119, and 120.
Required: To draw a three view orthographic sketch of the object shown in Fig. 118, 119, or 120, or any similar object with dimensions, as assigned by the instructor.

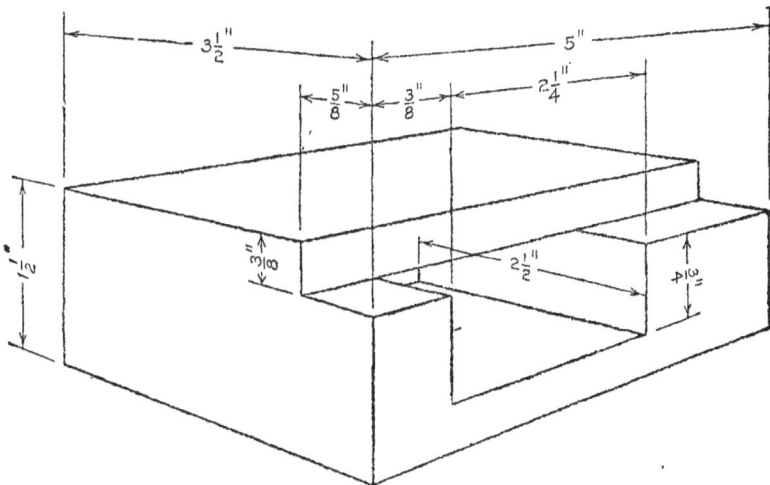

FIG. 118. MORTISE

Instructions:

1. This sketch should be drawn entirely freehand. Proportions and distances should be estimated—not measured.

2. Consider the form of the object carefully and select the views to be drawn.

3. Compare the over-all dimensions of each view and block in the view by drawing a rectangle the sides of which are in the proportions of the over-all dimensions of the object shown in this view. Example: In Fig. 117 the over-all dimensions of the object which are represented in the front view are 3¾″ and 1¾″. The length of the sides of the rectangle are therefore drawn in the

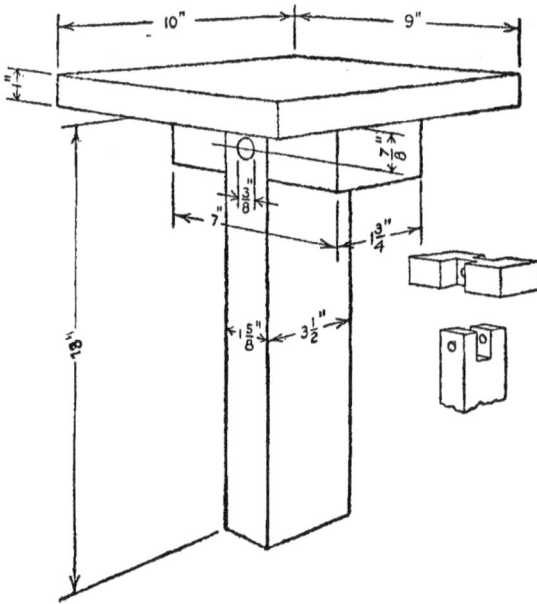

FIG. 119. MILK STOOL

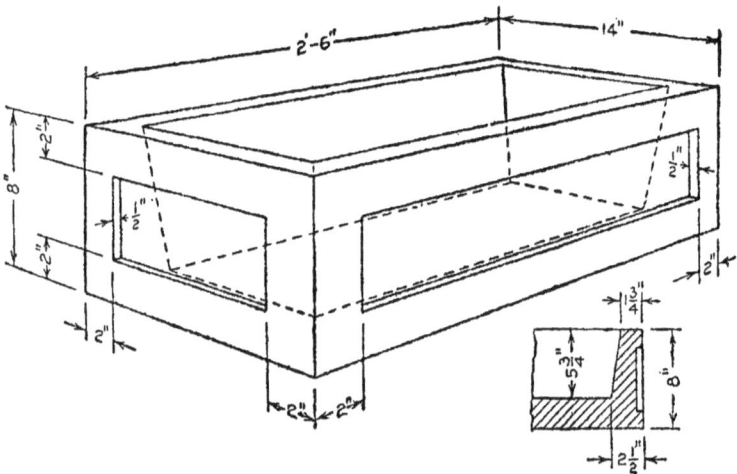

FIG. 120. CONCRETE FLOWER BOX

approximate ratio of $3\frac{1}{4}:1\frac{1}{4}=2:1$, or, in other words, the length of the rectangle is twice its width. The length of the top view is equal to the length of the front view. Its width is about one-fourth of its length. The height of the side view is equal to that of the front view. Its width is about one-half of its height.

4. The distances from the views to the border line above and below the views should be equal and the distances from the views to the border lines at the right and left should be equal. Use the same principle for placing the views as given for the problems of Chapter II, page 81.

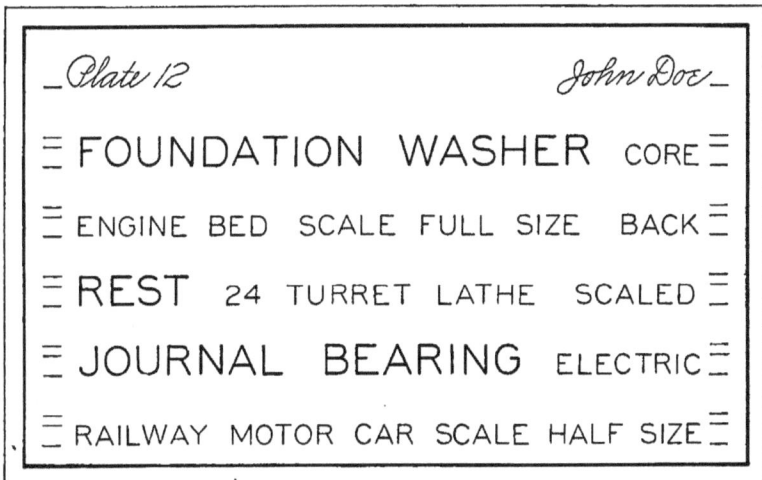

Plate 12   _John Doe_

≡ FOUNDATION WASHER CORE ≡

≡ ENGINE BED SCALE FULL SIZE BACK ≡

≡ REST 24 TURRET LATHE SCALED ≡

≡ JOURNAL BEARING ELECTRIC ≡

≡ RAILWAY MOTOR CAR SCALE HALF SIZE ≡

FIG 121. LETTERING PLATE 12

5. When the rectangles are properly located, sketch in lightly the details of each view, proportioning them by eye.

6. Complete the views by retracing the lines.

7. With the advice and suggestions of the instructor, select the necessary dimensions and then draw extension and dimension lines, arrowheads, and figures.

DATA FOR LETTERING PLATE 12

Given: Plate 12 to reduced size. Fig. 121.

Required: To make the plate to an enlarged scale.

PREPARATORY INSTRUCTIONS FOR DRAWING PLATE 13

The following is a list of materials needed in making the pencil mechanical drawing in this chapter:

1. Drawing board.
2. High-grade paper similar to Duplex or Cream, 11″ × 15″ sheets.
3. T-square.
4. 30°-60° and 45° triangles.
5. High-grade 5H pencil.
6. Scale.
7. Bow compass.
8. 4H compass lead.
9. Pencil pointer.
10. Eraser.
11. Erasing shield.

The Drawing Board. The best boards are designed to prevent warping, various means being used to accomplish this end. Some are built of narrow strips glued together; others have a series of saw cuts running lengthwise with the grain to reduce the transverse strength. Fig. 122. Such boards are made rigid by cleats of hard wood screwed through oblong slots fitted with metal bushings to the back of the board. This construction allows the board to expand and contract, the screws sliding in the slots.

For accurate work it is necessary that the edge of the board against which the head of the T-square is placed be perfectly straight and that the face of the board lie in a plane. *To test the edges* of the board, place on each a standard straightedge or the edge of a T-square blade which is known to be straight. An edge of the board is straight if, when held up to the light, the straightedge is in contact at all points.

The surface of the board may be tested in like manner by placing the straightedge upon it in various positions.

The edges and surface of the board should be kept free from cuts, scratches, and bruises. The board should not be subjected to extremes of temperature or moisture.

Drawing Paper. For mechanical drawing, where a sharp, fine line is to be produced with a hard pencil, a tough, hard paper should be used. It should stand considerable erasing without injury to the surface. It should not become brittle or discolored from reasonable exposure or age. If freehand lettering is to be done the surface must be reasonably smooth to secure the best results. If considerable time is to be spent on a drawing, a

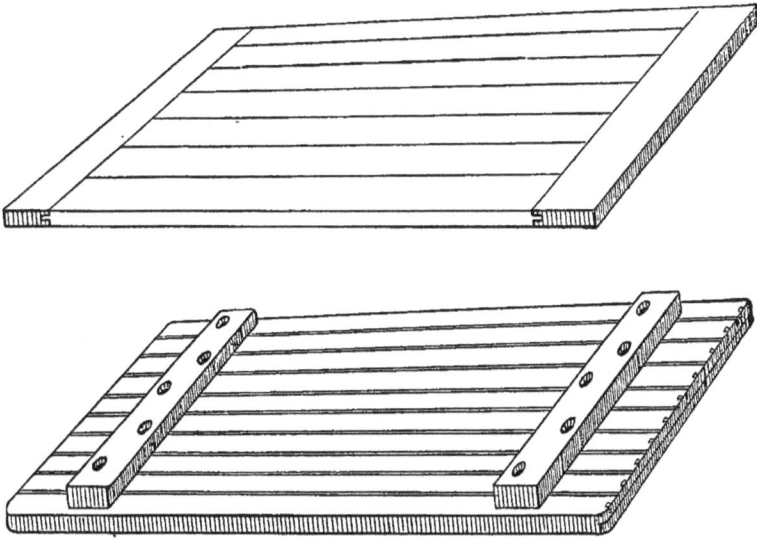

Fig. 122. Drawing Boards

paper should be selected which has an agreeable tint and which will not soil easily with handling. The paper used for the mechanical drawings of this course must fulfil these requirements.

The drawing paper should be fastened to the board in the upper left corner of the board as for sketching. After inserting the first tack, make the upper edge of the sheet horizontal by means of the T-square; stretch the sheet and insert the remaining thumb tacks in the usual manner.

The T-square is used to draw horizontal lines and to provide an edge against which the triangles are placed. It consists of a rule called the blade, attached to one end of which is a cross-piece called the head. Fig. 123. The head is sometimes fastened to the blade by means of a swivel, so that the blade may be set at any desired angle.

T-squares are made of steel, hard rubber, and wood. The steel blade is the most accurate but tends to soil the drawing. For ordinary work wooden blades are preferable. They are usually made of maple, mahogany, or pearwood, and their edges are often lined with hardwood or celluloid.

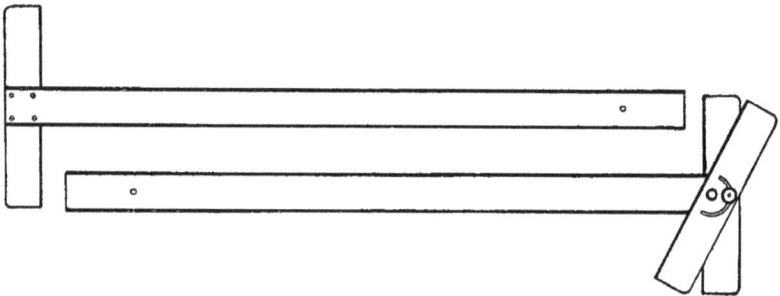

FIG. 123. T-SQUARE. PLAIN AND SWIVEL HEAD

The celluloid edges make it possible to see lines near the one to be drawn and are therefore quite convenient when joining lines at corners, etc.

The upper or working edge of the T-square and the edge of the head which rests against the drawing should be perfectly straight. The edge of the blade may be tested as follows:

1. Draw a long line along the edge of the blade.

2. Reverse the ends of the blade with respect to the ruled line, keeping the *same side up* and bringing the *same edge* against the ruled line.

3. Draw a second line along the edge of the blade. If the edge of the blade is straight the two lines will coincide. Both the head and the blade of the T-square may also be tested by means of a straightedge. Since the T-square is used only for ruling parallel lines, and as lines at other angles are drawn with the triangles in combination with the T-square, it is evident that the accuracy of the angles beween lines drawn with the T-square

and those drawn with the triangles does not depend upon the angle of the blade to the head of the T-square. It is not necessary, therefore, that the edge of the head and blade be exactly at an angle of 90° to each other.

Care should be taken to preserve the upper edge of the blade of the T-square from injury. It should never be used as a guide for the knife in cutting paper. When using the T-square the head is pressed firmly against the edge of the board with the left

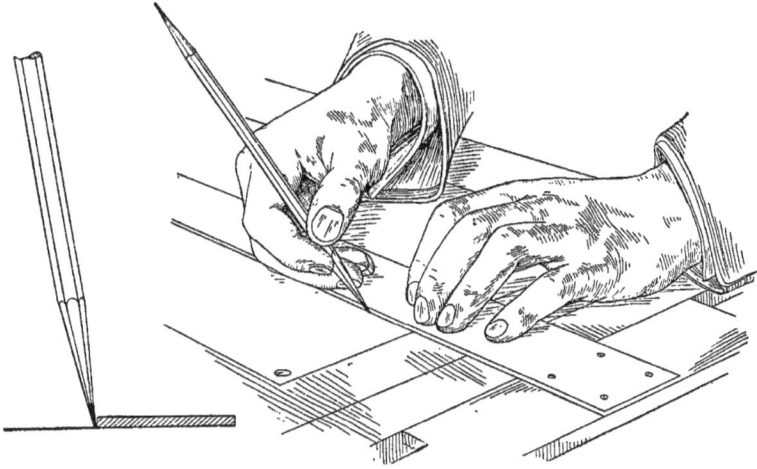

FIG. 124. RULING A HORIZONTAL LINE

hand as shown in Fig. 124. The lines are always drawn along its upper edge.

The triangles are used in combination with the T-square for drawing lines at certain angles to the horizontal. They are used in combination with each other for drawing lines at various angles with lines which are not horizontal.

Triangles are made of steel, wood, hard rubber, or celluloid. Steel triangles are used for the most accurate work. Triangles made of wood are easily injured and are likely to change their shape. Those made of celluloid have the advantage of being transparent and are more generally used. For accurate work it is necessary that the edges of the triangles be straight and that the angles be true. The edge may be tested by the method given for testing the T-square.

Assuming that the edge of the T-square has been found to be straight, the 90° angle of a triangle may be tested as follows:

1. Place the triangle in position D, as shown in Fig. 126, and draw the line A B.

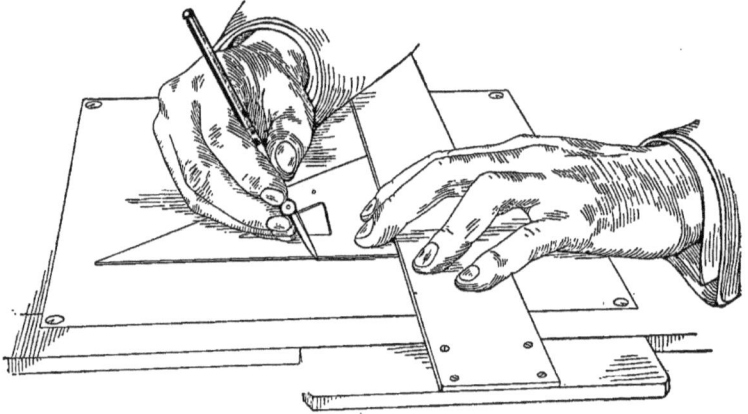

FIG. 125. RULING A VERTICAL LINE

2. If when the triangle is turned over into position C, the vertical edge coincides with the line A B, the angle is 90°

3. When the 90° angle of the 45° triangle has been found true, the 45° angles are true if equal.

Compare the two 45° angles as follows, Fig. 127:

1. Place the triangle against the T-square and draw a 45° line.

2. Turn the triangle over so that the other 45° angle comes into the position previously occupied by the first. If the edges of the triangle coincide with the line drawn, the 45° angles are equal.

The 60° angle of a 30°-60° triangle may be tested as follows:

1. Draw a horizontal line, A B, along the T-square. Fig. 128.

2. Draw a 60° line, B C, along the edge of the triangle crossing the horizontal line.

3. Turn the triangle over and draw a second 60° line, A C,

completing a triangle. If the triangle formed is equilateral, the 60° angle is true.

The lengths of the sides of the triangle may be compared by means of the dividers. When the edges are straight and the 90°

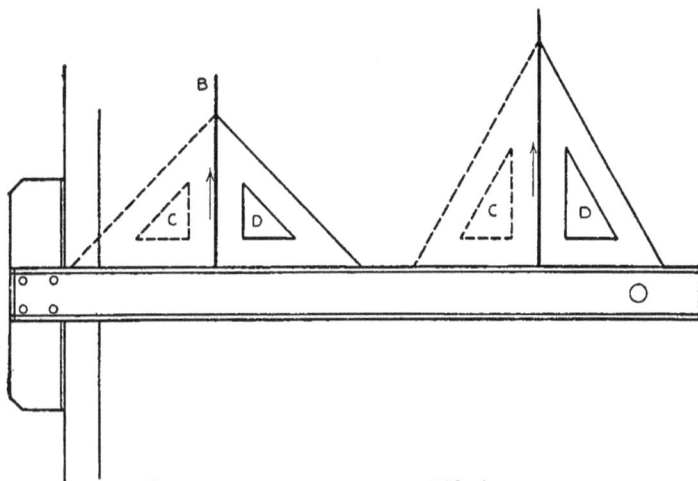

FIG. 126. TESTING THE 90° ANGLE

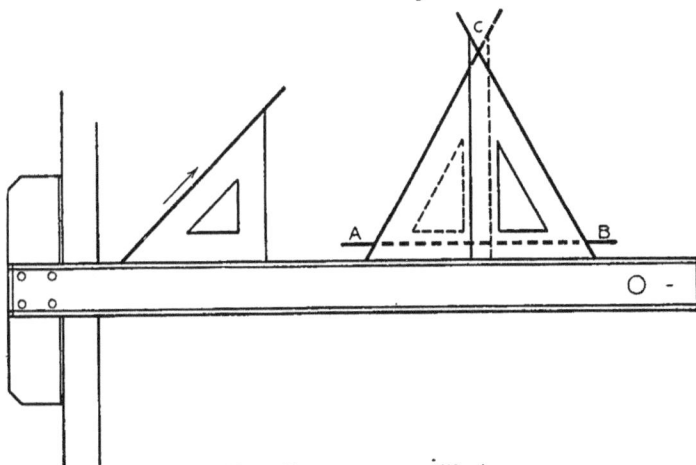

FIG. 127. TESTING THE 45° ANGLE
FIG. 128. TESTING THE 30° AND 60° ANGLES

and 60° angles are found to be true, the remaining angle, 30°, will also be true.

With the 30°-60° triangle, lines may be drawn at 90° to the

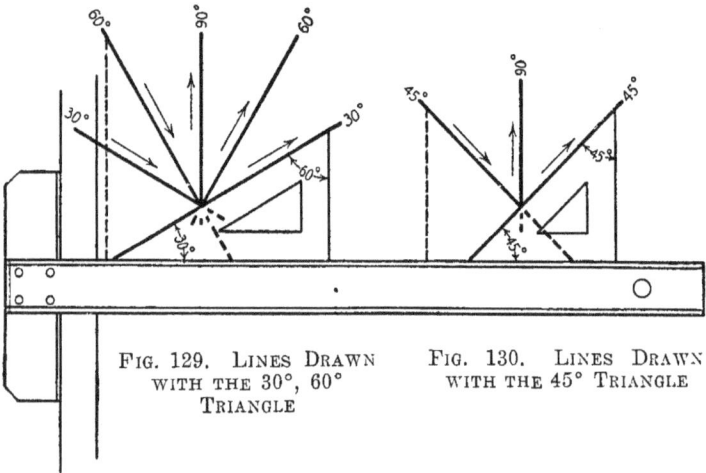

FIG. 129. LINES DRAWN
WITH THE 30°, 60°
TRIANGLE

FIG. 130. LINES DRAWN
WITH THE 45° TRIANGLE

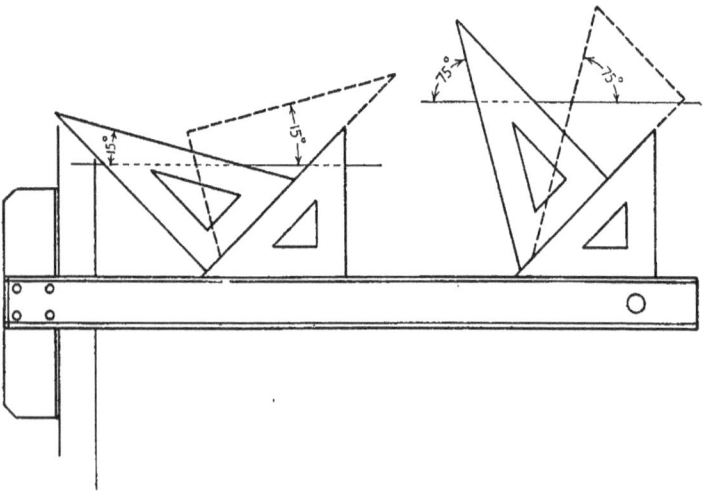

FIG. 131. LINES DRAWN WITH A COMBINATION OF THE 30°, 60° AND 45°
TRIANGLE

horizontal and at 30° or 60° with the horizontal to the right and
to the left. Fig. 129. With the 45° triangle, lines may be drawn

at 90° to the horizontal and at 45° with the horizontal to the right or to the left. Fig. 130.

By combining the two triangles, lines may be drawn at 15° or 75° with the horizontal to the right and to the left. Fig. 131. Lines parallel to any given line may be drawn by placing the two triangles in contact and sliding them as one tool until an edge of one coincides with the given line. Fig. 132. With triangle A held firmly in place, triangle B may be moved along it and lines drawn parallel to the given line.

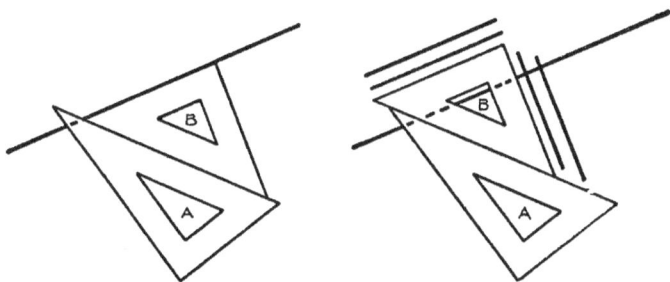

FIG. 132.   LINES DRAWN PARALLEL OR PERPENDICULAR TO ANY GIVEN LINE

Lines perpendicular to the given line may be drawn along the edge of triangle B which is at 90° to the given line.

The direction in which lines should be drawn along the triangles is shown in Figs. 129 and 130. The forearm should always make a right angle with the line being drawn.

The edges of the triangles should not be cut or bruised. If they are allowed to fall on the floor a corner may be blunted and as a result the angle will not be true. The celluloid triangles should not be allowed to remain bent for any length of time, as they will remain permanently so.

Pencils. For mechanical drawing where it is desired to produce fine sharp lines, a hard pencil should be used on a comparatively smooth, hard surfaced paper. The pencil should not be sharp enough or used with enough pressure to crease the paper. The 5H pencil should be used for drawing lines mechanically.

For convenience. in using the pencil for different purposes, it should be sharpened at both ends. One end is used for ruling lines and the other for laying off measurements. The ruling point is obtained by cutting away the wood to expose about $\frac{1}{4}''$ of lead and by rubbing opposite sides of the lead on a sandpaper pad or file to produce a wedge-shaped point. Fig. 133. This point is used for ruling continuous lines. The measuring point is similar to the conical point used in sketching except that the point is sharper in order that very accurate measurements may be laid off with it. It is used both for measuring and making dotted lines.

To insure accuracy in laying off measurements from the scale, the eye should be directly above the division on the scale from

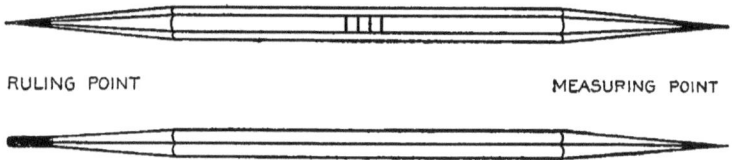

RULING POINT   MEASURING POINT

FIG. 133. RULING AND MEASURING JOINTS OF THE MECHANICAL DRAWING PENCIL

which the dimension is to be laid off. Care should be taken to place the point of the pencil on the paper exactly opposite the mark on the scale. The pencil should be revolved upon its axis while in this position without pressing the lead into the paper. The mark left by the pencil should be a small, round dot just visible to the eye.

Ruling Horizontal Lines. In ruling horizontal lines the position of the hand is the same as for sketching horizontal lines. In this case, however, the pencil is held leaning slightly forward with the point in the position shown in Fig. 124. The line is drawn with a continuous motion to the right with the tip of the fourth finger touching the T-square to steady the hand. Fig. 124. The forearm should always be at right angles to the line being drawn.

Vertical Lines are drawn along the edge of a triangle which is set against the T-square as shown in Fig. 125. Note that the triangle is to the right of the line. The line should be drawn away from the T-square with the hand and arm in the same relative position to the line being drawn as for horizontal lines.

Scales are used for taking measurements and laying off distances. They are made of paper, steel, and wood. Ordinarily scales are made of boxwood. There are two general forms, the flat and the triangular. The flat scale may have from one to four graduated faces and the triangular scale from four to six graduated faces. The graduations are placed directly on the wooden face of the scale or the face is coated with a white compound which makes the graduation easier to read.

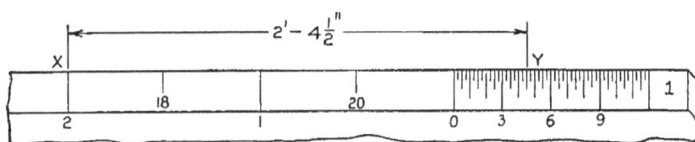

FIG. 134. READING THE ARCHITECT'S SCALE

The faces of the scales are graduated as follows:

The Engineer's Scale is divided to 10, 20, 30, 40, 50, and 60 parts to the inch. It is full divided, *i. e.*, the small divisions are marked off for the full length of the face.

The Architect's Scale is divided to $\frac{3}{32}$, $\frac{3}{16}$, $\frac{1}{8}$, $\frac{1}{4}$, $\frac{3}{8}$, $\frac{1}{2}$, $\frac{3}{4}$, 1, $1\frac{1}{2}$, and 3 inches to the foot. The edges on this scale are open divided, *i. e.*, only the portion of the face representing one foot is subdivided to read in smaller units. One face of the scale is usually divided into $\frac{1}{16}''$ for its full length.

To illustrate the reading of the architect's scale, consider the edge designated by a figure 1 at the end. Fig. 134. This indicates that one inch on this scale *represents* one foot. The inch to the right of the 0 at the right end of the scale is divided into forty-eight equal parts so that each of the smaller divisions *represents* $\frac{1}{4}''$ and the spaces between the 0, 3, 6, and 9 *represent* 3" each. To the left of the 0, the readings 1, 2, etc., are inches, and therefore represent feet. To measure off $2'-4\frac{1}{2}''$ to the right of point X, place the 2 opposite the point and read four and one-half inches to the right past the 0. In case it is desired to lay off this distance to the left of Y, place the four and one-half inch mark opposite Y and read past the 0 to the 2.

The Proportional Inch Scale is divided to read one-half or one-fourth inch to the inch and has one face divided to $\frac{1}{16}''$ for its full length. The open divided edges are read in the same manner as the architect's scale. The difference is that in this case the large divisions represent inches instead of feet. One of the large divisions is subdivided to read sixteenths.

Drawing Instruments. In beginning mechanical drawing it is important that the student have a good set of instruments. It is difficult to define a "good" set of instruments, for the better grades are extensively imitated. The student should be guided in his selection either by some experienced draftsman or by the trademark and the price charged by a reliable dealer.

A good set of instruments differs from a poor one, mainly, in the quality of materials used, correct tempering, and good workmanship. The steel of the pens must be properly tempered so that when once sharpened the points will remain in good condition for a reasonable time. The compass and dividers must be so made that they will retain their alignment and adjustment when handled with ordinary care. These qualities can only be definitely determined after the instruments have been given a fair trial.

To secure uniformly satisfactory results in drawing it is necessary to start with a good set of instruments and to keep them in good condition.

The Compass is used for drawing circles and arcs of circles. Fig. 135. The better grades are made of German silver. It is important that a compass be light yet rigid. The most important part of the compass is the head which, in the modern instruments, consists of two discs held in contact in a fork by means of pivot screws. By adjusting these screws the pressure between the discs is regulated. This pressure should be such that the legs of the compass may be opened or closed without springing them. On the other hand, the joint should be tight enough to retain the setting when the instrument is in use.

The thing of next importance is the socket joint of the removable pen and pencil parts. These are made in various forms.

They usually consist of a shank on the pen and pencil parts which fits into a corresponding socket in the compass leg. The proper position of the shank in the socket is insured by some device such as a feather or sharp corner on the shank which is matched by a corresponding slit or groove in the socket. These parts are made to clamp together with a thumb screw or else a bayonet fitting is used.

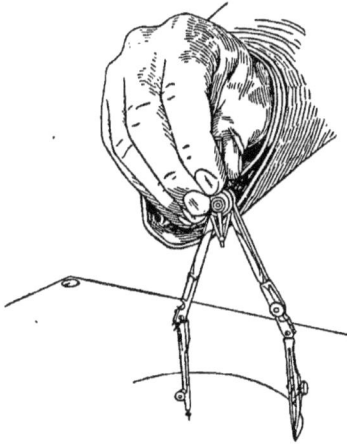

FIG. 135.   DRAWING A CIRCLE WITH THE COMPASS

The legs of the compass should move in the same plane. To test the compass for alignment:

1. Place the parts in the sockets.

2. Bend the legs out at the head, and

3. Bring the joints together, as shown in Fig. 136. If the points are exactly together the joints are true.

Before using a compass, the needle point and lead should be adjusted as follows:

1. Place the pen in the compass and adjust the needle point so that it projects slightly beyond the nibs of the pen.

2. Remove the pen.

3. Replace the pencil and adjust the head so that it is slightly shorter than the needle point. The pen and pencil parts are now interchangeable without adjusting the needle point.

In using the compass proceed in the following manner:

1. Place a 4H lead in the compass and sharpen it to a narrow wedge, in width about one-half the diameter of the lead.

2. Set the lead so that it projects about one-half the length of the needle point beyond the shoulder.

3. Draw the center lines of the circle to be drawn at right angles and lay off the radius on one of them.

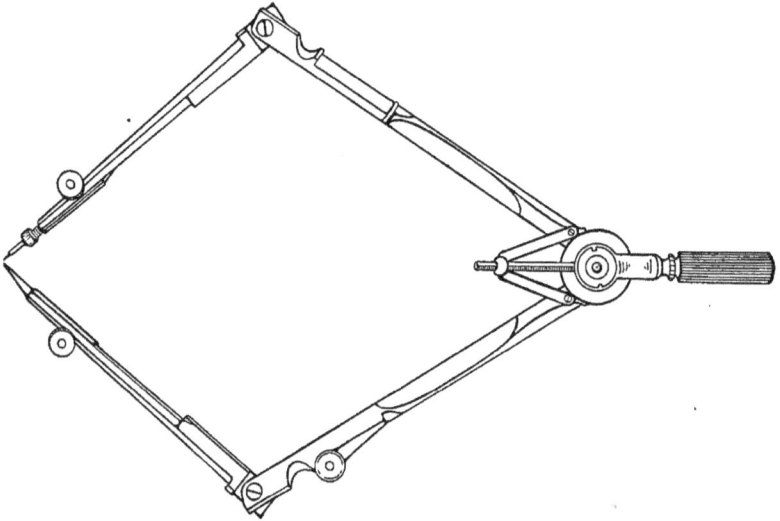

FIG. 136. TESTING THE COMPASS

4. Grasp the compass by the handle between the thumb and first finger of the right hand. Care should be taken to place the needle point *exactly* at the intersection of the center lines.

5. Adjust the lead to the exact radius and draw the circle, rolling the handle of the compass between the thumb and finger.

The large compass should not be used for circles of less than ¾″ radius. For very large circles the lengthening bar should be inserted between the leg of the compass and the pen or pencil point. When this does not suffice a beam compass should be used.

Dividers are similar to compasses in general appearance. The legs terminate in sharp steel points. The dividers are used for laying off distances from the scale, for transferring lengths, or for dividing straight or curved lines into any number of equal parts.

To divide a line into any number of equal parts with the dividers, proceed as follows: (Assume that the line is to be divided into three equal parts.)

1. Open the dividers to what is estimated to be one-third the length of the line.

2. Step off the estimated distance three times on the line.

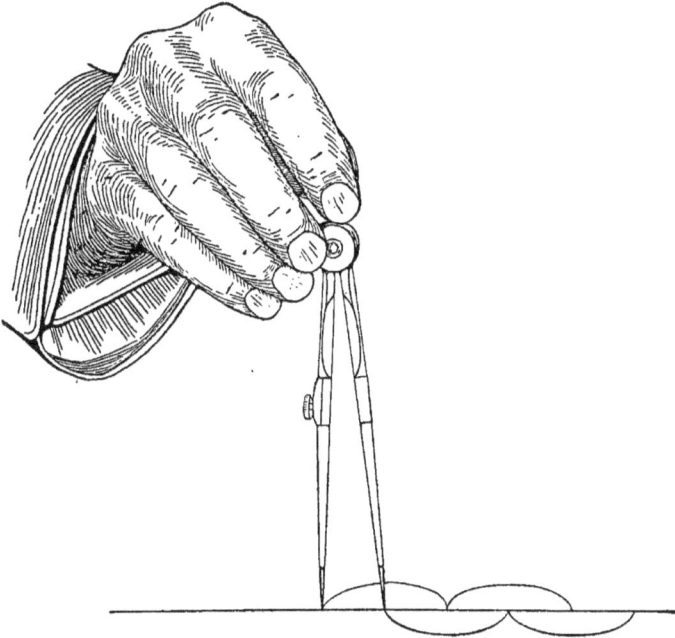

FIG. 137. STEPPING OFF WITH THE DIVIDERS

3. Adjust the dividers to one-third the error making the distance between the points larger or smaller as the case may require.

4. Repeat the process until the third step ends exactly at the end of the line. In taking the steps the dividers are held by the handle between the thumb and first finger and swung alternately first to one side of the line and then the other as shown in Fig. 137. This avoids rolling the handle in an awkward position between the thumb and finger.

The Bow Pen, Bow Pencil, and Bow Dividers. The bow pen and bow pencil are used to describe small circles, and the bow dividers to lay off small distances. They have the advantage over the larger instruments that they retain their adjustment. There are two forms of adjusting devices, as shown in Fig. 138. To make large adjustments in the instruments having side screws the pressure on the nut should be relieved by pressing the legs together with the left hand while the nut is made to spin with the first finger of the right hand.

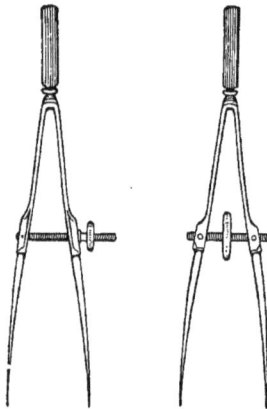

FIG. 138. CENTER AND SIDE SCREW ADJUSTMENT OF BOW INSTRUMENTS

The bow compass should be used in the same manner as the large compass, as described on page 129.

The Eraser. Ordinarily a medium hard eraser such as the ruby is used for removing pencil lines from a drawing. A soft flexible eraser is very satisfactory for cleaning a pencil drawing without erasing the lines. When erasing lines the paper near the lines to be erased should be held down with the thumb and first finger of the left hand to prevent it from crumpling.

The Erasing Shield is used to protect the parts of the drawing which are not to be erased. The opening in the shield is selected which is best suited to expose only the parts to be erased. The shield is held in position on the drawing with the thumb and first finger of the left hand, while the eraser is applied with the right. Fig. 139.

Steps in Making a Mechanical Drawing

The Border Rectangle. To draw the border rectangle, proceed as follows:

1 Lay off $\frac{1}{2}''$ from the upper and left hand edge of the sheet.

2. Through the points thus located draw the upper and left hand sides of the border rectangle.

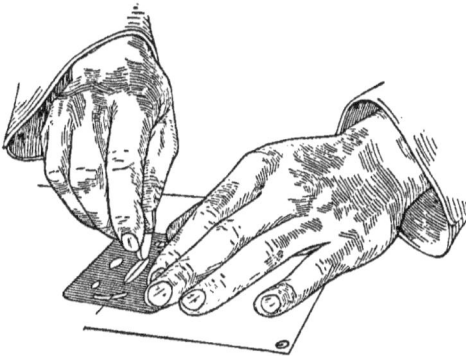

Fig. 139. Using the Eraser and Shield

3. On these lines, and from their intersection, lay off 14″ to the right and 10″ downward.

4. Through the points thus found draw the remaining sides of the border rectangle.

The Enclosing Rectangle. In mechanical drawing the views are located centrally by calculating the position of a rectangle in which they may be inscribed. In this course the distance between views should not be less than $\frac{3}{4}''$ or more than 1″. The student's calculation should be made as indicated in Fig. 140.

Accuracy. It is of prime importance that a mechanical drawing be accurate. Accuracy depends both on the quality and condition of the instruments and materials and upon the skill of the draftsman. All straightedges, angles, etc., should be tested as just described. When the tools are found to be in good condition the draftsman should take great care to lay off measure-

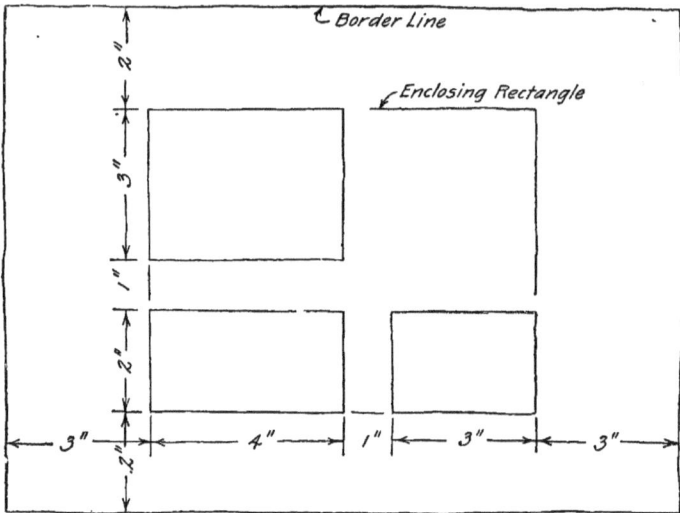

Scale of drawing = full size. Border rectangle = 10"x 14."
    4" = width of front view.
    3" = width of side view.
    1" = space between views.
    8" = width of enclosing rectangle.

    14" = width of border rectangle.
     8" = width of enclosing rectangle
$$2\overline{)6}\ ''$$
    3" = distance between the views
       and the border line at
       the right and left.

    2" = height of front view.
    3" = height of top view.
    1" = space between views.
    6" = height of enclosing rectangle.

    10" = height of border rectangle.
     6"
$$2\overline{)4}\ ''$$
    2" = distance between the views
       and the border line at
       top and bottom.

FIG. 140. CALCULATION FOR THE SIZE AND POSITION OF THE ENCLOSING
RECTANGLE
(134)

ments accurately, and draw the lines exactly through the points located. It is always well to place the point of the pencil in the located point, bring the straightedge up to the pencil, and then draw the line, being careful to maintain the same relationship throughout between the pencil and the straightedge.

Errors multiply with the number of operations involved, hence, other things being equal, the most direct construction is the most accurate one.

Constructive Stage. In this stage all measurements are laid off and lines drawn lightly and of indefinite length. This is what is sometimes known as the blocking-in stage. When using the scale make as many measurements as possible. Whenever practicable, consecutive measurements should be laid off with the scale in one position. Be as systematic as possible in making measurements and in using the T-square and triangles. It is a good plan to draw as many of the horizontal lines as may be drawn at one time, beginning at the top and moving the T-square downward. In like manner draw vertical lines, several at a time, moving the triangle from left to right. The lines should be drawn long enough so that there will be no need to extend them.

No distinction is made between visible and invisible edges in this stage. Fig. 116.

Finishing Stage. In this stage the drawing is completed as it will finally appear. First, erase all lines not needed in the finished drawing and, second, retrace all required lines with a carefully sharpened 5H pencil. All finished lines must be of uniform width and shade. They must be ended at the proper points. The lines should be drawn in the following order:

1. Horizontal lines beginning at the top of the sheet.

2. Vertical lines beginning at the left of the sheet.

The hidden edges of the object should now be represented by so-called dotted lines, which in reality are lines made up of short dashes and spaces. The dashes are $\frac{1}{8}''$ long, separated by $\frac{1}{32}''$ spaces. Fig. 76. The end of each dash can be made distinct by keeping the end of the pencil in contact with the paper until the end of the line is reached. The pencil should be placed upon the paper, drawn $\frac{1}{8}''$ on the paper, stopped, and then raised in making each dash.

Dimensioning. The draftsman's judgment is used more in dimensioning than in any other part of the drawing. To avoid mistakes and to facilitate the work of the mechanic, only necessary dimensions should be given. They should be placed in such a way as to make the drawing easily read. Cases are rare where it is advisable to repeat the same dimensions on different views. Repeating dimensions adds to the difficulty in checking them and when changes are made there is a possibility of making a change in one place and not in another. This leads to confusion. Placing dimensions in obscure and out of the way places should be avoided. Whenever possible, dimensions should be grouped in such a way as to make their relation obvious. It should not be necessary for the mechanic to do any calculating to obtain necessary dimensions.

No doubt, the best guide to follow, in placing dimensions on a drawing, is for the draftsman to imagine himself in the mechanic's place and to consider the operations through which the object must go to become a finished product. With this idea in mind most problems in dimensioning will be solved without difficulty. For example, when a machinist drills a hole he sets the point of the drill at its center; hence the hole should be dimensioned by referring its center to some surface, line, or point easily accessible.

In ordinary working drawings, dimensions are usually given in inches *up* to twenty-four inches. *Above* twenty-four inches they should be given in feet and inches. Examples: $23\frac{1}{2}''$, $2'-4\frac{1}{2}''$.

For all ordinary work, fractions in dimensions containing mixed numbers have the following denominators: 2, 4, 8, 16, 32, 64; such denominators as 6 or 19 are not used. When very small fractions of an inch are necessary, as in the case of special fits, etc., the fractional part of an inch may be expressed in decimals of three or four places. Example: $5.006''$ bore.

Extension and Dimension Lines. The extension lines and dimension lines should be drawn in the order suggested for the finishing stage of the pencil drawing, *i. e.*, draw all horizontal lines beginning at the top and moving downward, then draw all vertical lines beginning at the left and moving toward the right. As in orthographic sketching, the extension lines should begin

about $\frac{1}{32}''$ from the outline of the object and continue $\frac{1}{8}''$ beyond
the arrowheads. The space between the outline of the object and
the nearest dimension line, or between two parallel consecutive
dimension lines, should be about $\frac{1}{4}''$. The extension and dimension lines in pencil should be of the same width and shade as the

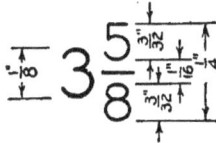

$$\stackrel{\mid}{\underset{\mid}{}}_{\frac{1}{8}} 3\frac{5}{8}$$

Fig. 141. Showing Actual Heights of Whole Number and Fraction

object lines. Center lines may be used as extension lines, but
not as dimension lines.

The Dimension Figures and Notes. If a drawing is to present
a neat appearance, a suitable type of letter and figure should be
used for all notes and dimensions. A very plain letter should be
selected; one that can be drawn with reasonable rapidity and
that will be in harmony with the remainder of the drawing. It
is essential that a standard height be adopted and adhered to for
all notes and figures on the drawing. For this course the standard height for the whole number is $\frac{1}{8}''$ and the total height of the
fraction $\frac{1}{4}''$ as shown in Fig. 141. Whenever possible, notes
should be lettered on horizontal guide lines. The letters should
be $\frac{3}{32}''$ high. To insure uniform heights for all notes the distance
between the ruled guide lines should be accurately laid off with
the scale or stepped off with the dividers.

As nearly as possible, place the dimension figures in the center of the dimension line, leaving a convenient free space for the
figures. Whenever a center line interferes with the dimension
figures, place it near the center line and either to the right or
left of it. In the case of consecutive parallel dimension lines
where the dimension figure in one line would naturally interfere
with the dimension figure in the other dimension line, the dimension figures should be "staggered"; that is, one dimension figure
should be placed a little to the right and the other to the left of
the center of the dimension line, so that they will not interfere.
Example: Fig. 147.

Information which the dimensioned drawing does not make clear is put into the form of notes. They usually relate to materials, finish, number of parts needed, etc. Example: See note below the view. Fig. 156.

The Title. The views of an object with their dimensions and notes do not convey all of the necessary information. A title which supplies the deficiency is therefore added. The title is

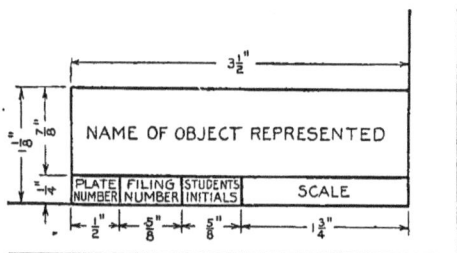

FIG. 142. DIMENSION AND POSITION OF ITEMS IN TITLE BLOCK

usually placed in the lower right-hand corner of the sheet so that it will be easily accessible when the drawing is filed.

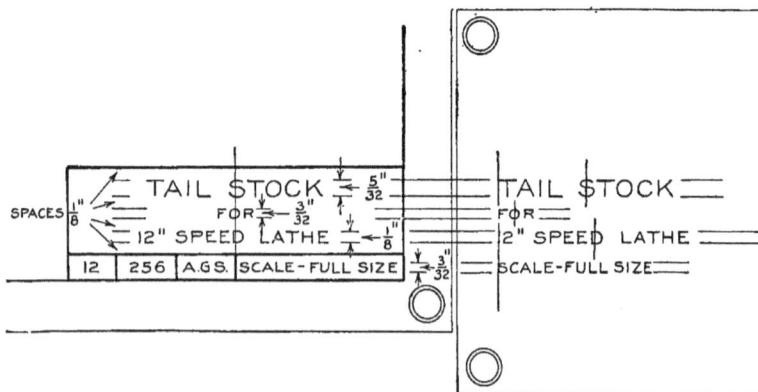

FIG. 143. SHOWING METHOD OF BALANCING THE LINES IN THE TITLE

Various elements may enter into the title, depending upon the character of the drawing and the use to be made of it. The following items are usually found in the titles of commercial drawings of machines or structures:

1. Name of part or parts of machine or structure.
2. Name of complete machine or structure.
3. Manufacturer's firm name and address.
4. Drawing number.
5. Date of finishing drawings.
6. Scale to which drawing is made.
7. Initials of draftsman, tracer, and checker.

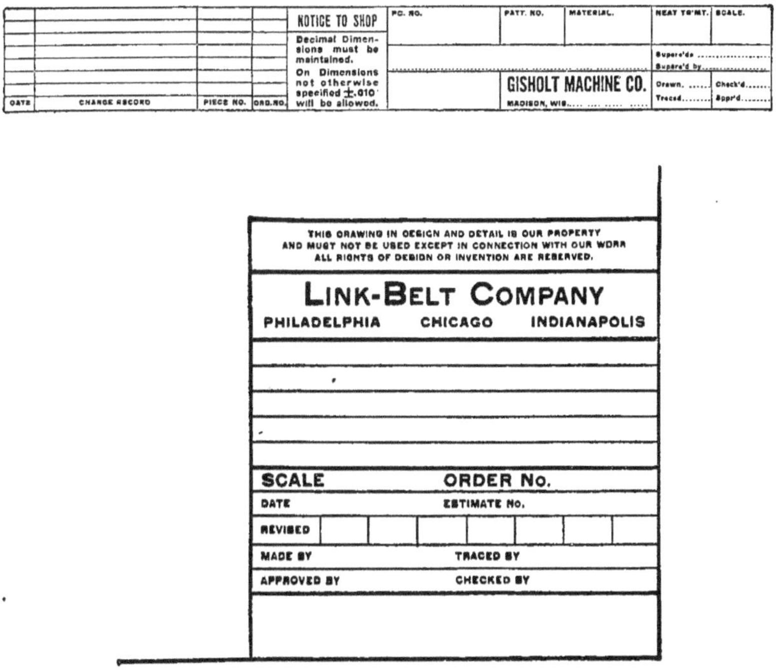

FIG. 144. COMMERCIAL TITLES

The titles of the plates given in this chapter will be much simpler than the ordinary commercial title. Figs. 142 and 143 show two forms of title suitable for this course. The words for each title in this chapter will be given below the figure from which the drawing is made.

The relative importance of the items of a title is shown by the varying heights or weight, or both, of the letters. In some drafting offices a rubber stamp is used on the pencil drawing to obtain the words and lines that are common to all drawings. The same words and lines are often printed on the tracing cloth in type. Example: Fig. 144. The style of letter used should be plain and dignified, whether printed in type or drawn freehand.

The Title Block. The title for each mechanical drawing plate in this course will be placed in a *title block*. The dimensions for this block with the names and position of the items are shown in Fig. 143. The height of the letters and spaces between lines of letters are shown in Fig. 143.

Balancing a Title. It is essential to the appearance of a title that the lines be symmetrical with respect to a vertical center line. Example: Fig. 143 shows a title properly balanced.

To balance the title proceed as follows:

1. Tack a piece of drawing paper to the board opposite the lower right hand corner of the sheet. Fig. 143. This will be referred to as the *trial sheet*.

2. Draw a line as a continuation of the lower border line on the trial sheet. This is a base line for measurements.

3. Lay off on the trial sheet the space for the letters as given in Fig. 143. Extreme accuracy in making these measurements is necessary, as the width of the letters varies with their height. A small error in height makes the letter appear much too large or too small.

4. Rule each guide line on the trial sheet and the drawing sheet with one setting of the T-square. Care should be taken to draw exactly through the points located. Check the heights of the spaces with the scale.

5. Letter each line of the title on the trial sheet, giving attention to the proportion of the letters and to spacing. Do not try to balance the lines on this sheet.

6. Locate the middle point of each line on the trial sheet.

7. Draw the vertical center line of the title through the center of the title rectangle.

8. Cut out each line of letters from the trial sheet and place it above the space in which it is to be lettered on the drawing sheet, with its middle point on the center line of the title.

9. Letter each line, following the spacing on the trial line. The result should be a perfectly balanced title.

DATA FOR DRAWING PLATE 13

Given: The orthographic sketch, Plate 12.
Required: To make a pencil mechanical drawing from Plate 12.

Instructions:

1. Test the drawing board, T-square, and triangles as explained under the corresponding headings in this chapter.

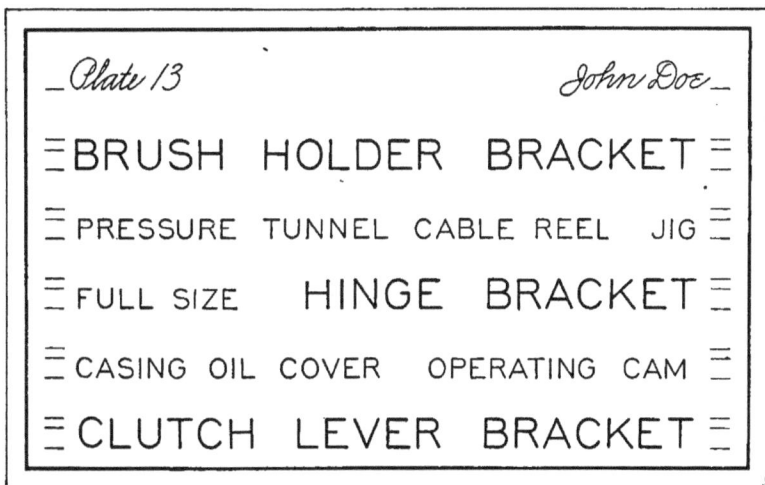

Plate 13   John Doe

≡BRUSH HOLDER BRACKET ≡

≡ PRESSURE TUNNEL CABLE REEL   JIG ≡

≡ FULL SIZE   HINGE BRACKET ≡

≡ CASING OIL COVER   OPERATING CAM ≡

≡ CLUTCH LEVER BRACKET ≡

FIG. 145. LETTERING PLATE 13

2. Draw the border line as previously explained.
3. Calculate the size of the enclosing rectangle.
4. Lay off as many of the dimensions of the object as possible at one time. Draw the lines lightly.
5. Check the drawing for accuracy.
6. Erase unnecessary lines and retrace the drawing, taking care to end the lines exactly at their intersections. Dot the lines representing hidden edges.
7. Draw extension and dimension lines and put in dimensions.

8. Letter a note, giving the number of parts required and the material from which they are to be made.

9. Letter the title, using the name of the object given below the figure from which the drawing was made.

DATA FOR LETTERING PLATE 13

Given: Plate 13 to reduced size. Fig. 145.

Required: To make the plate in ink to an enlarged scale.

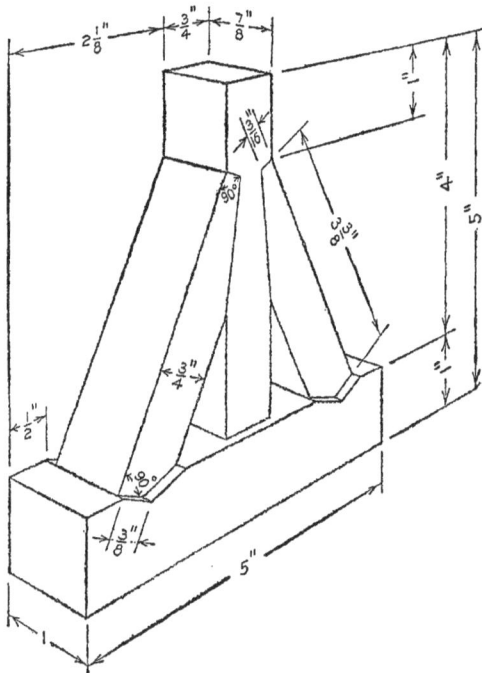

Fig. 146. Type Problem. Perspective of Brace

PREPARATORY INSTRUCTIONS FOR DRAWING PLATE 14

In this plate the student will need to decide for himself the number of views necessary, and their arrangement, to show the form of the object. It is suggested that for this purpose he review the principles given on pages 110 and 111.

The methods of dimensioning, particularly those relating to the dimensioning of inclined lines, should also be reviewed. Pages 136 and 137.

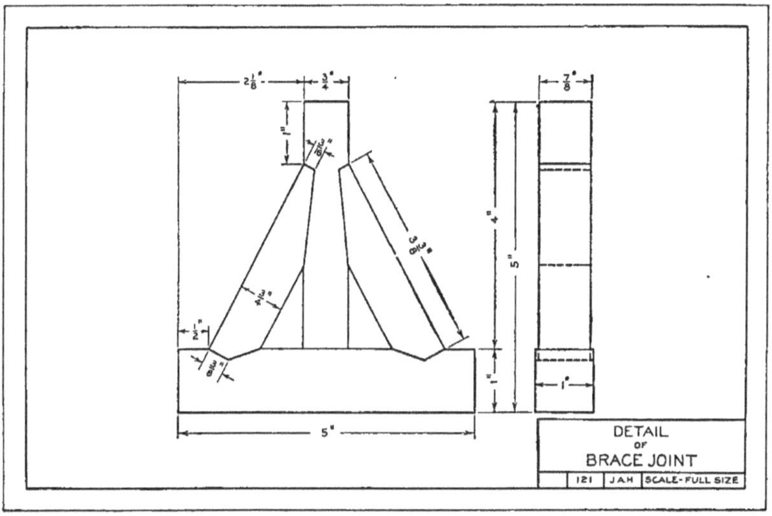

FIG. 147. TYPE PROBLEM. DETAIL OF BRACE JOINT

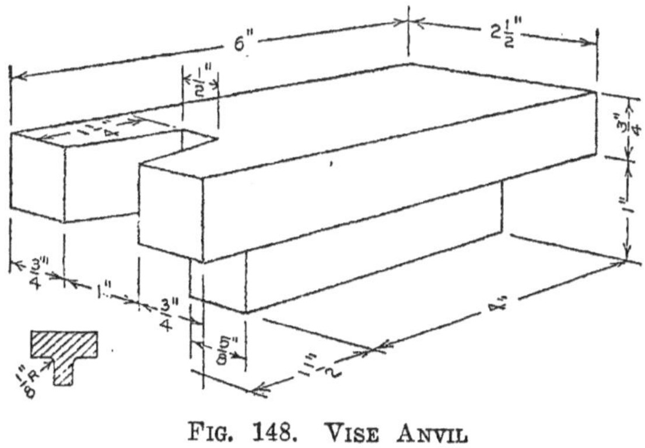

FIG. 148. VISE ANVIL

DATA FOR DRAWING PLATE 14

Given: Perspective sketches, Figs. 148, 149, 150, and 151.

Required: To draw an orthographic sketch of the object

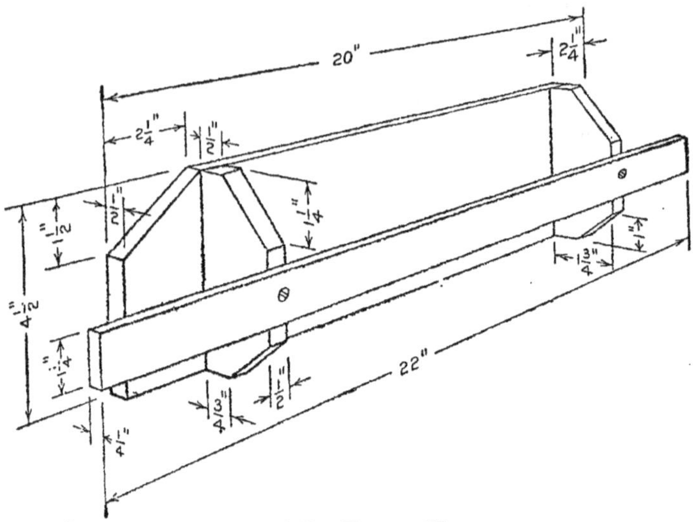

FIG. 149. TOWEL HANGER

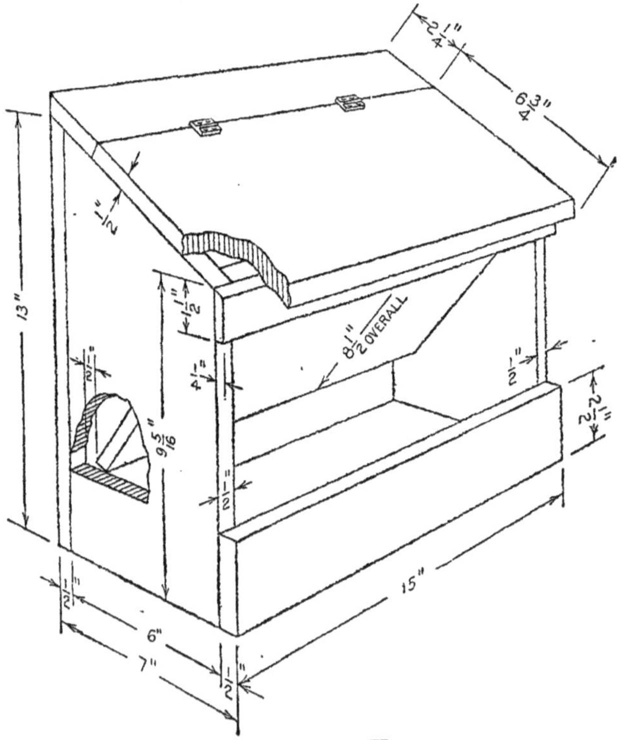

FIG. 150. FEED HOPPER

(144)

shown in Fig. 148, 149, 150, or 151, or any similar object, with dimensions, as assigned by the instructor. *Instructions:* Proceed as for previous orthographic sketches.

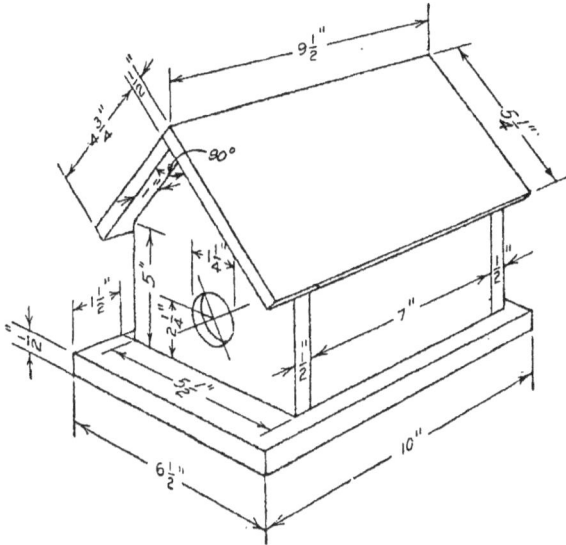

FIG. 151. BIRD HOUSE

DATA FOR LETTERING PLATE 14

Given: Plate 14 to reduced size. Fig. 152.
Required: To make the plate in ink to an enlarged scale.

PREPARATORY INSTRUCTIONS FOR DRAWING PLATE 15

Inclined Lines. Lines at such angles as 15°, 30°, 45°, 60°, and 75° may be drawn with the T-square and triangles described on pages 123 and 124. When the inclination of a line is not given in degrees at least two points on it must be located. The line is then drawn by placing an edge of a triangle or the T-square so that it passes through the two points. *Scale.* When an object is too large to be drawn full size on the sheet, it may be drawn to some fraction of the actual size.

Half and quarter sizes are common scales for shop drawings. The edge of the scale, graduated to read half or quarter size, should be used instead of dividing the dimensions by 2 or 4. See page 127 for a description of the method of using the scale.

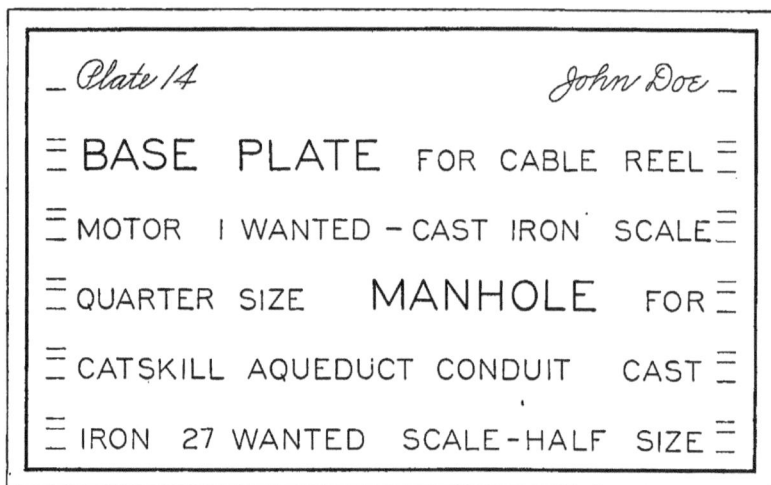

_ *Plate 14* *John Doe* _

≡ BASE PLATE FOR CABLE REEL ≡

≡ MOTOR I WANTED – CAST IRON SCALE ≡

≡ QUARTER SIZE MANHOLE FOR ≡

≡ CATSKILL AQUEDUCT CONDUIT CAST ≡

≡ IRON 27 WANTED SCALE - HALF SIZE ≡

FIG. 152. LETTERING PLATE 14

DATA FOR DRAWING PLATE 15

Given: The orthographic sketch, Plate 14.

Required: To make a pencil mechanical drawing from Plate 14.

Instructions:

1. Draw the border line and calculate the size of the enclosing rectangle as in plate 13.

2. Lay off the dimensions of the object and complete the constructive stage.

3. Check carefully each dimension for accuracy.

4. Retrace the object lines, drawing (1) horizontal lines, beginning at the top; (2) vertical lines beginning at the left; (3) inclined lines.

5. Draw extension and dimension lines and put in dimensions.

6. Letter a note, giving the number of parts required and the material from which they are to be made.

7. Letter the title, using the name of the object given below the figure from which the drawing was made.

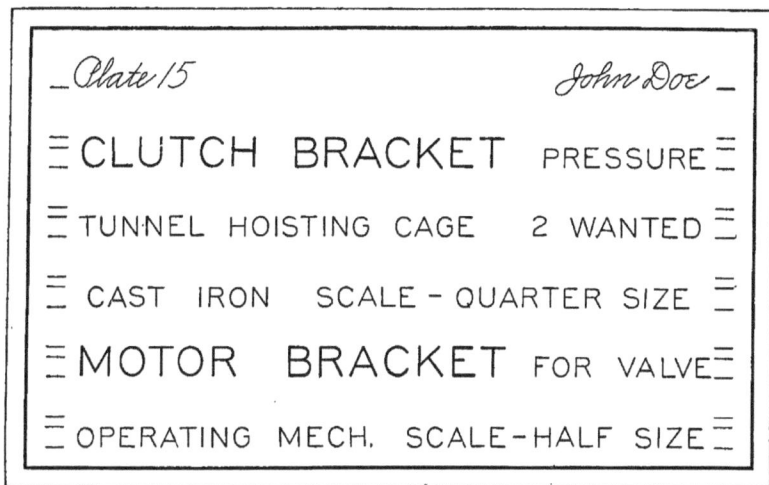

Plate 15   _John Doe_ _

≡CLUTCH BRACKET PRESSURE≡

≡TUNNEL HOISTING CAGE   2 WANTED ≡

≡ CAST IRON   SCALE – QUARTER SIZE ≡

≡MOTOR BRACKET FOR VALVE≡

≡OPERATING MECH. SCALE–HALF SIZE≡

FIG. 153. LETTERING PLATE 15

DATA FOR LETTERING PLATE 15

Given: Plate 15 to a reduced size. Fig. 153.
Required: To make the plate in ink to an enlarged scale.

PREPARATORY INSTRUCTIONS FOR DRAWING PLATE 16

In a freehand or mechanical drawing where a straight line is tangent to an arc, the arc should be drawn first. In the constructive stage the arc should be drawn long enough so that it will extend beyond the point of tangency when the line is drawn. A straightedge may then be laid tangent to the arc and the straight line drawn in. Before the drawing is finished the unnecessary part of the arc is erased. Fig. 155.

Centers for rounded corners, fillets and other arcs of circles, which do not have their centers on any line of the drawing, are located by what is called the "trial and error" method. The compass should be first adjusted to the proper radius. To locate the center of the arc, set the lead on the tangent line at A, Fig.

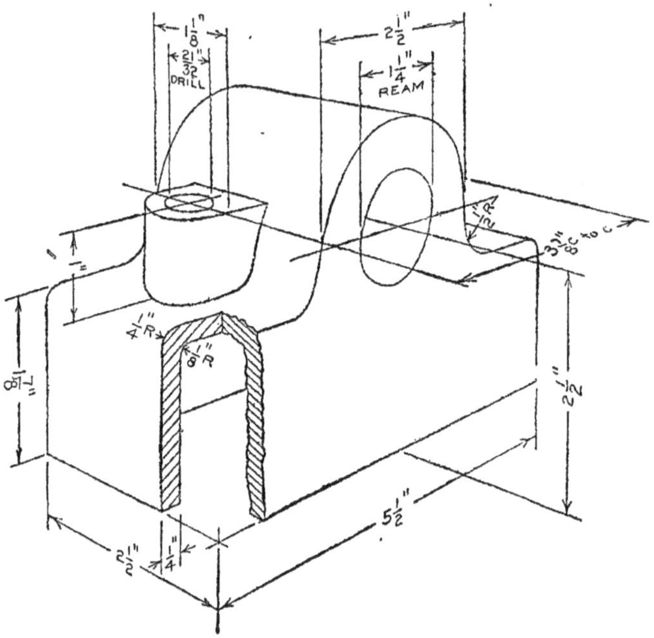

FIG. 154. TYPE PROBLEM. PERSPECTIVE OF BEARING

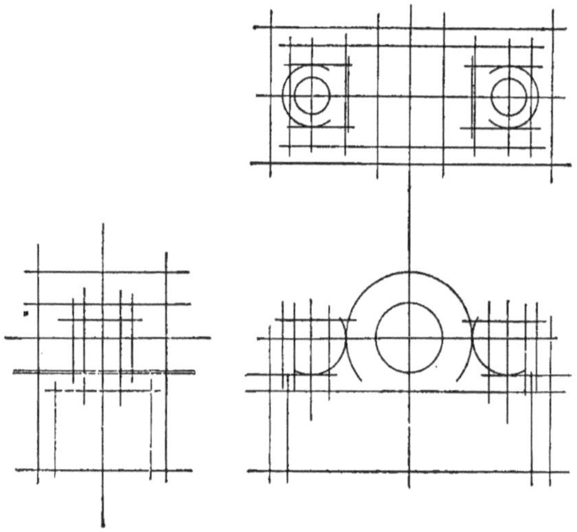

FIG. 155. TYPE PROBLEM. CONSTRUCTIVE STAGE OF MECHANICAL DRAWING

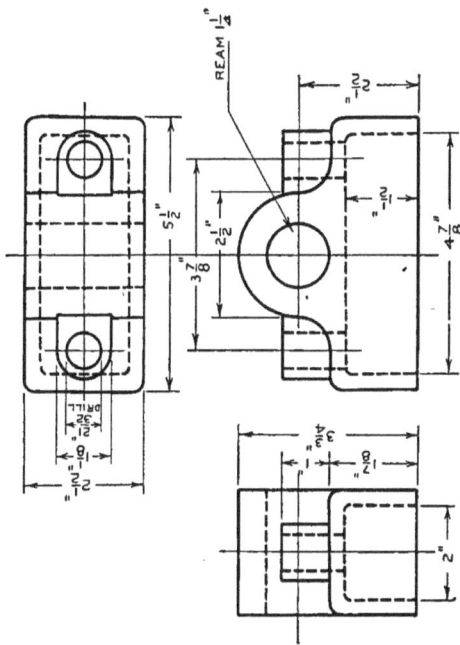

ONE WANTED   CAST IRON

SPARKER BEARING

| 17 | 101 | DAT. | SCALE-HALF SIZE |

FIG. 156.   TYPE PROBLEM.   FINISHED DRAWING

157, estimating A C as nearly as possible equal to the radius of
the arc. Set the needle point at B opposite A and bring the lead
around to D. Move the needle point parallel to A C an amount
equal to the error. The compass should then be again rotated
back to A to test for accuracy, and if necessary further adjustment should be made before drawing the arc.

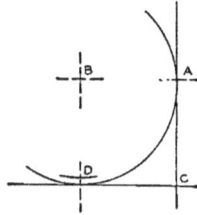

FIG. 157. TRIAL AND ERROR METHOD OF LOCATING CENTERS

Radius Dimensions. The dimension form for radius dimensions is shown in Fig. 189. When the distance between the arc
and its center is great enough to admit the figures and arrowheads the form is as shown in Fig. 189. Sometimes a small circle is drawn around the center in place of an arrowhead. This
circle should be made freehand and about $\frac{1}{16}''$ in diameter. When
the distance between the arc and its center is short the center,
as shown by the $\frac{1}{8}''$ radius, is not indicated. Fig. 189.

DATA FOR DRAWING PLATE 16

Given: Perspective sketches, Figs. 158, 159, and 160.

Required: To draw an orthographic sketch of the object
shown in Fig. 158, 159 or 160, or any similar object, with dimensions, as assigned by the instructor. The student should decide
what views are necessary to show the form of the object.

Instructions: In drawing the circles and arcs, sketch in the
center lines and lay off the radii on each, as in Plate 10.

PREPARATORY INSTRUCTIONS FOR LETTERING PLATE 16

Composition. This and the following plates will be devoted
to the practice of notes which frequently appear on the drawing
to give information not shown by the views.

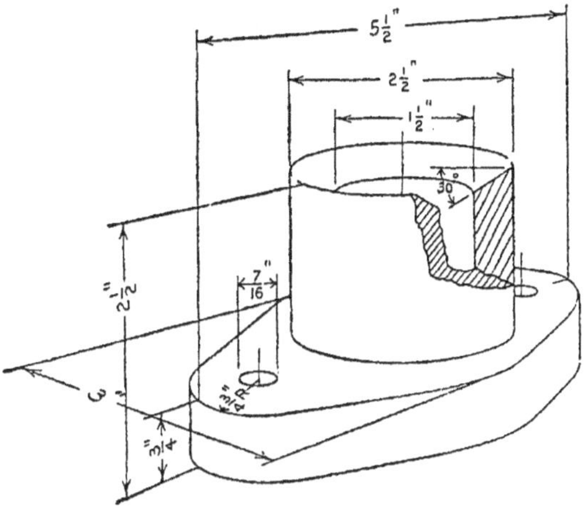

FIG. 158. STUFFING BOX GLAND

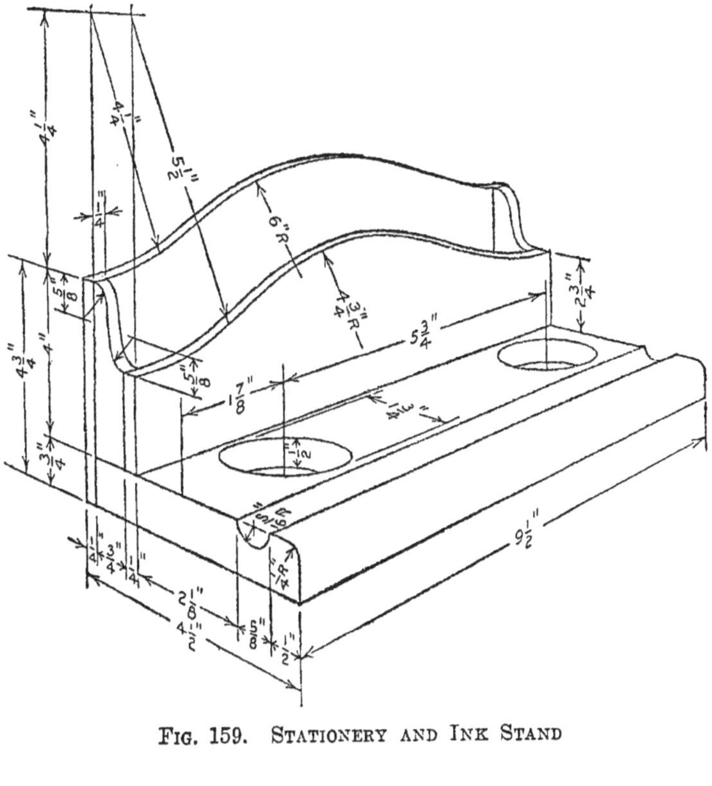

FIG. 159. STATIONERY AND INK STAND

DATA FOR LETTERING PLATE 16

Given: Plate 16 to reduced size.   Fig. 161.

Required: To make the plate in ink to an enlarged scale.

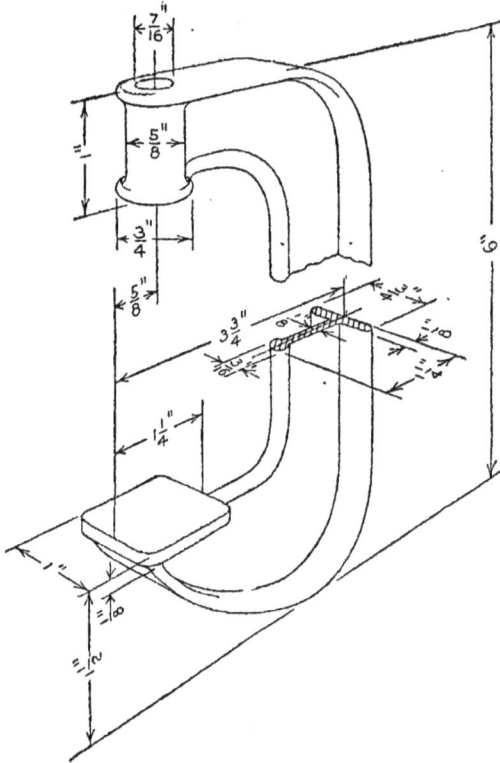

FIG. 160.   CLAMP

DATA FOR DRAWING PLATE 17

Given: The orthographic sketch, Plate 16.

Required: To make a pencil mechanical drawing from Plate 16.

Instructions:

1. Draw the border line and enclosing rectangle.

2. Locate and draw two center lines at right angles to each other for each arc or circle.

3. Draw the arcs of indefinite length so they extend beyond the points of tangency.

4. Draw the straight lines tangent to the arcs.

5. When the constructive stage is complete retrace the lines in the following order: (1) circles and arcs; (2) horizontal lines, beginning at the top of the sheet; (3) vertical lines, beginning at the left of the sheet; (4) inclined lines.

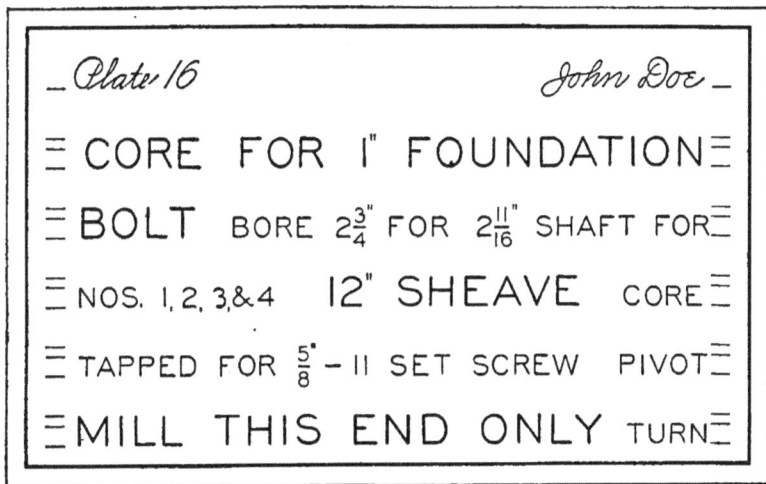

FIG. 161. LETTERING PLATE 16

6. The center line may be produced and used as an extension line where appropriate.

7. Letter a note, giving the number of parts required and the materials from which they are to be made.

8. Letter the title.

DATA FOR LETTERING PLATE 17

Given: Plate 17 to reduced size. Fig. 162.

Required: To make the plate in ink to an enlarged scale.

REVIEW QUESTIONS

1. (a) What determines the number of views of an object?
(b) When are more than two views necessary?

2. Where is the front surface of an object represented in the side view?

3. (a) What dimension is common to the top and side views?
(b) If only the top and side views were drawn how should they be related?

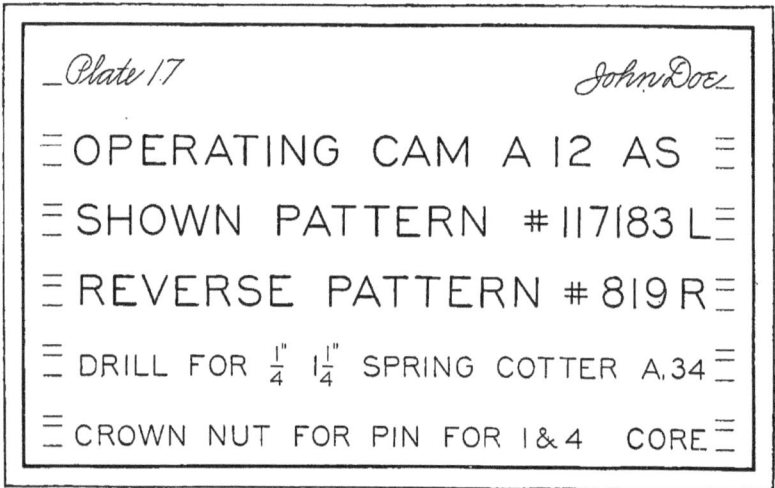

FIG. 162. LETTERING PLATE 17

4. What are the requisites of a good drawing board?

5. (a) Describe a method for testing the surface and working edge of a drawing board. (b) What care should be taken of the surface and working edge of a drawing board?

6. Give requisites of a good T-square and explain its uses.

7. Is it necessary for the head of a T-square to be at right angles to the blade? Why?

8. Describe a method for testing the working edge of a T-square for straightness.

9. (a) Describe the position of the T-square for drawing horizontal lines. (b) How is it held? (c) Illustrate by a sketch the position of the pencil in ruling a line along the T-square.

10. Describe the process of squaring and fastening the paper on the board for a mechanical drawing.

11. What is the advantage of the celluloid triangle over triangles made of other materials?

12. (a) For what are triangles used? (b) For what angles are they usually cut?

13. Describe a method of testing the accuracy of a 90° angle of a triangle.

14. Describe a method of testing the accuracy of a 45° angle of a triangle.

15. Describe a method of testing the accuracy of a 30° and 60° angle of a triangle.

16. (a) Show by a sketch how to construct an angle of 15° with a horizontal line by means of the T-square and triangles. (b) What angle does this make with a vertical line?

17. (a) Show by a sketch how to construct an angle of 75° with a horizontal line by means of the T-square and triangles. (b) What angle does this make with a vertical line?

18. When using the triangle against the T-square in which direction should the line be drawn? (b) Show different cases by sketching.

19. Show by a sketch how to draw a line parallel to any given line using only two triangles. Perpendicular.

20. (a) Describe the positions of the T-square and triangle for drawing a vertical line. (b) In which direction is the line always drawn?

21. (a) What is the shape of the ruling point of the pencil? (b) How is it obtained? (c) How does the measuring point of the mechanical drawing pencil differ from the point of the sketching pencil?

22. What are the uses of a scale in laying out a drawing?

23. Show by a sketch how to lay off a distance of $16' - 3\frac{1}{2}''$, using the architect's scale $\frac{1}{2}''$ to $1' - 0''$.

24. How are the legs of the compass set for describing circles?

25. (a) What is the shape of the point of the lead used in the bow compass? (b) How should it be set with reference to the needle point?

26. (a) What is the range of the bow compass? (b) How

are the circles drawn which are too large for the ordinary compass?

27. Show by a sketch how to divide a line into five equal parts by means of the dividers.

28. Illustrate by a sketch and show calculations for determining the size of an enclosing rectangle.

29. Describe the process of drawing the border rectangle for a mechanical drawing sheet.

30. (a) Define the constructive stage of the mechanical drawing. (b) How are hidden edges shown in this stage?

31. In what order are the lines drawn in the finishing stage?

32. (a) What space is left between the outline of the object and the end of the extension line? (b) How far should the extension line run beyond the arrowhead? (c) How far should the nearest dimension line be from the outline of the object? (d) How far apart should dimension lines be placed?

33. (a) What is the height of the whole number in a dimension? (b) The total height of the fraction?

34. What is the purpose of notes on a drawing?

35. What is the title block?

36. Describe the steps taken in balancing two or more lines in a title.

37. What dimension forms are used in showing the inclination of a line?

38. In what order are the lines drawn when an arc and a straight line are tangent to each other?

39. (a) Show two ways of dimensioning a radius. (b) Under what condition is each used?

DATA FOR REVIEW PROBLEMS

Given: A perspective sketch, Fig. 163.

Required:

1. To make an orthographic sketch of the object shown in Fig. 163.

2. To make a pencil mechanical drawing from the orthographic sketch.

Given: A perspective sketch, Fig. 164.

Required:

1. To make an orthographic sketch of the object shown in Fig. 164.

2. To make a pencil mechanical drawing from the orthographic sketch

Given: A perspective sketch, Fig. 165.

Required:

1. To make an orthographic sketch of the object shown in Fig. 165.

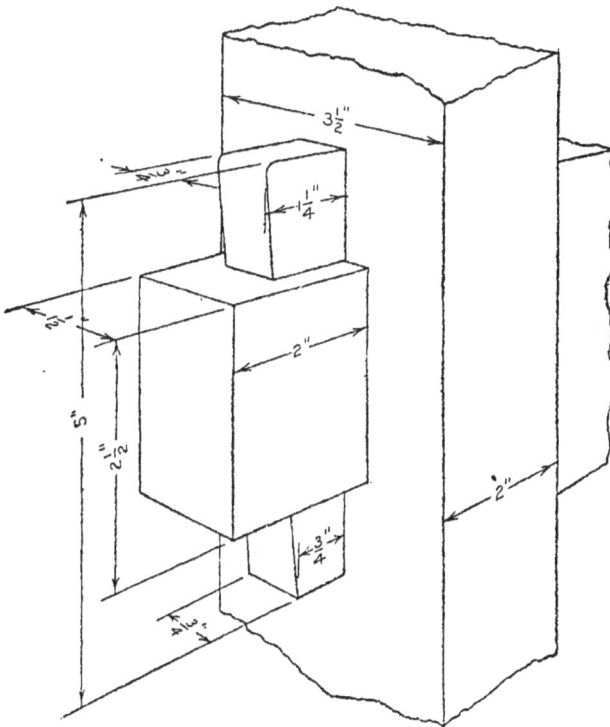

FIG. 163. KEYED MORTISE AND TENON

2. To make a pencil mechanical drawing from the orthographic sketch.

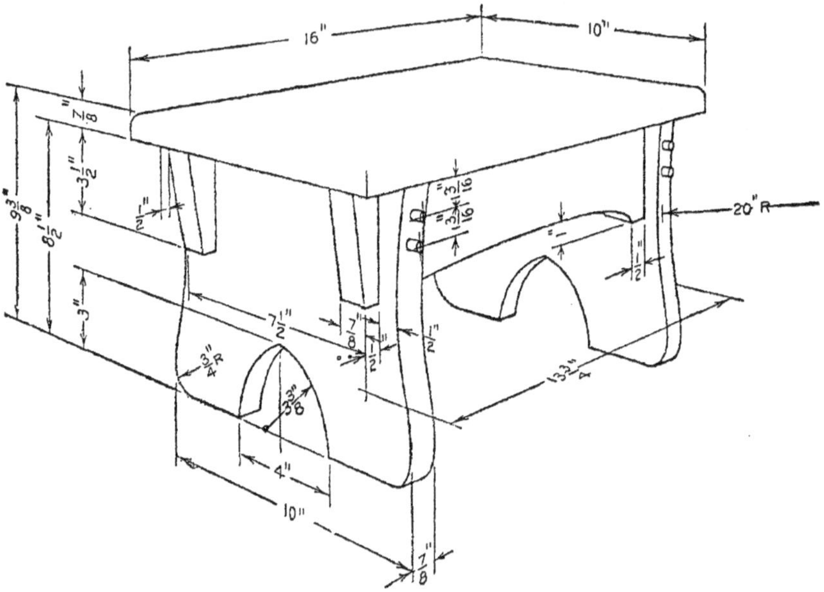

FIG. 164. FOOT STOOL

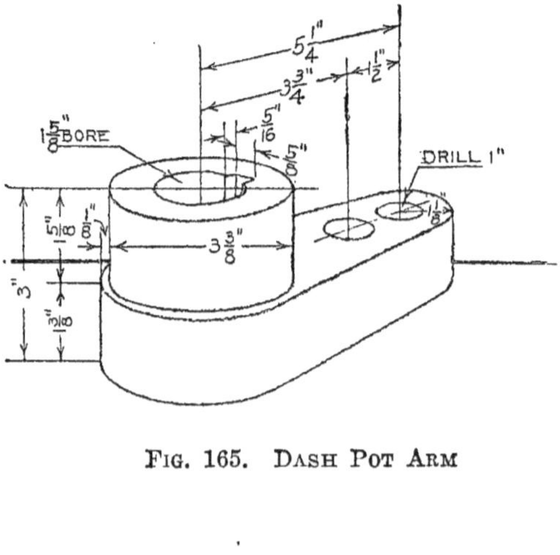

FIG. 165. DASH POT ARM

CHAPTER IV

TRACING AND BLUEPRINTING

PROSPECTUS

The pencil mechanical drawing of Chapter III is continued in this chapter to develop further skill in the use of instruments and to improve the technique in both the mechanical and freehand elements of the drawing. It is the chief aim of this chap-

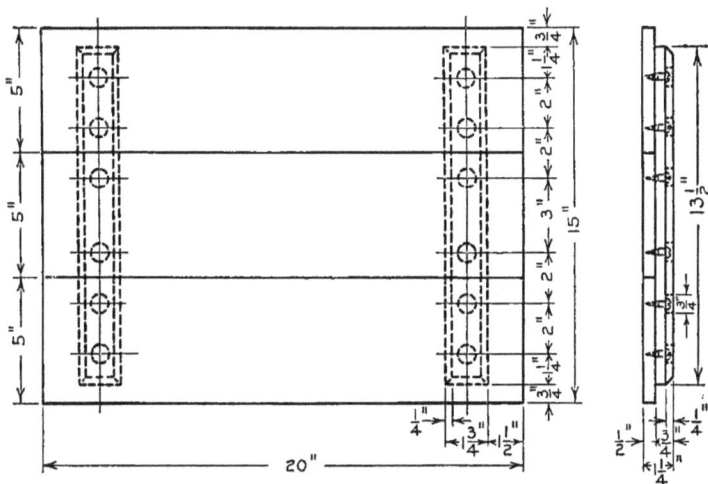

FIG. 166. TYPE PROBLEM. DRAWING BOARD

ter to familiarize the student with the instruments, materials, and methods used in inking and to fix a standard for the ink drawing. As a result of the work of this chapter the student should be able to make neat tracings with proper width of lines, good joints, and uniform spacing in crosshatching. The technique of the lettering, arrowheads, and figures should be comparable with that secured in the mechanical line work.

159

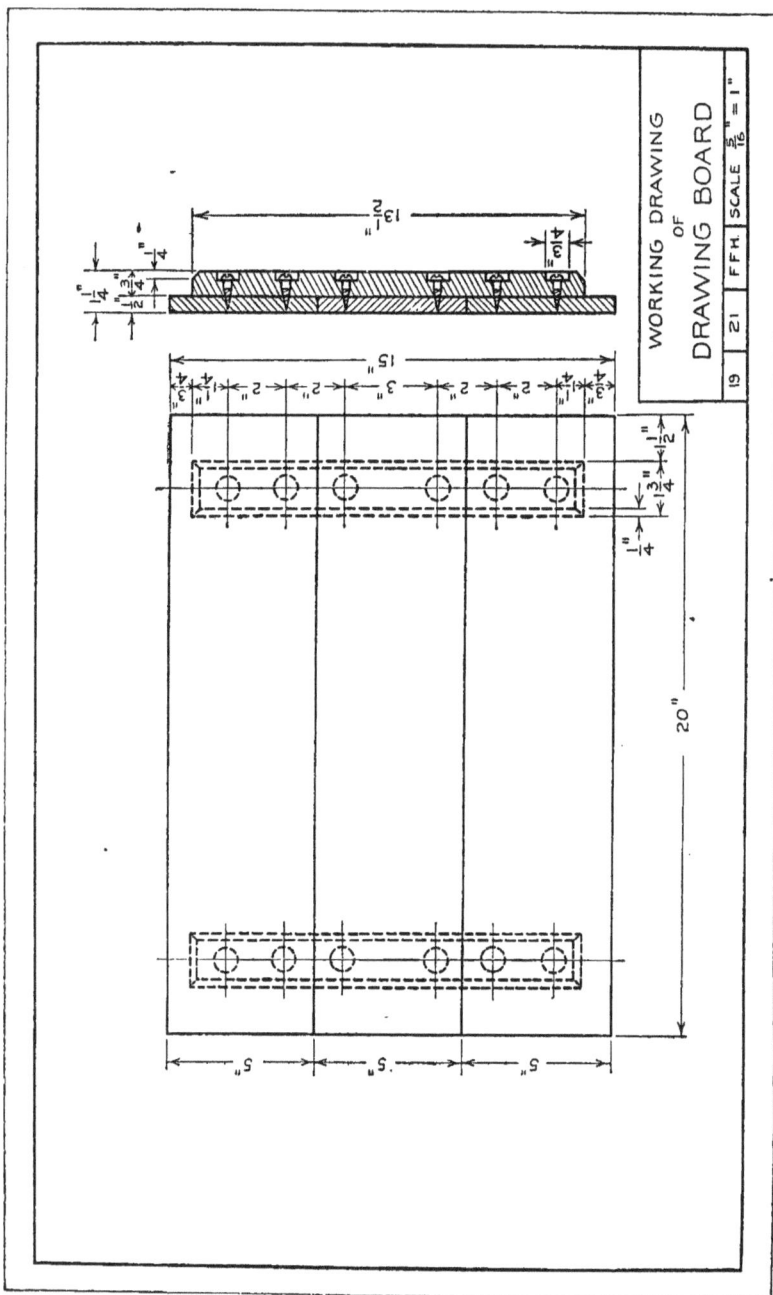

WORKING DRAWING
OF
DRAWING BOARD

| 19 | 21 | F F H | SCALE $\frac{5}{16}" = 1"$ |

FIG. 167.   TYPE PROBLEM.   DRAWING BOARD.   HALF SECTION

The tracing of the pencil mechanical drawing on tracing cloth with ink is usually the last step in the production of a drawing for the shop or for other purposes where a number of copies of the drawing are desired. The tracing is made on a transparent cloth or paper in order that the blueprints may be made from it as described later. The use of the blueprint makes it possible to have several copies of the drawing and at the same time preserve the original tracing from which other copies may be made at any time.

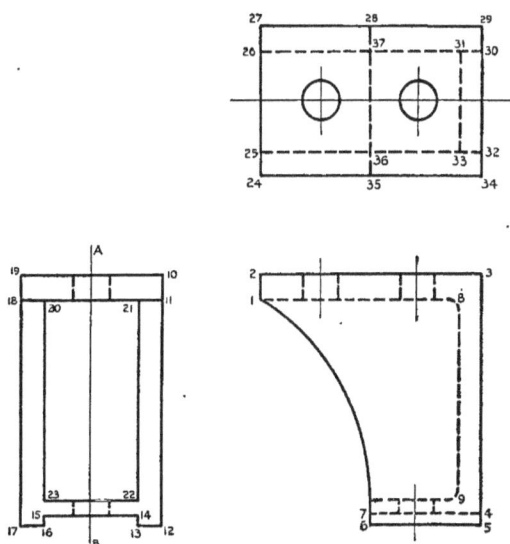

FIG. 168.   REVIEW PROBLEM

PREPARATORY INSTRUCTIONS FOR DRAWING PLATE 18

Inasmuch as the drawings of many objects require the use of sections, the student should review both the half and the quarter sections discussed in Chapter III, pages 98 and 99, and test his knowledge of the orthographic principles involved in making sectional views by answering the following questions. See Figs. 99 and 168.

1. Where is the surface 1, 6, shown in the side view?

2. (a) Does the rectangle 20, 21, 22, 23 in the side view represent an opening or a solid part of the object? (b) Why?

3. Where is the surface 7, 4 shown in the side and top views?

4. Make a front view of the object when cut on A B.

5. (a) Is the side view affected by the section? (b) Top view?

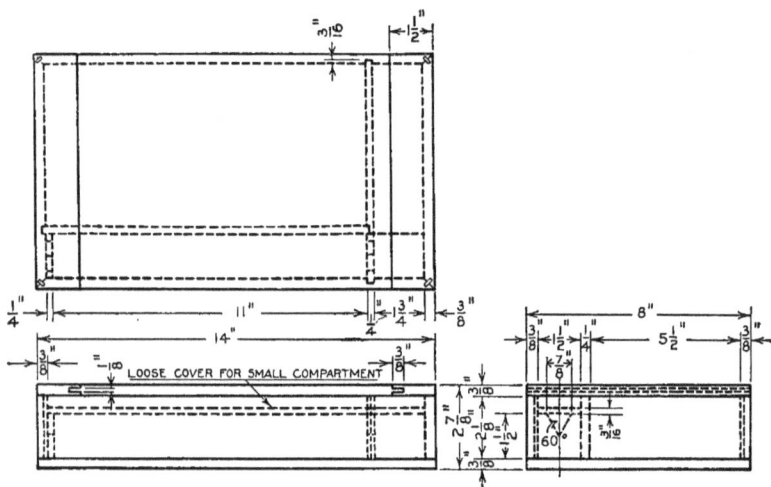

FIG. 169. INSTRUMENT CASE

Each tracing in this chapter will be preceded by a pencil mechanical drawing. The pencil drawing is made to give the student additional practice in the handling of the instruments already used, to introduce the use of new instruments, and to provide drawings for the tracings.

DATA FOR DRAWING PLATE 18

Given: Orthographic drawings, Figs. 169, 170, 171.

Required: To make a pencil mechanical drawing of the object shown in Fig. 169, 170, or 171 as assigned by the instructor. The views given and required may be obtained from the following statements. Any similar problems may be substituted by the instructor.

Given: Fig. 169. The front, top, and right side views.

Required: To draw the front, top, and left side half section views.

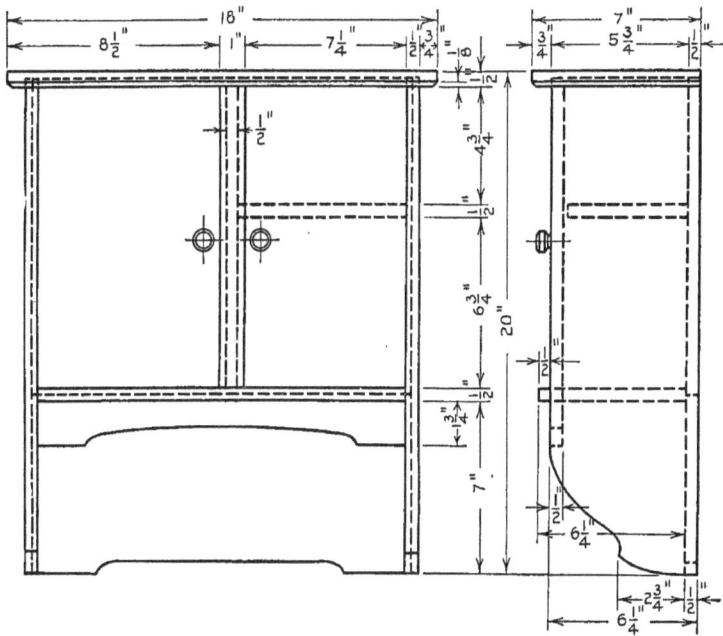

FIG. 170. MEDICINE CABINET

Given: Fig. 170. The front and right side views.

Required: To draw the front and left side half section views.

Given: Fig. 171. The front and right side views.

Required: To draw the front, and left side half section.

Instructions:

Proceed as for the mechanical drawing plates of Chapter III.

DATA FOR LETTERING PLATE 18

Given: Plate 18 to reduced size. Fig. 172.
Required: To make the plate to an enlarged scale.

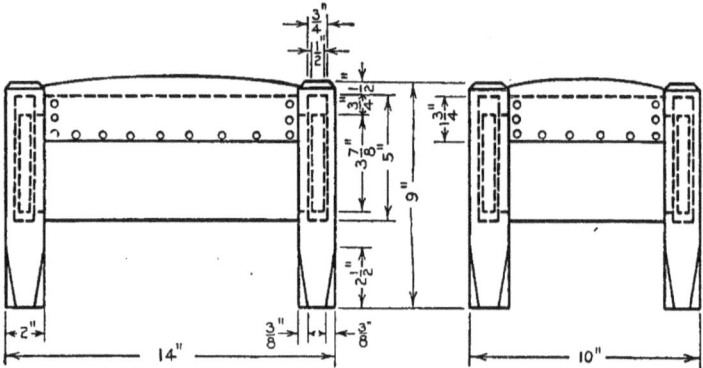

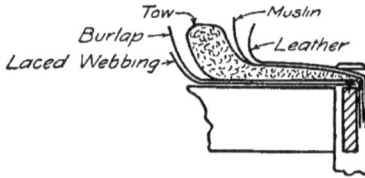

Section showing construction
of cushion

FIG. 171. FOOT STOOL

PREPARATORY INSTRUCTIONS FOR DRAWING PLATE 19

The following is a list of the instruments and materials needed
to make tracings and reproductions of mechanical drawings:

1. Tracing cloth.
2. Tracing paper.
3. Blueprint paper.
4. Black waterproof ink.
5. Ruling pen.
6. Compass.
7. Bow pen.

Tracing Cloth is a thin, firm cloth sized to hold ink and to make the cloth transparent. It is generally used when drawings are to be reproduced by the blue, black, or brown printing process. Drawings made on tracing cloth may be kept indefinitely if the cloth is kept dry and handled carefully. Changes may be made on the drawings and new prints made from time to time.

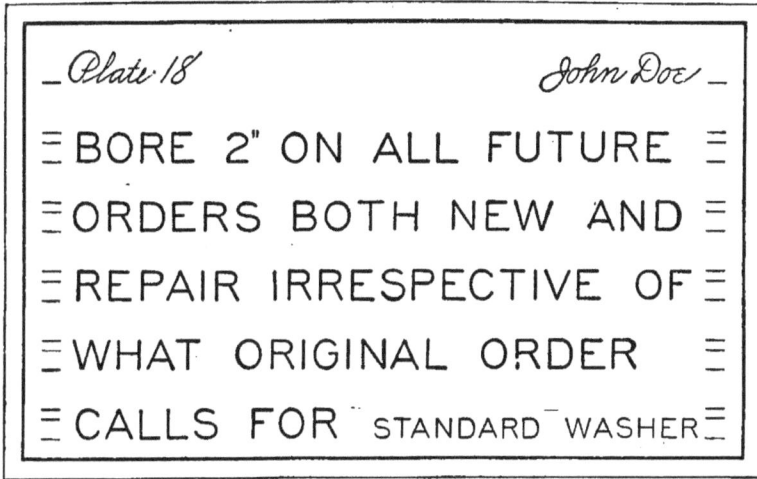

Plate 18   _John Doe_ _

≡ BORE 2" ON ALL FUTURE ≡

≡ ORDERS BOTH NEW AND ≡

≡ REPAIR IRRESPECTIVE OF ≡

≡ WHAT ORIGINAL ORDER ≡

≡ CALLS FOR STANDARD WASHER ≡

FIG. 172. LETTERING PLATE 18

One side of the cloth is glazed and the other is dull. Either side may be used for inking. The glazed side will admit of the most erasing, but when inking is done on this side the cloth will curl. For work where penciling is to be done on the cloth, for drawings to be used for photographic reproduction, and for tinting, the dull side should be used. For the tracings of this chapter use the dull side.

Sometimes the ink does not adhere readily to the surface of the cloth, particularly when the glazed side is used. To overcome this difficulty powdered chalk may be rubbed into the surface with a soft cloth. The chalk should be thoroughly removed before beginning inking.

The cloth is fastened to the board with the same thumb tacks used to hold the pencil drawing. In order to avoid shifting this drawing, the cloth should be spread over the sheet and one tack at a time removed and inserted through the cloth into the hole from which it came.

Tracing Paper. For temporary drawings, especially where some portion of a drawing already made can be traced and used as part of a new drawing, a thin, transparent paper called *tracing paper* may be used with considerable saving of time. It should not be used for a permanent drawing or one which requires much handling.

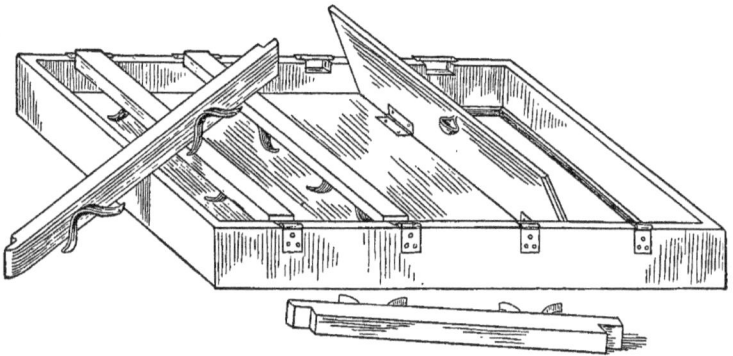

FIG. 173. BLUEPRINTING FRAME

Blueprint Paper. Instead of sending the tracing into the shop where it would soon be injured or worn out, prints are made, usually on blueprint paper. This is a white paper covered with a solution which after being exposed to light, turns blue.

The Blueprinting Process. To make prints, the inked side of the tracing is placed against the glass of a printing frame. The sensitized side of the blueprint paper is then placed against the tracing cloth and held firmly in contact with it. The contact is secured by means of clamps attached to the back of the board of the printing frame which holds both the tracing and the blueprint paper in place. Fig. 173.

The printing frame should be placed in a direct light. If sunlight is used the exposure should be made during the middle of the day. The length of exposure to the light depends on the intensity of the sunlight or electric light and upon the "speed" of the blueprint paper. After removing the paper from the frame it should be washed by turning it over several times in a basin of water. This removes the chemical on the sensitized side of the paper which was covered by the lines of the drawing on the tracing cloth and leaves the white paper exposed, forming the outline of the blueprint drawing. The result is a reproduction of the drawing in white lines with a blue background. After the blueprint has been washed it should be hung vertically by one edge or over a horizontal stick to drain, and allowed to remain until it is dry.

Black Ink will be used for all lines on the plates of this chapter. Black drawing ink is composed of finely divided carbon held in suspension in a liquid. When a line is drawn with this ink the liquid dries and leaves the carbon deposited on the paper or cloth. It is important that enough ink be left on the line so that when the ink is dry the amount of carbon deposited will be sufficient to make the line black. Thin ink gives brown lines. The liquid used in drawing ink evaporates quickly. The carbon therefore dries quickly, permitting one to work rapidly while tracing. The rapid evaporation of the ink necessitates keeping the stopper always in the bottle to prevent the ink from becoming too thick.

The Ruling Pen is used more than any other instrument in the draftsman's outfit and should therefore be carefully selected. The steel of which the pen is made should be properly tempered and of such quality as to retain a smooth sharp edge. The blades should be of the same length, the inner one sufficiently stiff to resist a light pressure against the ruling edge. The nibs should be of the same width, equally rounded and directly opposite one another. The ends of the nibs should be narrow enough to give control in starting and ending lines, but broad enough to hold a reasonable amount of ink. When the nibs are too narrow the ink is drawn from the points by capillary attraction, making it difficult to start the ink at the beginning of a line.

Filling and Using the Pen. The ruling pen should be adjusted, filled, and used in the following manner:

1. Adjust the pen by turning the thumb screw to approximately the proper width of line.

2. Fill the pen by inserting the quill, attached to the stopper of the ink bottle, between the nibs of the pen. The pen should be filled to a height of about $\frac{1}{4}''$. Care should be taken to avoid getting ink on the outside surfaces of the nibs.

3. Set the pen to give the exact width of line required, testing it on the margin of the drawing or on a separate sheet. It should be tested on the same kind of surface as that on which it is to be used and by ruling along a straightedge—*not freehand.*

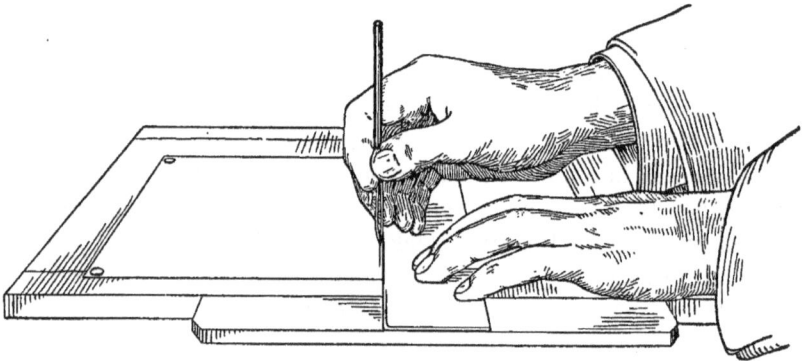

Fig. 174.   Ruling a Horizontal Line

4. Hold the pen in the hand, as shown in Fig. 174, with the first finger above the thumb screw and the second finger against the right side of the pen. It should be held in a vertical plane, but may be allowed to lean slightly in the direction of motion. In this position both nibs will touch the cloth with equal pressure, which is essential to the production of smooth, sharply defined lines.

5. Draw rather slowly with a movement of the hand and arm, the forearm remaining perpendicular to the line being drawn. There should be no wrist movement, as the pen must not be rotated upon its axis. The tips of the third and fourth fingers should slide on the surface of the T-square or triangle to steady the hand. As the end of the line is approached the motion of the hand and arm should cease and the line should be completed with a finger movement. The center of the ink line on the tracing should be directly over the pencil line on the drawing being

traced. Care must be taken to set the pen exactly at the beginning of a line. At the end of a line the pen should be lifted
vertically in order that the ink will not run out and cause the
line to overrun. In drawing dotted lines, the pen must be set
down vertically, the dash drawn, and the pen then lifted vertically so as to make both ends of the dash square.

The spacing of section lines is done entirely by eye. In order
to avoid varying the spaces the pen should be placed against the
ruling edge and the perpendicular distance from the point of
the pen to the last line drawn made equal to the perpendicular
distance between any two sequential preceding lines.

When starting the crosshatching in a corner, there is a ten-
dency to space the lines too closely, the spaces increasing as the
lines become longer. The student should practice crosshatching
rectangular areas on a scrap of tracing cloth before attempting to
work on the drawing.

Cleaning the Pen. The pen should be cleaned frequently by
inserting a cloth at the side and pulling it out between the nibs.
This should be done frequently while the pen is in use. The pen
should not be laid away until the surfaces are thoroughly cleaned,
as ink will corrode steel. If the ink does not start readily at the
beginning of a line, squeeze the nibs of the pen together slightly
to draw the ink down to the point. If the ink has been allowed
to stand for some time, the pen should be cleaned and refilled.
Do not touch the pen to the hand or a cloth to start the ink.

Sharpening the Pen. The nibs of the pen should be as sharp
as they can be made without producing the sensation of cutting
when the pen is in use. They should not scratch the paper when
drawing a line. This occurs if they are sharpened to a point
instead of a rounded edge, or if the point is rough or notched.
The length and condition of the points may be tested by holding
the pen up to the light and bringing the nibs together slowly.

In case the pen becomes broken or dull from use it should be
sharpened as follows:

1. Provide a close grained oilstone.

2. Close the nibs until they just touch each other.

3. Hold the pen on the stone as in drawing a line and move
it back and forth, revolving it slowly in the plane of motion until

the nibs are evenly rounded and of the same length. Fig. 175. This will dull the nibs.

4. Separate the nibs and sharpen them by rubbing the *outside* on the oilstone, giving at the same time a slight rotary motion

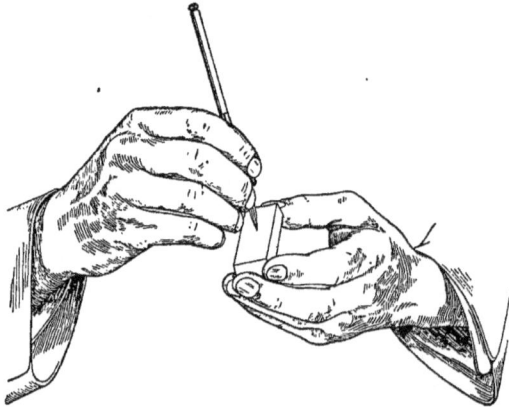

FIG. 175. SHARPENING THE PEN. EVENING THE NIBS

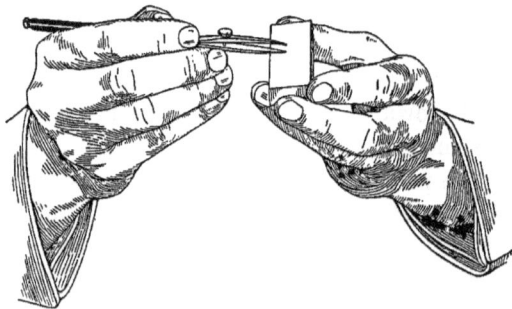

FIG. 176. SHARPENING THE PEN. GRINDING THE NIBS

to the handle, which is held at a small angle with the face of the stone. Fig. 176. The point of the pen should be examined frequently and the process continued until the nibs are sharp. If a burr is produced on the inside of a nib it may be removed by placing the inside surface flat against the oilstone and rubbing it lightly.

The Compass. When using the compass for either penciling **or** inking, the legs should be adjusted so that the pen or pencil part and the needle point are *perpendicular* to the drawing board. With the legs in this position, the compass revolves about the needle point as an axis and the two nibs of the pen bear with equal pressure, thus producing sharply defined lines. The compass is held by the handle, between the thumb and first finger of the right hand. It is rotated by rolling the handle between the thumb and finger. Fig. 177. If the compass is allowed to lean very slightly in the direction of motion, sufficient pressure may be put on the pen or pencil to hold it firmly in contact with the paper or cloth without danger of the needle point being lifted from the center.

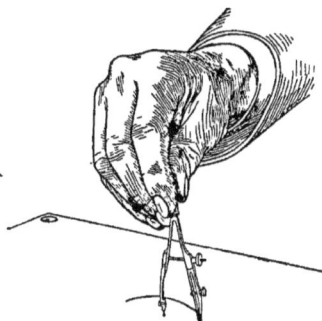

FIG. 177. DRAWING A CIRCLE WITH THE COMPASS

The pen of the compass is filled, adjusted, cleaned, and sharpened in the same manner as the ruling pen.

The Bow Pen should be used for all circles and arcs of $\frac{3}{4}''$ radius or less. The pen should be filled and adjusted in the same manner as the ruling pen.

Line Notation. In inking, the object lines are drawn noticeably heavier than all other lines except the border line. The difference in width produces a sharp contrast between classes of lines which makes the drawing easy to read and gives it a good appearance. In small drawings or those containing intricate detail the width of the object lines is slightly reduced.

No system of line notation has ever been universally adopted. In this course a simple one conforming to average commercial drafting room practice will be used. All lines except the dotted line used to represent invisible edges are solid. The widths of lines to be used in this course are given in Fig. 178. All widths as indicated in Fig. 178 should be estimated by the student. As far as possible all lines of the same width should be drawn while the pen is set for that width. Before starting to ink a group of lines of the same width a sample line should be drawn on the edge of the sheet. This may be used as a guide in setting another instrument to give the same width of line. For instance, when the compass is used in drawing circles a sample of the width of line should be drawn to aid in estimating the setting of the ruling pen which will be used later. In case the pen must be reset for a particular line the estimated width should be determined by drawing lines near the sample until the proper width of line is secured.

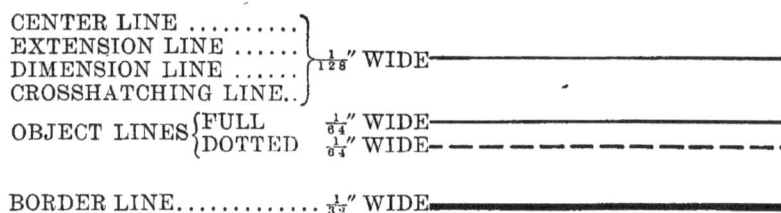

CENTER LINE ....⎫
EXTENSION LINE ...... ⎬ $\frac{1}{128}$″ WIDE————————————
DIMENSION LINE ......
CROSSHATCHING LINE..⎭

OBJECT LINES {FULL $\frac{1}{64}$″ WIDE————————————
            {DOTTED $\frac{1}{64}$″ WIDE— — — — — — — — — — .

BORDER LINE............. $\frac{1}{32}$″ WIDE━━━━━━━━

FIG. 178. LINE NOTATION FOR INK DRAWINGS

Order of Inking the Drawing. The drawing should be inked in the order given below to secure economy of time and effort.
  1. Object lines.
      a. Circles and arcs of circles.
      b. Horizontal lines (beginning at the top).
      c. Vertical lines (beginning at the left).
      d. Inclined lines.
  2. Center lines (same order as object lines).
  3. Extension and dimension lines (same order as object lines).
  4. Arrowheads.
  5. Dimension figures and notes.
  6. Crosshatching lines.
  7. Title.
  8. Border line
In inking the title and notes, pencil guide lines on the tracing

cloth will be found of great assistance in keeping the letters uniform in height. In no case should letters and figures be penciled on the tracing cloth over those which appear on the pencil drawing, before they are inked. The pencil drawing should serve as a copy for inking figures, letters, and arrowheads as it does for all mechanical lines All freehand inking should be done with the writing pen as described under, "Preparatory Instruction for Lettering Plate 11," page 102.

Erasure. On ink drawings erasures must be carefully made, especially if inking is to be done over the erased areas. It will be found that if the ruby eraser is used for removing ink lines the drawing surface will be left in good condition for re-inking. In case a blot occurs the ink should not be allowed to soak into the tracing cloth. As much of the ink as possible should be taken up with a blotter or cloth and the remainder allowed to dry before erasing. The erasing shield should be used to protect the parts of the drawing which are not to be erased, as described on page 132.

Trimming the Tracing Cloth. When the tracing is finished lay off one-half an inch from each corner of the border rectangle to make a one-half inch margin. Place the tracing on the back of the drawing board. With a sharp knife running along the edge of the T-square blade *not used for ruling,* trim the sheet to the rectangle determined by the eight pencil points. In this process the T-square blade should be placed over the finished portion of the sheet. The drawing will then be held firmly and will be protected from the knife in case it should slip.

DATA FOR DRAWING PLATE 19

Given: The pencil mechanical drawing, Plate 18.
Required: To make a tracing of Plate 18.

Instructions:

1. Fasten the tracing cloth over the mechanical drawing and prepare the surface for inking as previously described under, "Tracing Cloth," page 165.

2. Ink the drawing, following the steps outlined under, "Order of Inking the Drawing," page 172.

DATA FOR LETTERING PLATE 19

Given: Plate 19 to reduced size. Fig. 179.
Required: To make the plate to an enlarged scale.

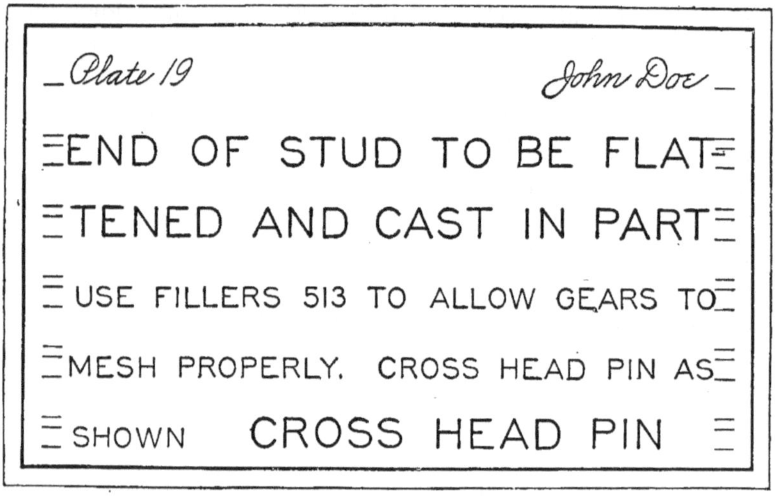

_Plate 19   John Doe _

=END OF STUD TO BE FLAT=

=TENED AND CAST IN PART=

= USE FILLERS 513 TO ALLOW GEARS TO=

=MESH PROPERLY. CROSS HEAD PIN AS=

= SHOWN   CROSS HEAD PIN   =

FIG. 179. LETTERING PLATE 19

DATA FOR DRAWING PLATE 20

Given: Orthographic drawings, Figs. 182, 183, 184, and 185.
Required: To make a pencil mechanical drawing of the object shown in Fig. 182, 183, 184, or 185, as assigned by the instructor. The views given and required may be obtained from the following statements. Any similar problem may be substituted by the instructor.

Given: Fig. 182. The front and left side views.
Required: To draw the front and right side views.
Given: Fig. 183. The front and left side views.
Required: To draw the front and right side views.
Given: Fig. 184. The front and left side views.
Required: To draw the front and right side views.
Given: Fig. 185. The front and left side views.
Required: To draw the front and right side views.

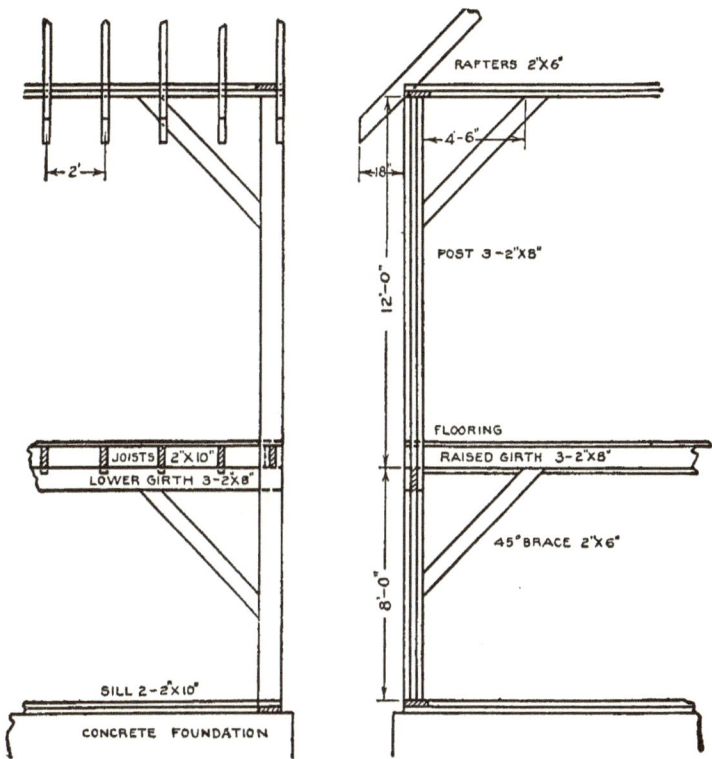

FIG. 180. TYPE PROBLEM. BARN FRAMING. GIVEN VIEWS

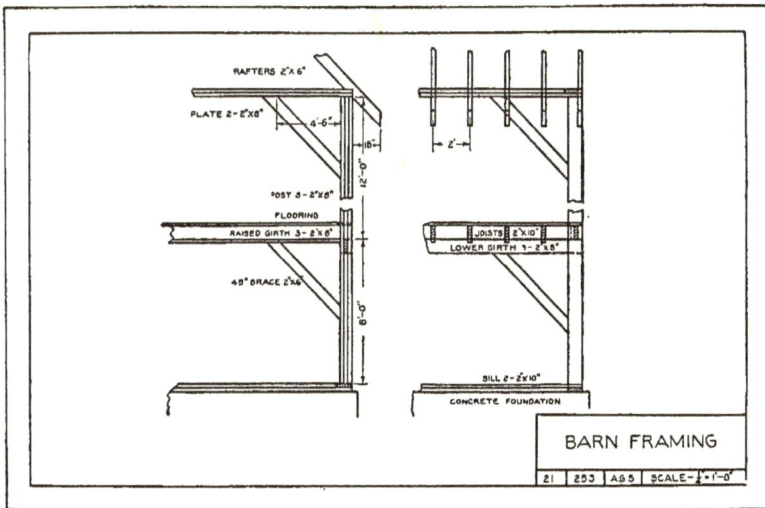

FIG. 181. TYPE PROBLEM. BARN FRAMING. FINISHED DRAWING

(175)

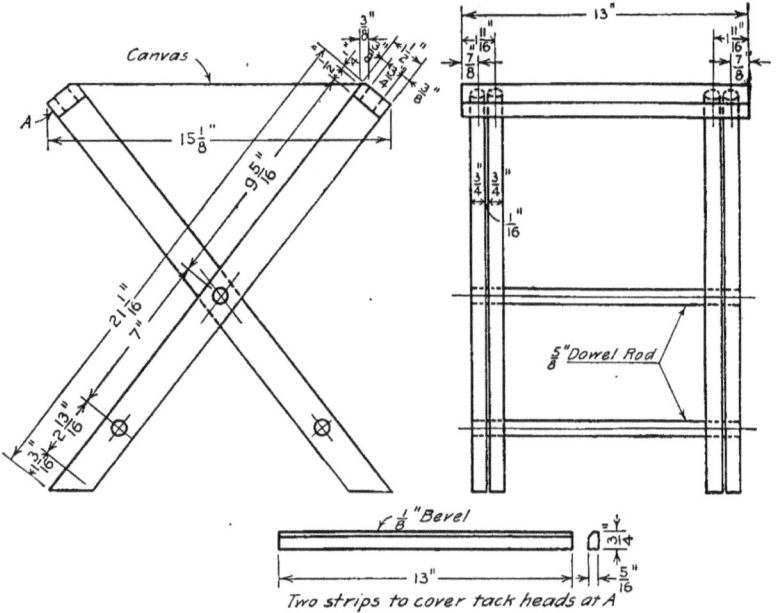

FIG. 182. CAMP STOOL

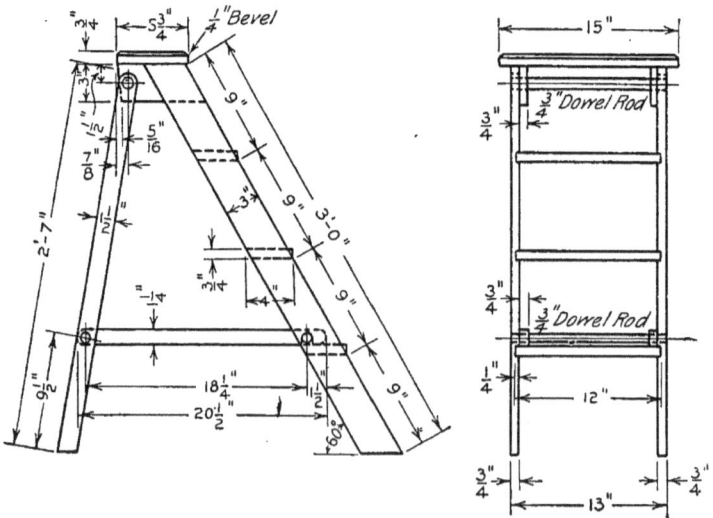

FIG. 183. STEP LADDER

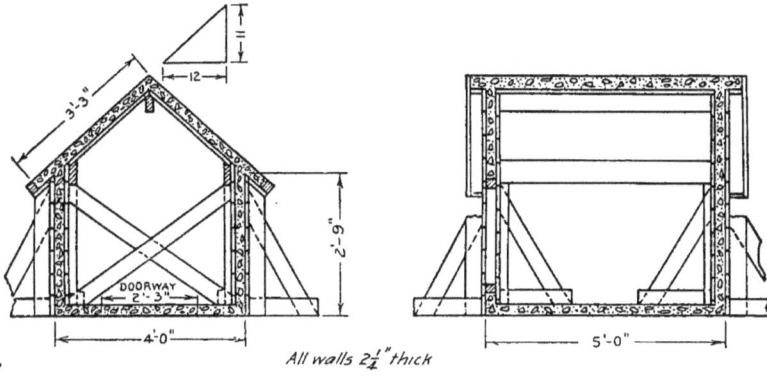

All walls $2\frac{1}{4}$" thick

FIG. 184. FORMS FOR CONCRETE DOG KENNEL

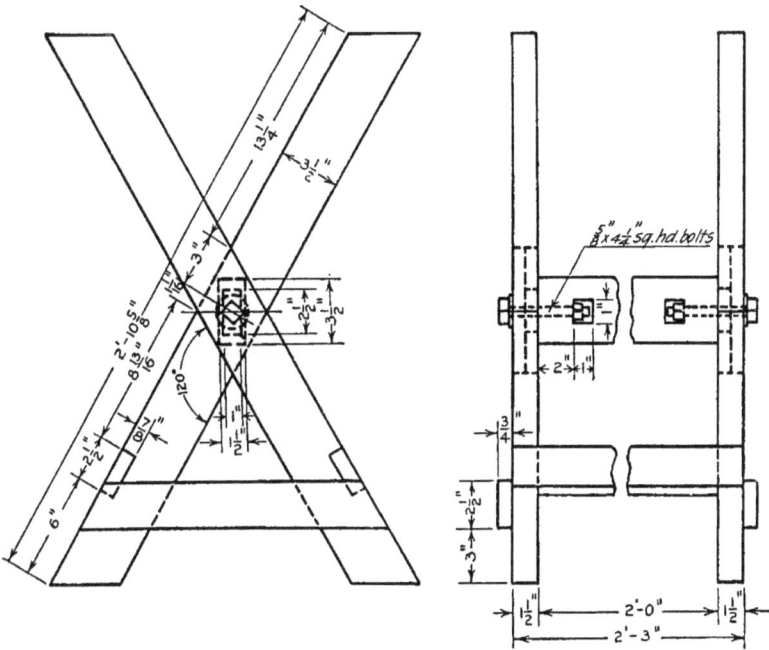

FIG. 185. SAW BUCK

DATA FOR LETTERING PLATE 20

Given: Plate 20 to reduced size. Fig. 186.
Required: To make the plate to an enlarged scale.

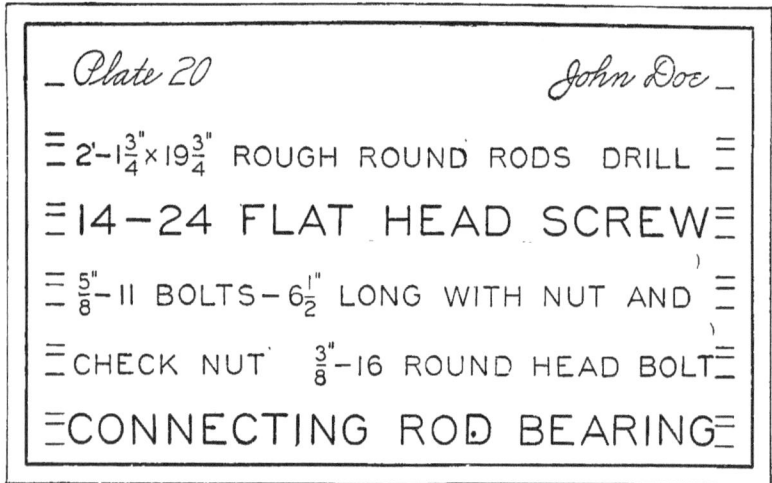

_ Plate 20   John Doe _

$2'-1\frac{3}{4}'' \times 19\frac{3}{4}''$ ROUGH ROUND RODS DRILL

14-24 FLAT HEAD SCREW

$\frac{5}{8}''$-11 BOLTS-$6\frac{1}{2}''$ LONG WITH NUT AND

CHECK NUT $\frac{3}{8}''$-16 ROUND HEAD BOLT

CONNECTING ROD BEARING

FIG. 186. LETTERING PLATE 20

DATA FOR DRAWING PLATE 21

Given: The pencil mechanical drawing, Plate 20.
Required: To make a tracing of Plate 20.

Instructions:

1. Fasten the tracing cloth over the mechanical drawing and prepare the surface for inking.

2. Ink the drawing, following the steps outlined under "Order of Inking the Drawing," page 172.

3. Trim the sheet and press the cloth back into the tack holes.

DATA FOR LETTERING PLATE 21

Given: Plate 21 to reduced size. Fig. 187.
Required: To make the plate to an enlarged scale.

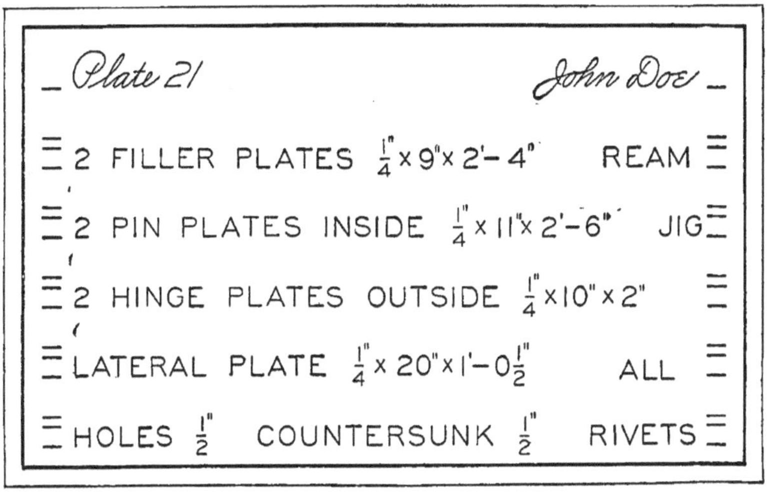
Plate 21
John Doe
2 FILLER PLATES 1/4" x 9" x 2'-4" REAM
2 PIN PLATES INSIDE 1/4" x 11" x 2'-6" JIG
2 HINGE PLATES OUTSIDE 1/4" x 10" x 2"
LATERAL PLATE 1/4" x 20" x 1'-0 1/2" ALL
HOLES 1/2" COUNTERSUNK 1/2" RIVETS

FIG. 187. LETTERING PLATE 21

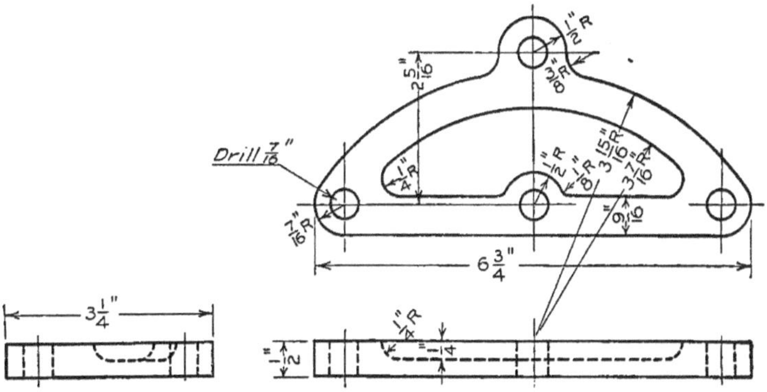
Drill 7/16"
6 3/4"
3 1/4"

FIG. 188. TYPE PROBLEM. RESERVOIR CAP. GIVEN VIEWS

PREPARATORY INSTRUCTIONS FOR DRAWING PLATE 22

Before starting to draw this plate the student should review the method of drawing tangent lines as described on page 147.

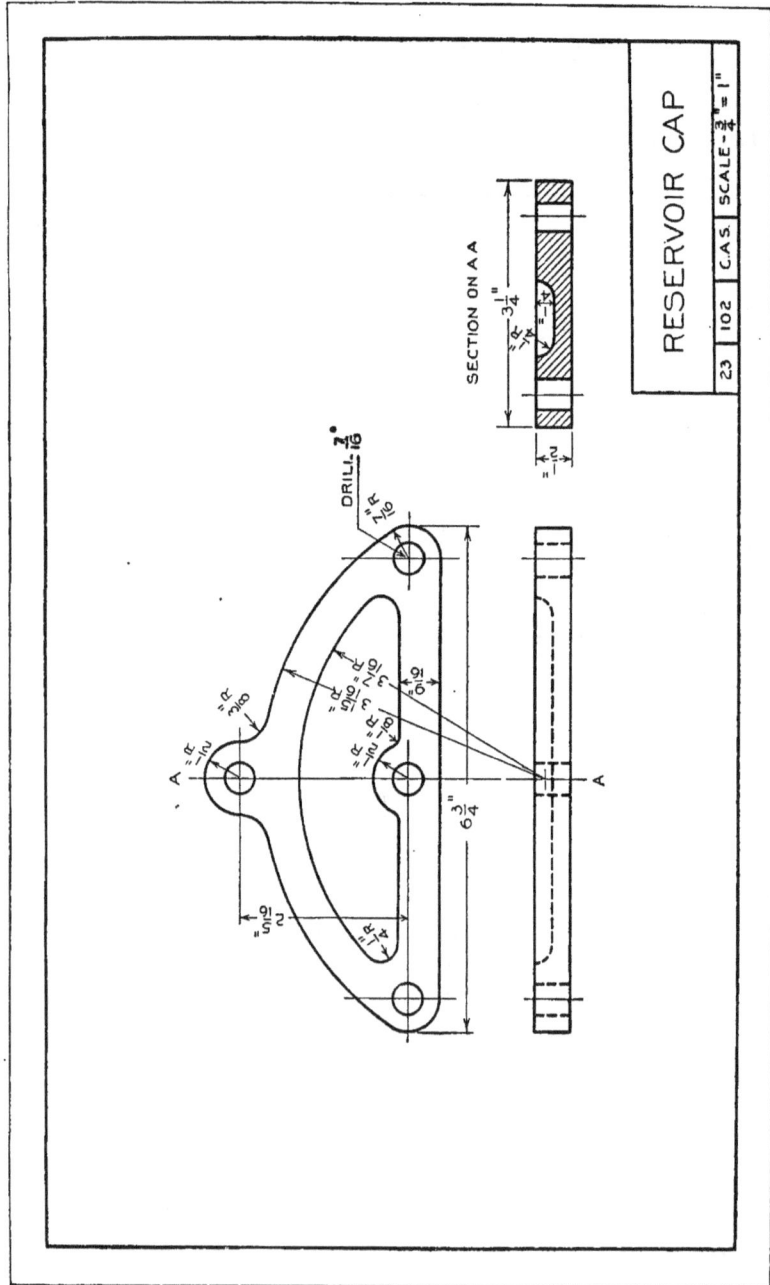

FIG. 189. TYPE PROBLEM. RESERVOIR CAP. FINISHED DRAWING

(180)

DATA FOR DRAWING PLATE 22

Given: Orthographic drawings, Figs. 190, 191, 192, and 193.
Required: To draw the views of the object shown in Fig. 190, 191, 192, or 193, as assigned by the instructor, from the following statements. Any similar problem may be substituted by the instructor.

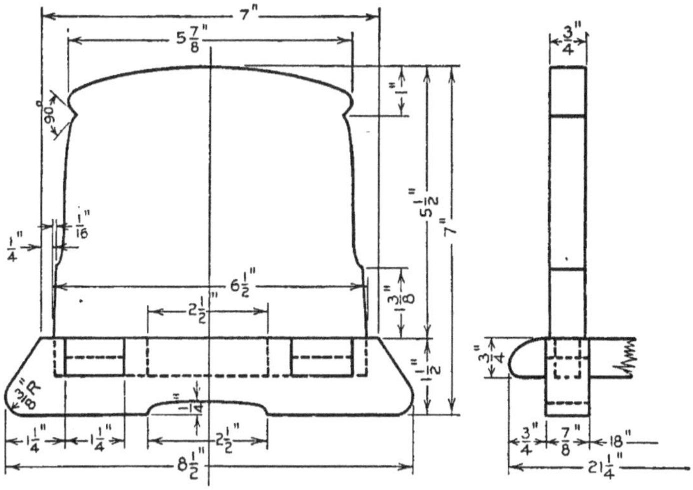

FIG. 190.   BOOK RACK

Given: Fig. 190.   The front and right side views.
Required: To draw the front and left side views.

Given: Fig. 191.   The front and left side views.
Required: To draw the front half section and right side views.

Given: Fig. 192.   The front and left side views.
Required: To draw the front half section and right side views.

Given: Fig. 193.   The front and left side views.
Required: The front half section and left side views.

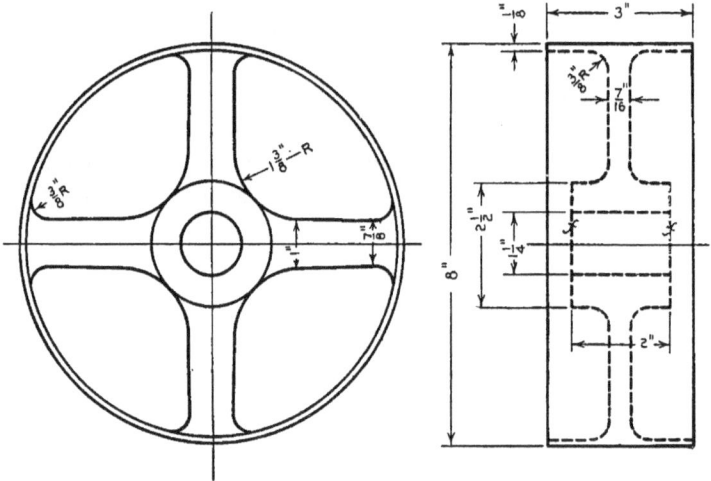

FIG. 191.   COUNTER SHAFT PULLEY FOR 12″ WOOD LATHE

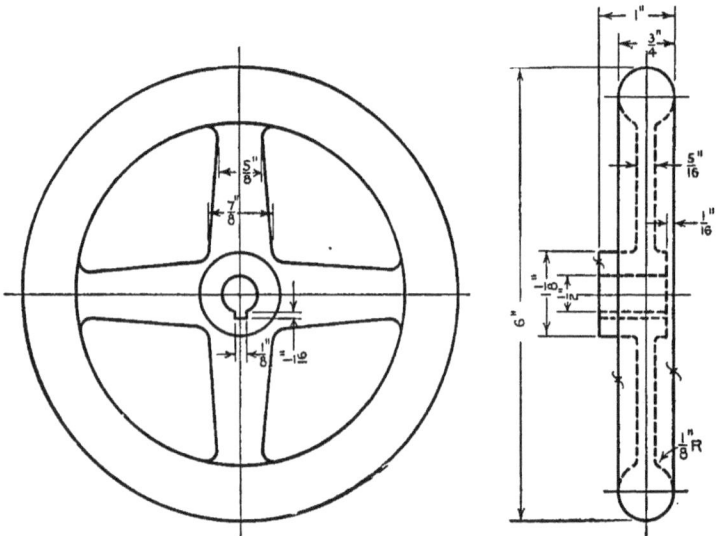

FIG. 192.   HAND WHEEL FOR 12″ WOOD LATHE

DATA FOR LETTERING PLATE 22

Given: Plate 22 to reduced size.   Fig. 194.
Required: To make the plate to an enlarged scale.

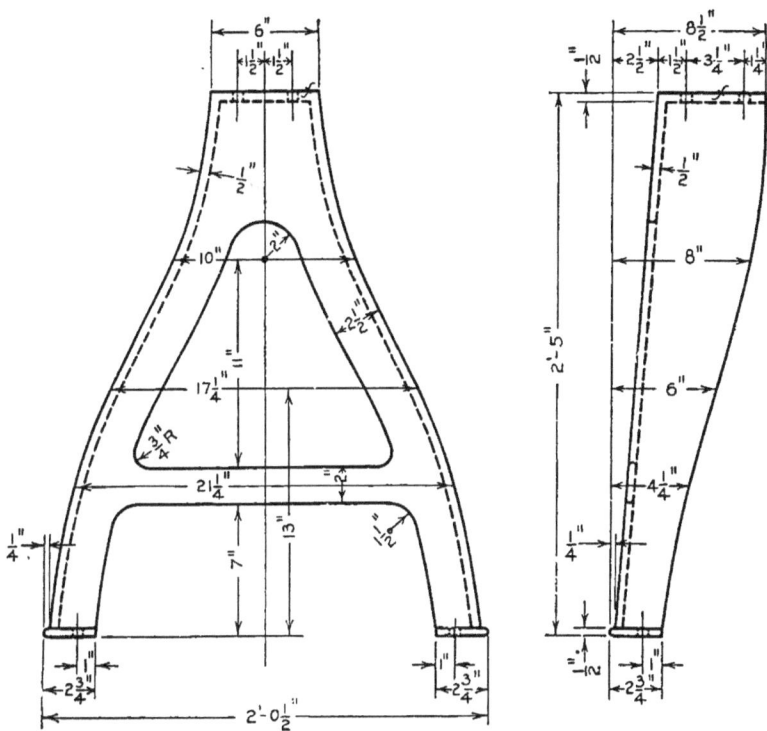

FIG. 193. LEG FOR 12″ WOOD LATHE

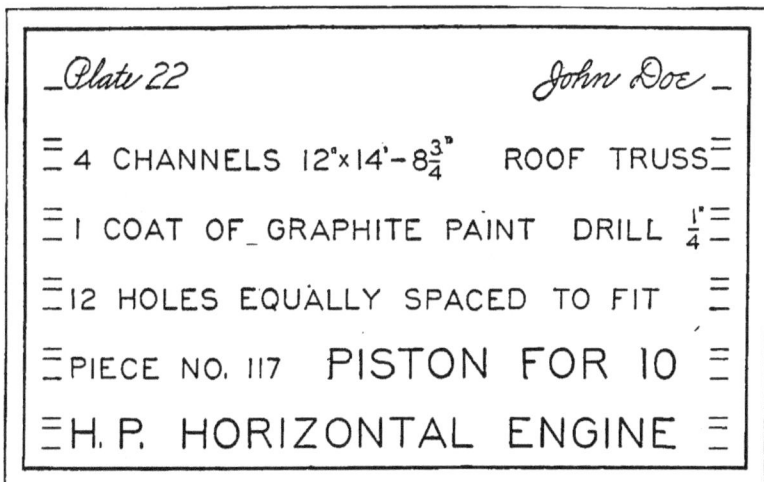

Plate 22   John Doe _

4 CHANNELS 12″×14′-8¾″   ROOF TRUSS

1 COAT OF_ GRAPHITE PAINT   DRILL ¼″

12 HOLES EQUALLY SPACED TO FIT

PIECE NO. 117   PISTON FOR 10

H. P. HORIZONTAL ENGINE

FIG. 194. LETTERING PLATE 22

(183)

Locating Points of Tangency. To secure perfect joints where lines are tangent in the tracings, the exact points of tangency should be located and marked in pencil on the tracing cloth. The method of locating the tangent points depends upon the geo-

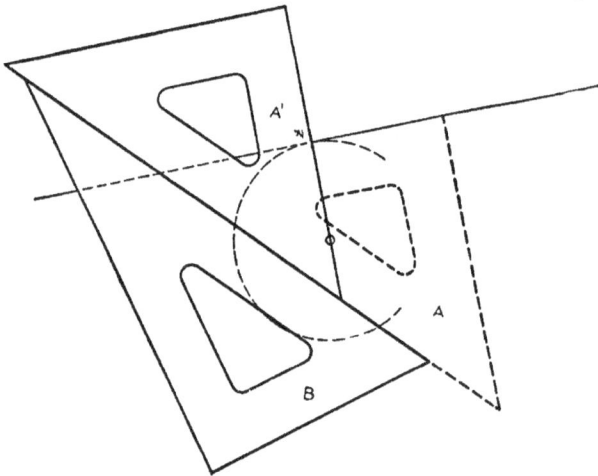

FIG. 195. METHOD OF LOCATING POINTS OF TANGENCY

metrical principle that a line perpendicular to a tangent at its point of contact passes through the center of the circle.

To locate a point of tangency, place the hypotenuse of either triangle against any edge of the other triangle, as shown in Fig. 195. Move both triangles as one tool until a side of the triangle A is coincident with the tangent line. With triangle B held firmly in place, slide triangle A into the position marked A′ where the side at right angles to the tangent line passes through the center of the arc. A short dash should be drawn across the tangent line to mark the point of tangency.

The point of tangency between two arcs may be located by drawing the straight line joining their centers. This line passes through their point of contact. Fig. 195.

DATA FOR DRAWING PLATE 23

Given: The pencil mechanical drawing, Plate 22.
Required: To make a tracing from Plate 22.

Instructions:

1. Fasten the tracing cloth and prepare it for inking.
2. Locate the points of tangency.
3. Ink the drawing in the usual order.
4. Trim the sheet and press the cloth back into the tack holes.

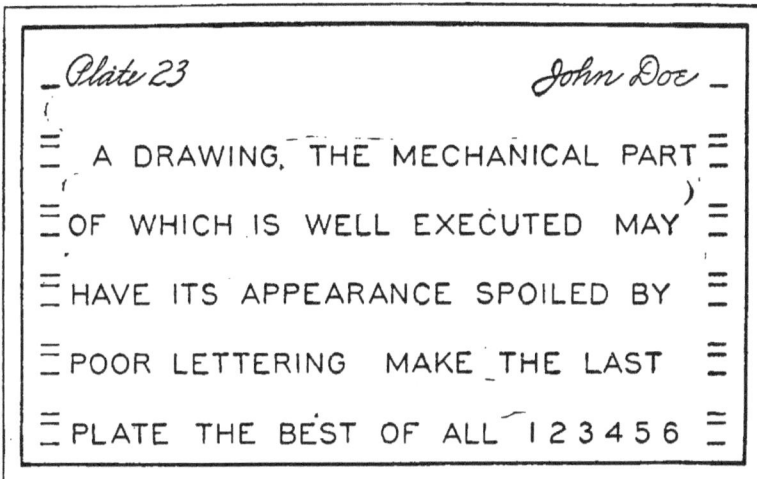

FIG. 196. LETTERING PLATE 23

DATA FOR LETTERING PLATE 23

Given: Plate 23 to reduced size. Fig. 196.
Required: To make the plate to an enlarged scale.

REVIEW QUESTIONS

1. (a) What is the difference between the two sides of the tracing cloth? (b) Which side is used in this course?

2. (a) Describe the process of fastening the cloth over the pencil drawing. (b) How is the cloth prepared for inking?

3. Describe the process of making a blueprint from a tracing.

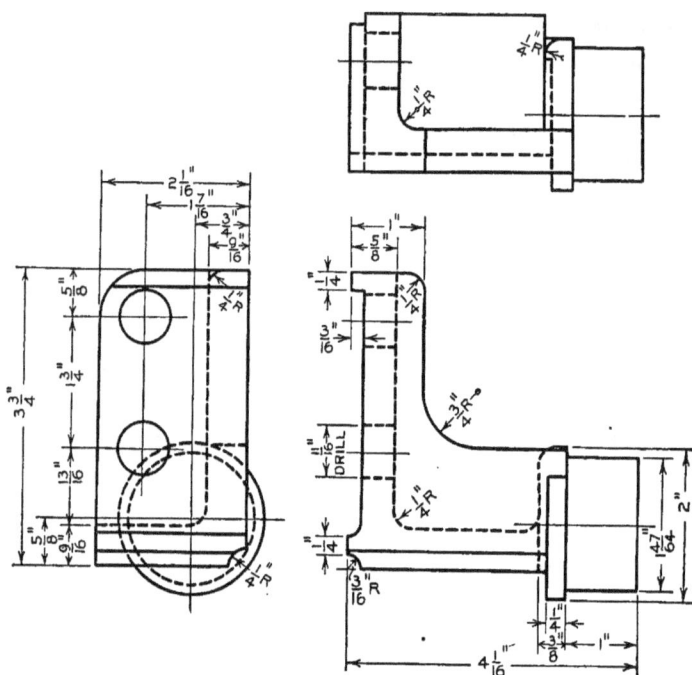

FIG. 197. GOVERNOR SUPPORT

4. (a) How is the ruling pen held for ruling lines? (b) How is it adjusted to the proper width of line? (c) How is it filled? (d) How cleaned?

5. (a) What precautions are taken in beginning and ending a line? (b) How does the pen approach and leave the paper in drawing dotted lines?

6. How are the spaces between crosshatching lines estimated?

7. (a) Why are the needle point and the pen and pencil points of the compass set at right angles to the plane of the drawing paper? (b) How is the compass held when drawing a circle? (c) How is it rotated?

8. (a) In inking, why are the object lines made wider than the other lines? (b) Give the standard width of inked object, extension, dimension, and center lines, and the border line.

9. In what order are the different kinds of lines inked?

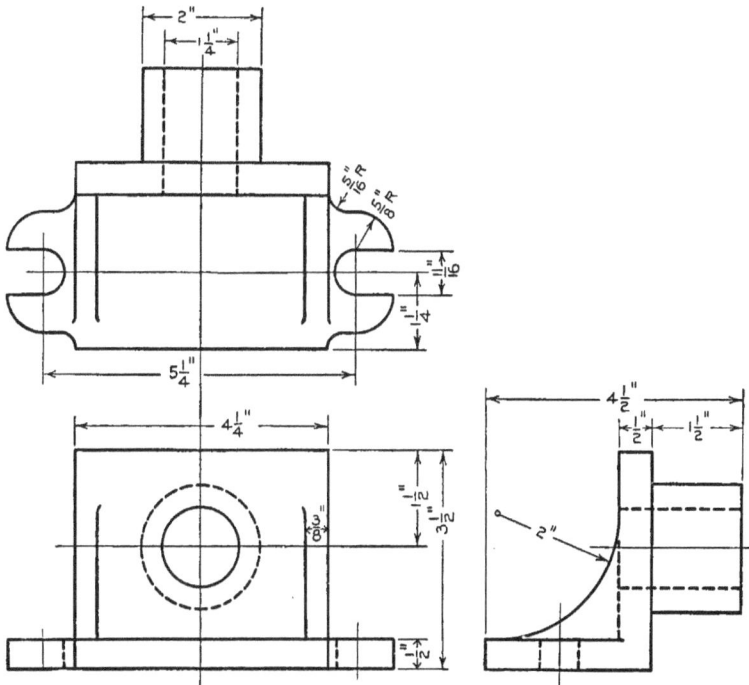

FIG. 198. PLANING JIG FOR ROD BRASSES

10. (a) In what order are the object lines inked? (b) Center lines? (c) Extension and dimension lines?

11. How is ink removed from a drawing?

12. How is the tracing trimmed to the required size?

13. (a) Upon what geometrical principle does the method of finding the point of tangency between an arc and a straight line depend? (b) Give the steps in the construction necessary to locate a point of tangency.

DATA FOR REVIEW PROBLEMS

Given: The top, front, and right side views of an object. Fig. 197.

Required: To draw the top, front half section, and right side views of the object shown in Fig. 197. Scale, full size.

Given: The top, front, and right side views of an object. Fig. 198.

Required: To draw the top, front, and left side views of the object shown in Fig. 198. Scale, half size.

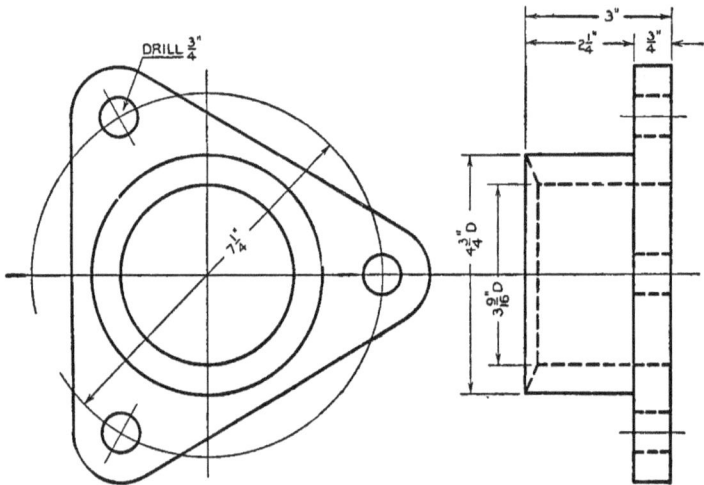

Fig. 199. Stuffing Box Gland

Given: The top and right side views of an object. Fig. 199.

Required: To draw the top and front half section views of the object shown in Fig. 199. Scale, half size.

CHAPTER V

ADVANCED DRAWING

PROSPECTUS

The first year's work outlined in Chapters I, II, III, and IV are intended to give opportunity for a thorough grounding in the fundamentals of the theory and practice of drawing. The second year's work outlined in this and the succeeding chapter assumes a knowledge of, and skill in, the work of the preceding year. With this knowledge and skill as a foundation the aim of this chapter is to furnish applications of principles in a broader and more general way and to introduce various details such as conventional sections, screw threads, etc.

SHEET METAL PATTERNS

PREPARATORY INSTRUCTIONS FOR DRAWING PLATE 24

Development of a Surface. The student is familiar with many articles made of sheet metal—tin, zinc, galvanized iron, etc. An examination of these objects, such as pails, cups, pans, etc., will make it clear that some of them were made from metal cut from flat sheets and rolled or bent into particular forms. The student will recognize the geometrical solids—prism, cylinder, cone, etc., as the bases for many of these forms. For example, an ordinary tomato can is in the form of a cylinder. Before an object of this kind can be cut from a sheet of metal a pattern must be made which, when rolled up, will give the correct form.

189

To construct this pattern the object is imagined rolled on a flat surface, such as that of the drawing board, until the entire surface of the solid has come in contact with the plane surface. Example: In Fig. 200 the prism was rolled until each of its faces

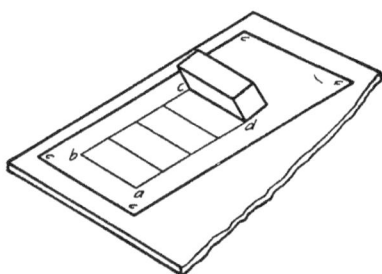

FIG. 200. ROLLING A PRISM TO OBTAIN THE DEVELOPMENT OF ITS LATERAL SURFACE

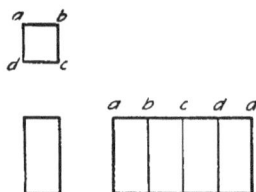

FIG. 201. ORTHOGRAPHIC VIEWS AND DEVELOPMENT OF THE LATERAL SURFACE OF A PRISM

came into contact with the board. The prints of these faces are shown. If the whole figure a, b, c, d were cut out of the paper and folded up on the lines representing the edges of the prism the result would be a prism like the original.

Fig. 201 shows two views of a square prism with the development of the lateral surface. It is evident from the orthographic views that the edges of the bases are at right angles to the lateral edges. When this is the case each face is a rectangle which is represented in the development by a rectangle equal to that of one side of the prism. When these rectangles are joined together as they are when the surface is imagined unrolled, the edges of the bases will form straight lines. Example: Line a a. Fig. 201.

The steps in the construction of the pattern are as follows:

1. Draw two parallel lines at a distance apart equal to the length of the prism. It is preferable to project these lines from the orthographic view as in Fig. 201.

2. Lay off with the dividers on one of these lines distances equal to the widths of the sides of the prism, taken in consecutive

order, as a b, b c, c d, d a. Fig. 201. For the square prism these distances are all equal. For a rectangular prism these distances would not be equal; hence care must be taken to lay off the distances *in consecutive order*

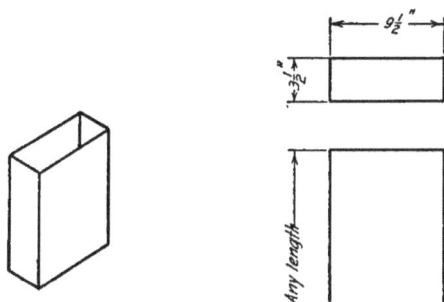

FIG. 202. FURNACE HEAT PIPE

DATA FOR DRAWING PLATE 24

Given: Two orthographic views of a rectangular furnace heat pipe. Fig. 202.

Required: To draw a pattern, quarter size, from which this pipe could be made, or any similar object assigned by the instructor.

Instructions:

1. Draw the two given orthographic views.
2. Make a construction similar to that shown in Fig. 201. In this case the widths of adjacent faces are not equal. The width of each face should be transferred to the pattern with the dividers from the top view starting at one corner and continuing around the top view until the same corner is reached.

Beginning at this point in the course the lettering plates are made up of lower case letters and numerals.

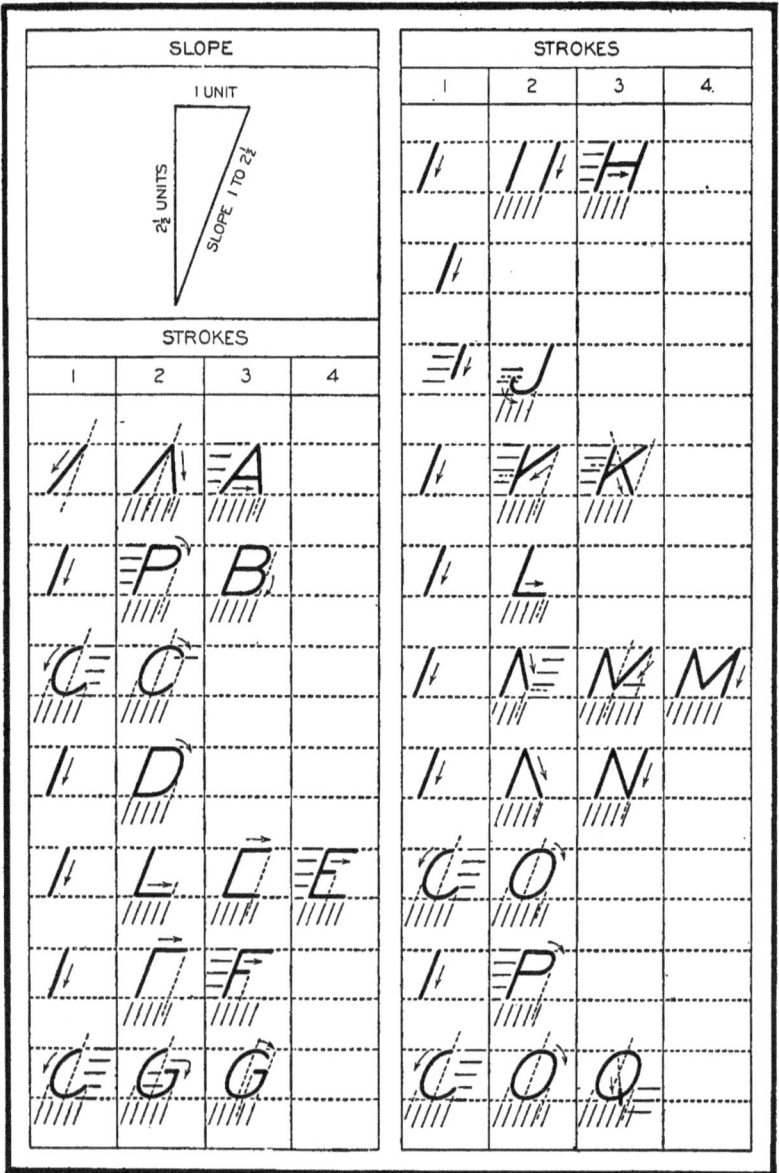

FIG. 203. INCLINED CAPITAL LETTERS

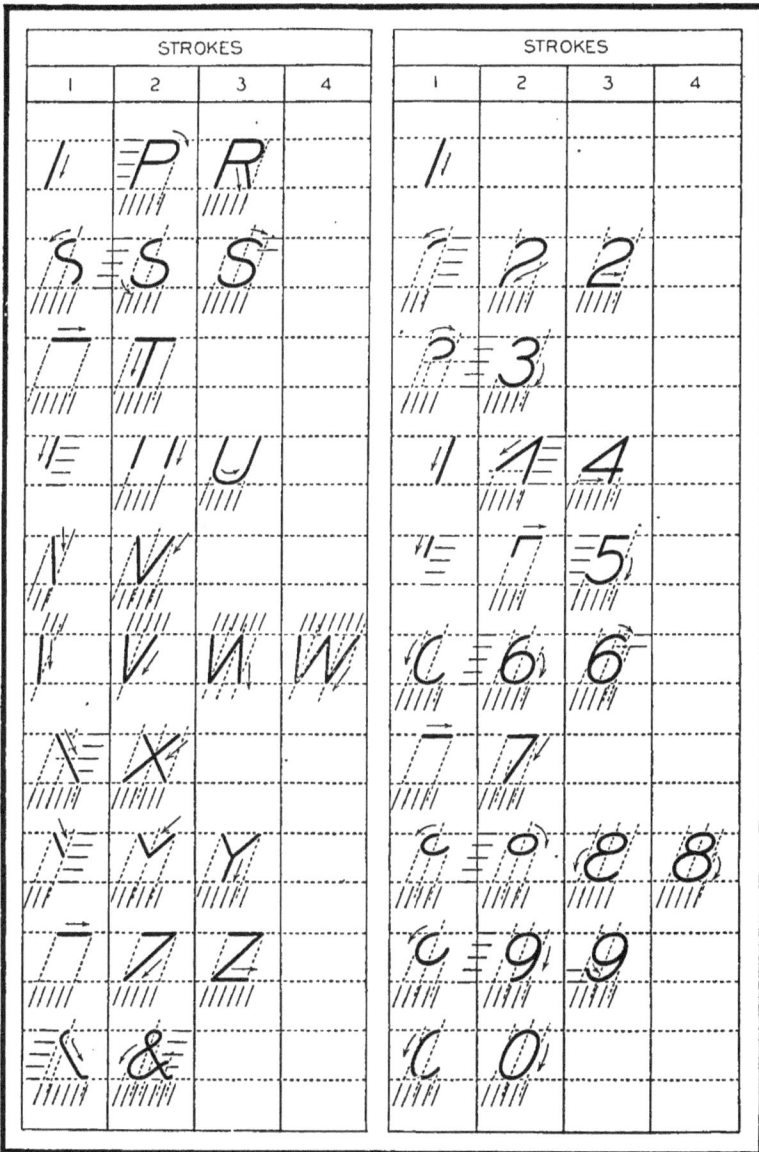

FIG. 204. INCLINED CAPITAL LETTERS

PREPARATORY INSTRUCTIONS FOR LETTERING PLATE 24

The Slope of the inclined letters is equal to that of the hypotenuse of a right triangle, the vertical leg of which is two and one-half units long and the horizontal leg one unit long. Fig. 203.

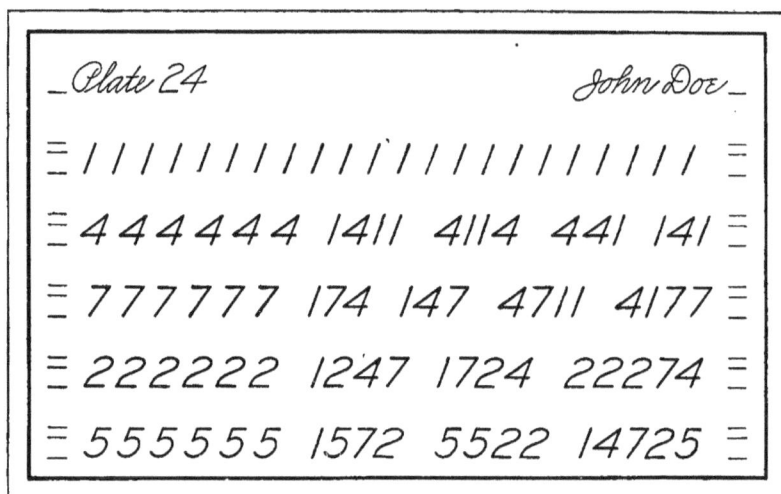

FIG. 205. LETTERING PLATE

Lettering in Ink. The following list of plates will be made in ink directly on tracing cloth. A list of materials needed and directions for lettering in ink are given for Plate 11 of the vertical Gothic letters.

DATA FOR LETTERING PLATE 24

Given: Plate 24 to reduced size. Fig. 205.
Required: To make the plate to an enlarged scale.

PREPARATORY INSTRUCTIONS FOR DRAWING PLATE 25

A somewhat more difficult pattern to lay out than the one just drawn is illustrated by a square pipe cut away at an angle to meet another pipe to form an elbow. Fig. 206. The edges of the lower base are at right angles to the vertical edges and will therefore unfold into a straight line. The lengths of these lines may be taken from the end view of the pipe and laid off on the pattern. It is evident in this case that not all of the vertical edges are of

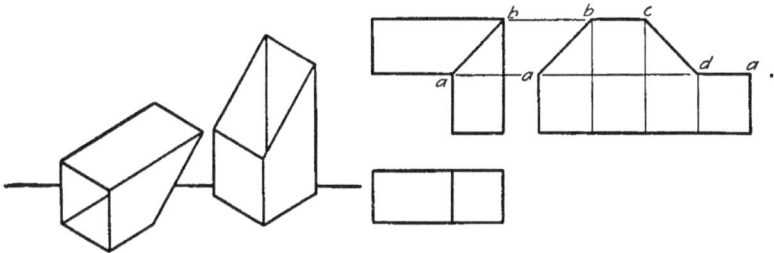

FIG. 206. TYPE PROBLEM. DEVELOPMENT OF SQUARE PIPE CUT AT AN ANGLE

the same length. Their true lengths are shown in the front view and may be transferred to the pattern by drawing horizontal lines from it to the pattern as shown in Fig. 206. The orthographic views, also, show that two of the edges of the slanting base are horizontal and the other two inclined. The lines a b, b c, c d, and d a are drawn, connecting points a, b, c, d, which should be located in consecutive order.

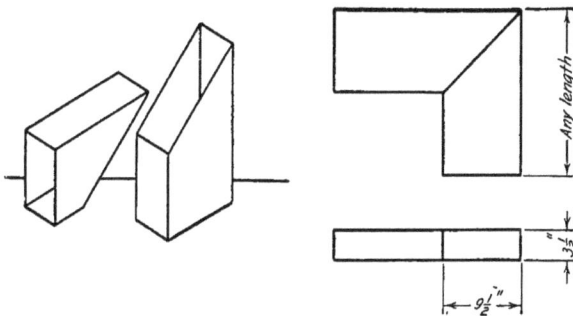

FIG. 207. RECTANGULAR PIPE

DATA FOR DRAWING PLATE 25

Given: Two orthographic views of an elbow for a rectangular pipe. Fig. 207.

Required: To draw a pattern, quarter size, from which the pipe could be made, or any similar object assigned by the instructor.

1. Draw two given orthographic views and make a construction similar to that shown in Fig. 206.

PREPARATORY INSTRUCTIONS FOR LETTERING PLATE 25

Curved Strokes. The 6 and 9 have the same oval outline as the 0. This form should be kept in mind while drawing the 6 and 9.

Given: Plate 25 to reduced size. Fig. 208.

Required: To make the plate to an enlarged scale.

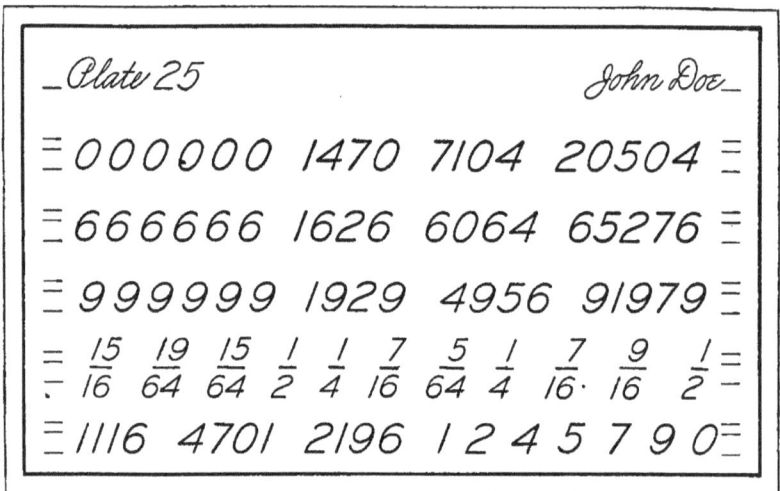

Plate 25   John Doe

≡ 000000 1470 7104 20504 ≡

≡ 666666 1626 6064 65276 ≡

≡ 999999 1929 4956 91979 ≡

= 15/16 19/64 15/64 1/2 1/4 7/16 5/64 1/4 7/16· 9/16 1/2 =

≡ 1116 4701 2196 1 2 4 5 7 9 0≡

FIG. 208. LETTERING PLATE

PREPARATORY INSTRUCTIONS FOR DRAWING PLATE 26

The cylinder is a very common form in sheet metal work. Many cans, pails, pipes, etc., are cylindrical in form. When the base of a cylinder is at right angles to its axis, the base will unroll into a straight line. Fig. 209. The length of this line must be found by dividing the circle representing the end view of the cylinder into a number of small parts and stepping off these lengths with the dividers on a straight line. Each distance transferred is the chord of the arc between two points. These divisions must, therefore, be small enough so that the straight

line distance between consecutive points is not greatly different
from the distance measured between these points on the circle,
or the arc distance. These divisions are usually made equal so
that one setting of the dividers is sufficient for stepping off all
the lengths.

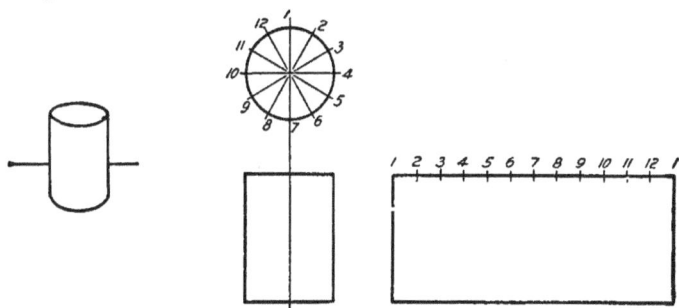

FIG. 209. TYPE PROBLEM. DEVELOPMENT OF CYLINDRICAL SURFACE

Eight points equally spaced on the circumference are very
easily obtained with the 45° triangle, or twelve points may be
obtained with the 30°-60° triangle as shown in Fig. 210.

Sixteen points equally spaced may be obtained by subdividing
each of the eight divisions with the dividers.

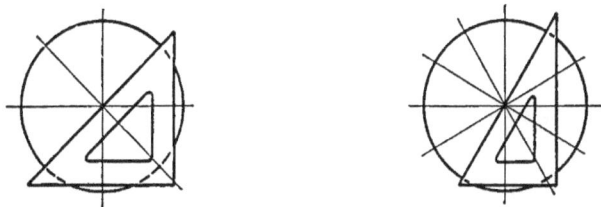

FIG. 210. DIVIDING A CIRCLE INTO 8 OR 12 PARTS WITH TRIANGLES

DATA FOR PLATE 26

Given: Two orthographic views of a bench oil-waste **cup.**
Fig. 211.

Required: To draw a pattern from which the cup can be
made, or any similar object assigned by the instructor.

Instructions: Draw the two orthographic views and make a
construction similar to that shown in Fig. 209.

PREPARATORY INSTRUCTIONS AND DATA FOR LETTERING
PLATE 26

The two ovals of the 8 have their major axes at 45°. The same combination of ovals is the basic form for the 3.

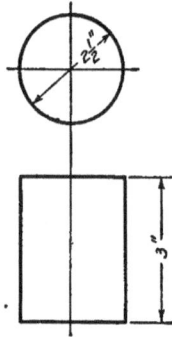

FIG. 211. OIL WASTE CUP

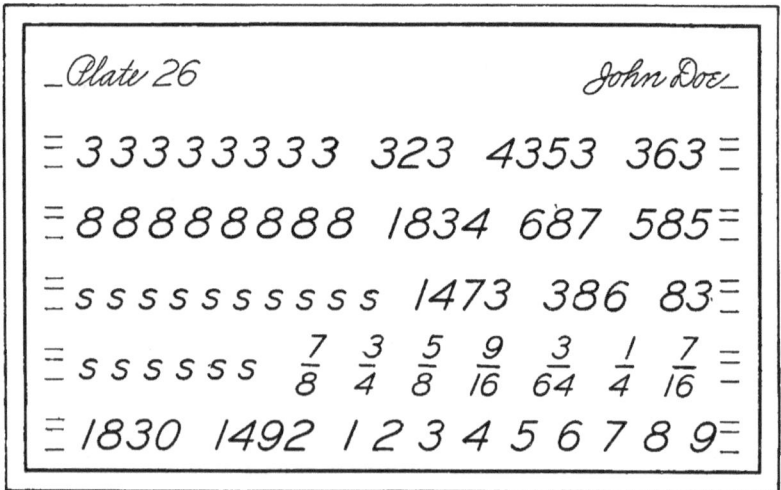

FIG. 212. LETTERING PLATE 26

Given: Plate 26 to reduced size. Fig. 212.
Required: To make the plate to an enlarged scale.
The strokes for the S are given in Fig. 238.

PREPARATORY INSTRUCTIONS FOR DRAWING PLATE 27

If a cylindrical object is cut away at an angle, as in the case of the elbow in Fig. 213, the end which is cut will not roll out into a straight line. The curved line into which it will unroll must be determined by locating a number of points through which it passes. This may be done by drawing lines in the orthographic view which represent elements or lines in the cylindrical surface parallel to the axis of the cylinder. For convenience these lines should be drawn perpendicularly up from the points located in the base circle. The length of each of these

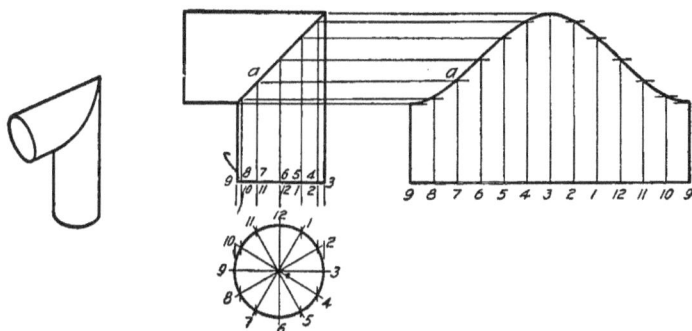

Fig. 213. Development of Cylindrical Surface Cut at an Angle

lines from the base up to the inclined line is the length of a corresponding line to be located in the pattern, because it is shown in its true length.

Vertical lines are drawn in the pattern from the points stepped off in the base. These lines represent the ones drawn in the surface of the cylinder. The length of these lines may be transferred from the orthographic view to the pattern with the dividers or by drawing horizontal lines across from the orthographic view as in Fig. 213. Example: Line a 7 in the orthographic view is equal to a 7 in the pattern. Here again it is necessary that these lengths be transferred *in consecutive order*.

When both ends of the cylinder are cut at an angle as in Fig. 214, neither end will develop into a straight line. For the purpose of determining the length of the pattern a line such as a b must be drawn in to represent an imaginary base which is at right angles to the axis of the cylinder. The lengths of the division from the circular view, A, are then stepped off on a

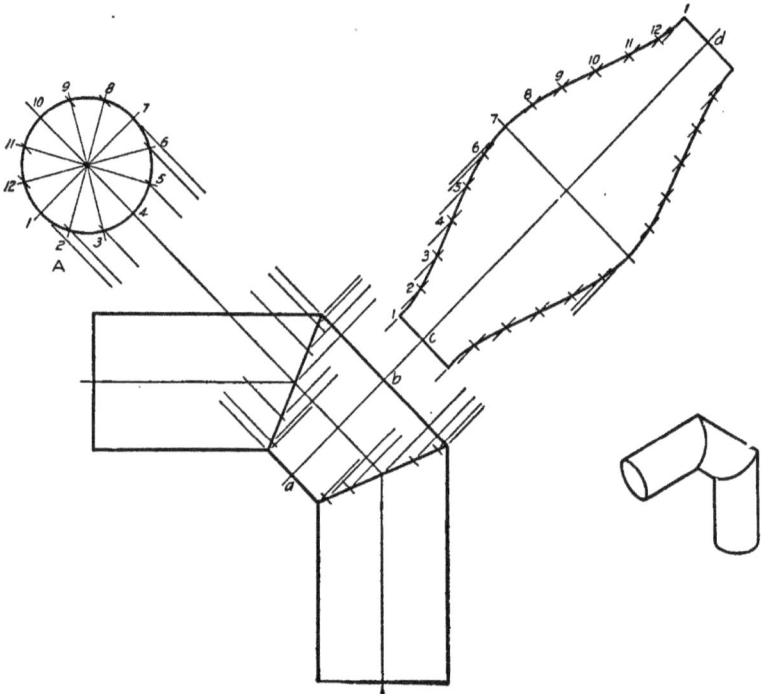

FIG. 214. TYPE PROBLEM. DEVELOPMENT OF CYLINDRICAL SURFACE CUT AT BOTH ENDS

line representing this imaginary base, c d, Fig. 214. The points on the curved lines in the development are located by using this imaginary base to measure from. If the base is centrally located between the two cut ends, points on both curves, on any one element, may be located with one measurement, as in the case shown in Fig. 214.

DATA FOR DRAWING PLATE 27

Given: Two orthographic views of the objects shown in Figs. 215, 216, and 217.

Required: To draw a pattern for one of the objects shown in Fig. 215, 216, or 217, or any similar object as assigned by the instructor.

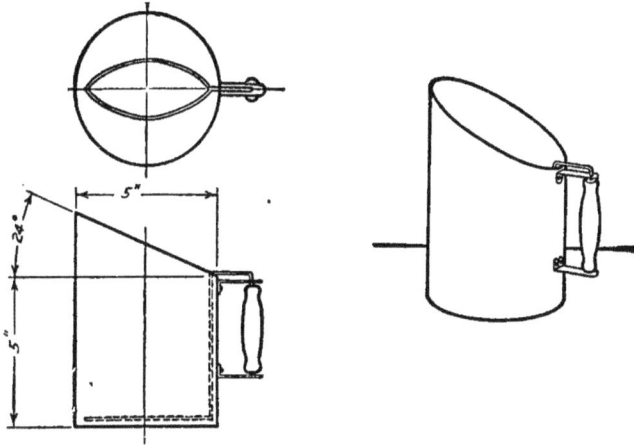

FIG. 215. FLOUR SIFTER

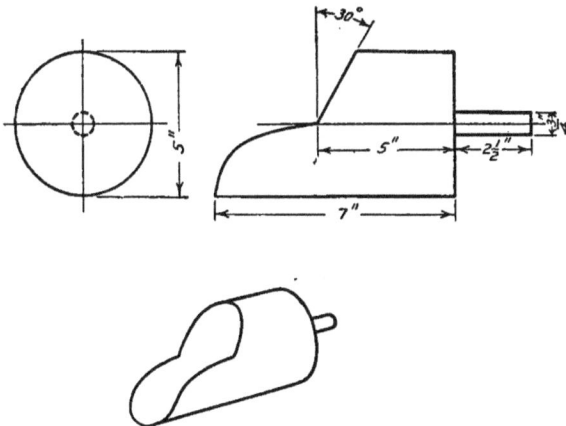

FIG. 216. SCOOP

Instructions: Draw the orthographic views and make a construction similar to that shown in Fig. 215, 216, or 217.

PREPARATORY INSTRUCTIONS AND DATA FOR LETTERING.
PLATE 27

Spacing of Letters. Observe carefully the spacing of the

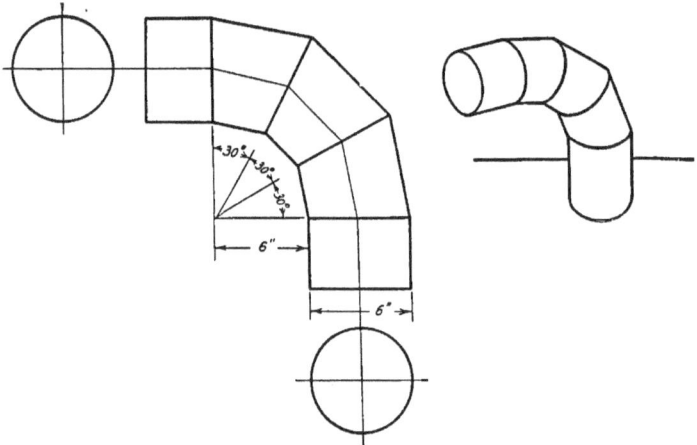

FIG. 217. FIVE PIECE ELBOW

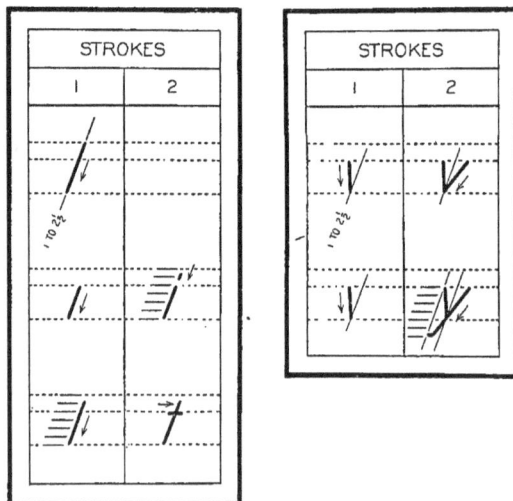

FIG. 218. LETTERING PLATE. l, i, t, v, y

letters in the words. Correct spacing is as essential as correct forms.

·**Given:** Plate 27 to reduced size. Fig. 219.

Required: To make the plate to an enlarged scale.

PREPARATORY INSTRUCTIONS FOR DRAWING PLATE 28

Another geometrical form commonly found in sheet metal work is the cone. If the cone were rolled on a flat surface its surface would come in contact with an area as shown in Fig. 220.

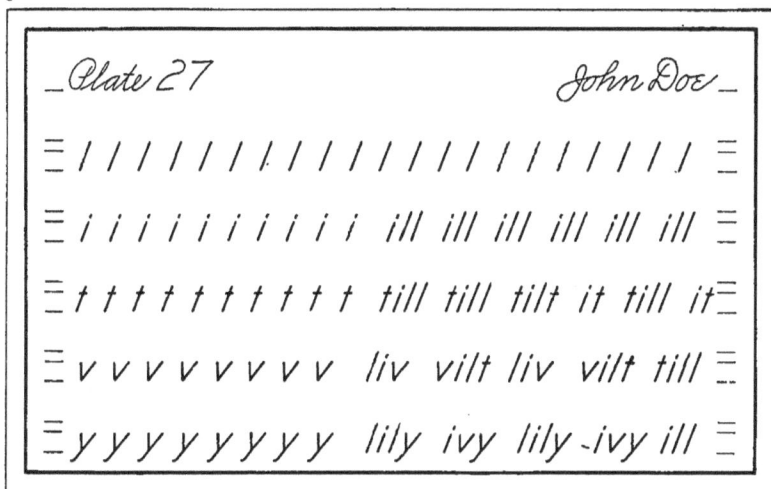

FIG. 219. LETTERING PLATE 27. l, i, t, v, y

The vertex would remain at a fixed point. Since all straight lines drawn in the surface of the cone from the vertex to the base circle are equal in length, the base circle will unroll into an arc with a radius equal to the true distance from the vertex to a point in the base circle. Fig. 220. The length of this arc is equal to the circumference of the base circle of the cone and may be laid off by dividing the orthographic view of the base circle into a number of small divisions as described for the cylinder, page 197. These lengths are then transferred to the arc with the dividers.

The pail shown in Fig. 221 is not a complete cone. It is therefore necessary to draw a second arc to represent the circular bottom of the pail.

Given: Two orthographic views of the objects shown in Figs. 222 and 223.

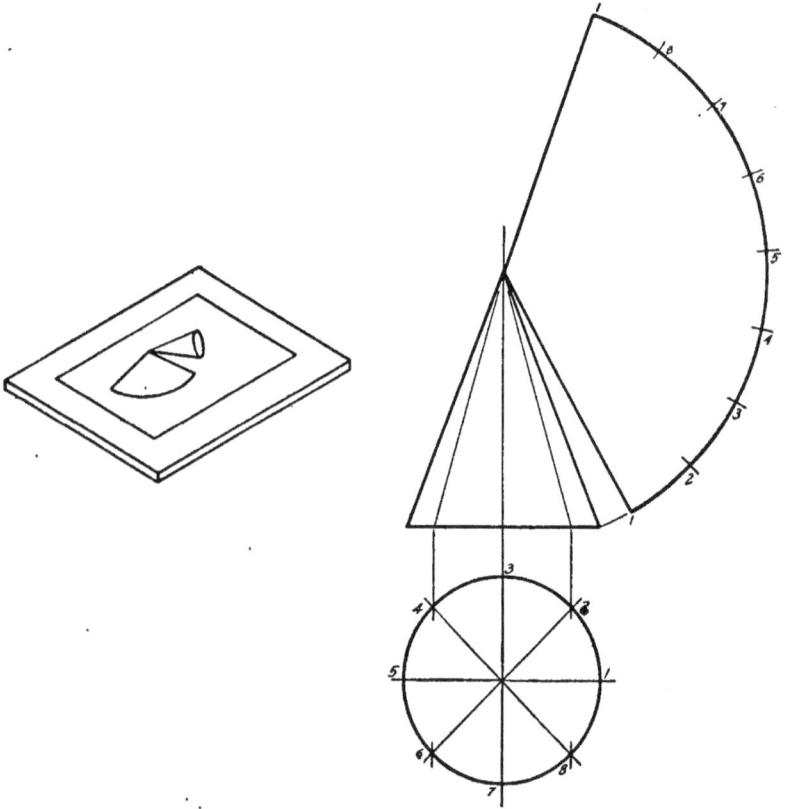

FIG. 220. ROLLING A CONE TO OBTAIN THE DEVELOPMENT OF ITS SURFACE

Required: To draw a pattern for one of the objects shown in Fig. 222 or 223, or any similar object as assigned by the instructor.

Instructions: Draw the orthographic views and make a construction for the pattern similar to that shown in Fig. 221.

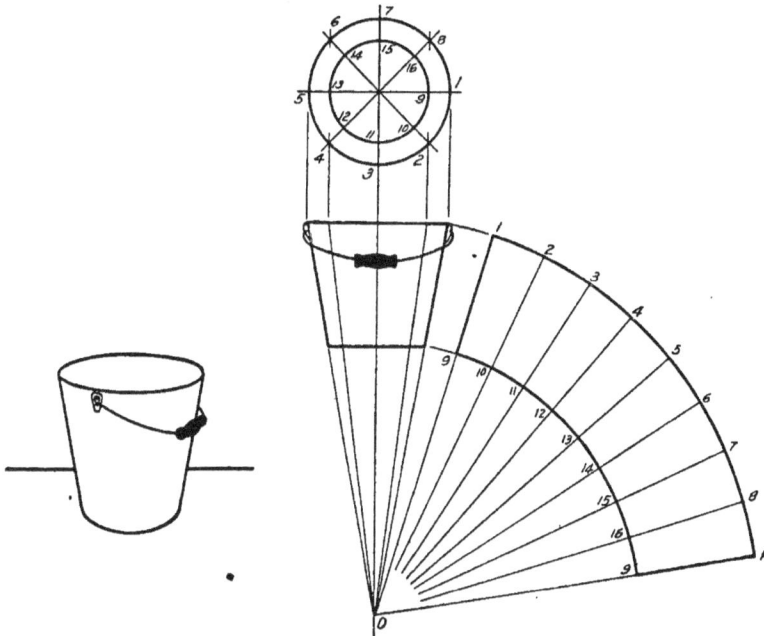

FIG. 221. TYPE PROBLEM. PAIL

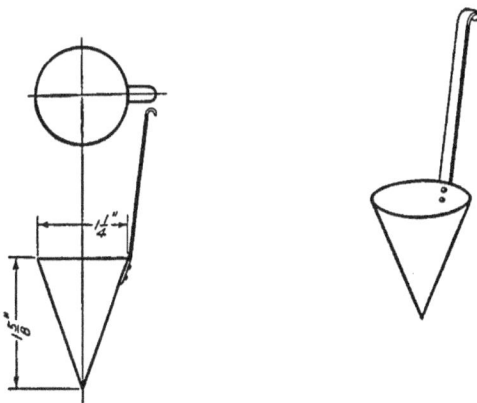

FIG. 222. CREAM DIPPER

DATA FOR LETTERING PLATE 28

Given: Plate 28 to reduced size. Fig. 225.
Required: To make the plate to an enlarged scale.

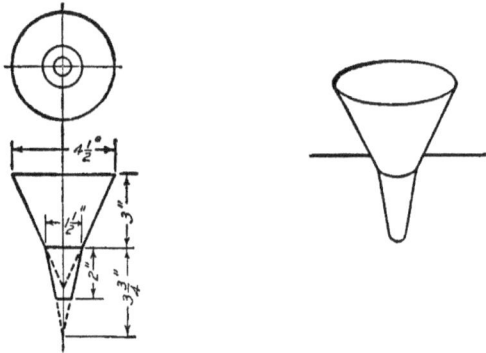

FIG. 223. FUNNEL

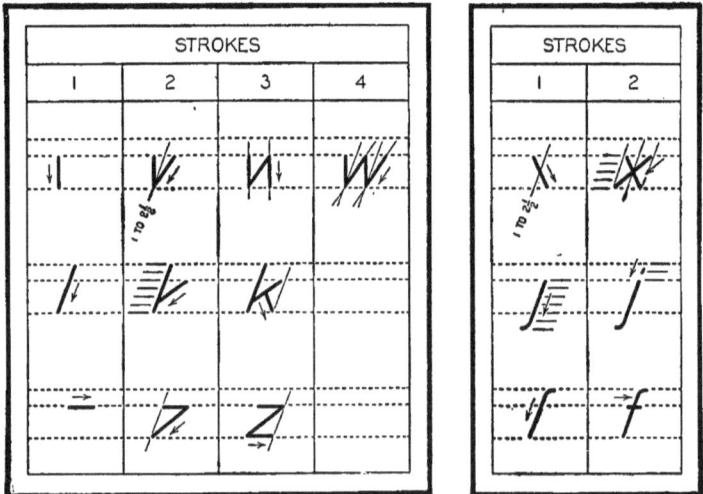

FIG. 224. LETTERING PLATE. w, k, z, x, j, f

PREPARATORY INSTRUCTIONS FOR DRAWING PLATE 29

The section of a right conical surface cut at right angles to the axis of the cone will develop into the arc of a circle as shown

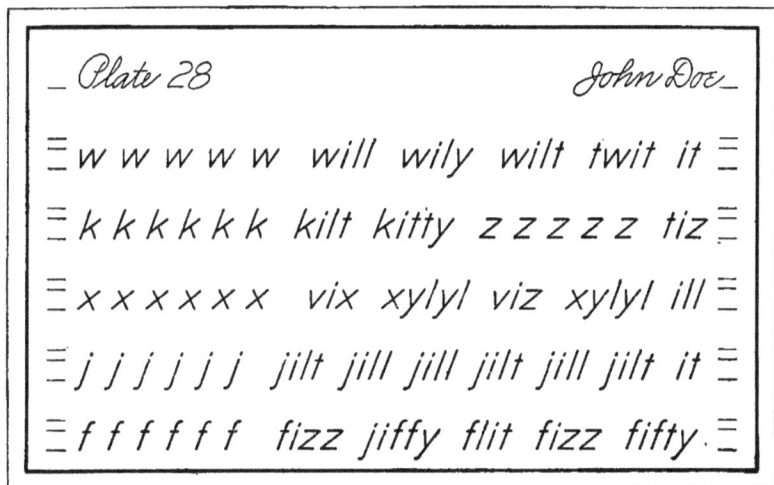

FIG. 225. LETTERING PLATE. w, k, x, j, f

in the case of the bottom of the pail, Fig. 221. When the conical surface is cut on a slant as shown in Fig. 226, the distances from the vertex to points on the cut are not equal and therefore the cut edge will not roll out into an arc of a circle. Points on the curved line into which this cut edge will develop may be located by drawing in the elements or lines in the conical surface from the vertex to points in the base circle. These elements may be located in the pattern by joining the vertex with points on the arc of the base corresponding to the points in which these lines meet the base circle of the cone.

This construction is illustrated in Fig. 227, where the points in which the elements pass through the cut edge of the cone are located in the development by finding their *true distances* on the elements from the vertex (see horizontal lines in front view) and transferring these lengths to the corresponding lines in the development. This may be done with the dividers or by swinging arcs with the compass as shown in the figure.

The true lengths of the contour elements of the cone are shown in the front view, Fig. 227, in o e and o d. None of the

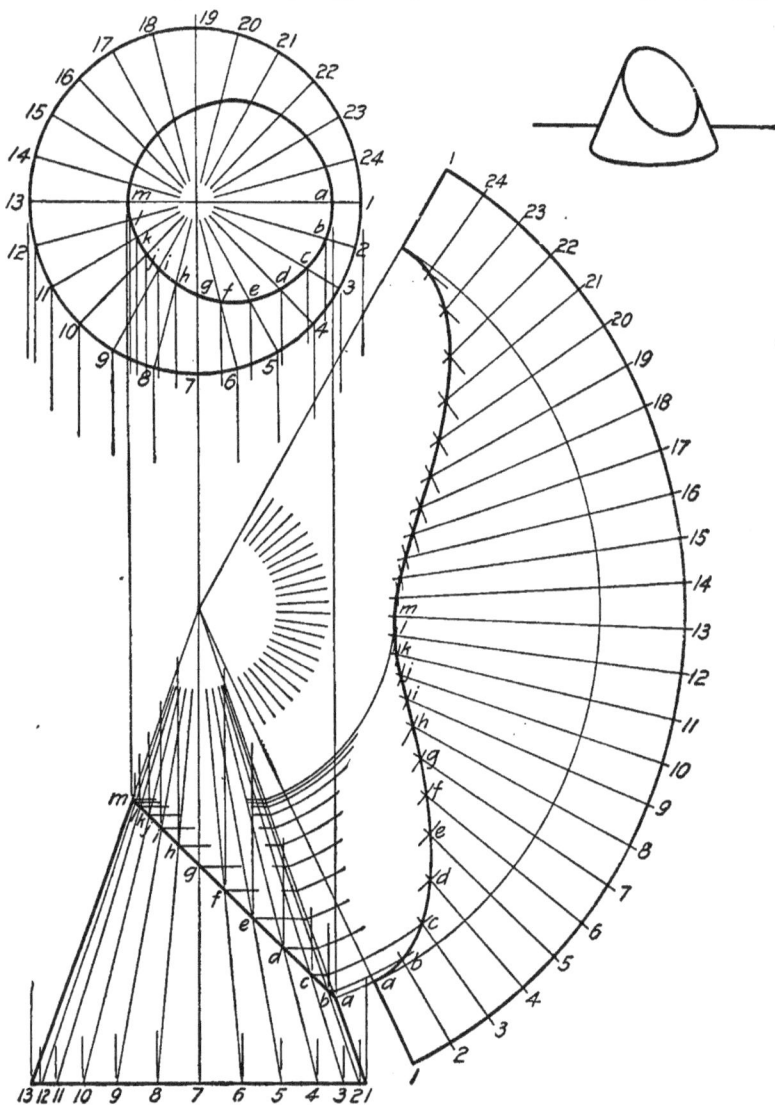

FIG. 226. DEVELOPMENT OF A CONE CUT AT AN ANGLE TO ITS AXIS

other elements show in their true length, although they are known to be equal in length to o e and o d.

In order to find the distance from the vertex to a point such as **a** on an element a construction is necessary. The student should try to fix in mind the principle on which this construction is made, which is as follows. Since any element o c is equal in length to o d it may be imagined turned around into coincidence with o d by keeping the end c always in the base circle. Thus the

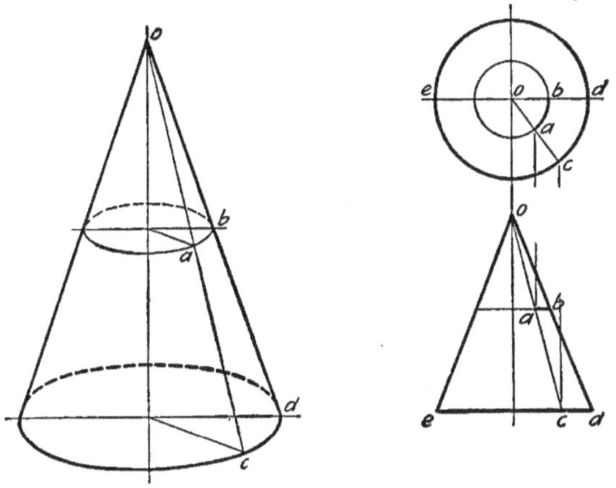

FIG. 227. ILLUSTRATING METHOD OF OBTAINING TRUE LENGTHS OF ELEMENTS

point c moves through an arc c d. The point at the other end of the line o c remains fixed, but any point, such as a between o and c will move on an arc a b. Thus when o c is brought into coincidence with o d the true length of o c is shown in o d and the true length of any part of it, such as o a, is shown in o b. Notice that the base circle of the cone appears as a straight horizontal line e d, Fig. 227, and also that the arc a b appears as a *straight horizontal line* in the front view of the cone. With this point clearly in mind it will be evident that to find the length of o a it is only necessary to draw *the horizontal line a b as construction.*

DATA FOR DRAWING PLATE 29

Given: The orthographic views of the objects shown in Figs. 228 and 229.

Required: To draw a pattern for one of the objects shown in Fig. 228 or 229, or any similar object assigned by the instructor.

Instructions: Draw the orthographic views and make a construction for the pattern similar to that shown in Fig. 226.

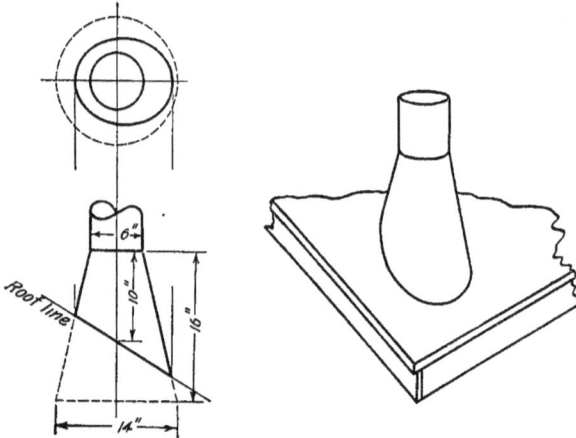

FIG. 228.   VENTILATOR PIPE

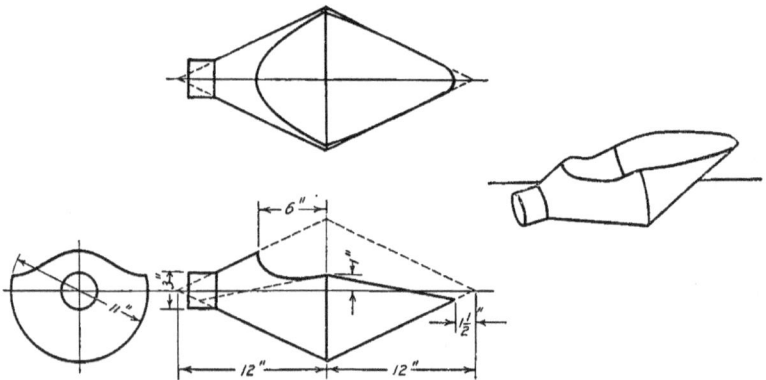

FIG. 229.   SCALE SCOOP

DATA FOR LETTERING PLATE 29

Given: Plate 29 to reduced size. Fig. 231.

Required: To make the plate to an enlarged scale.

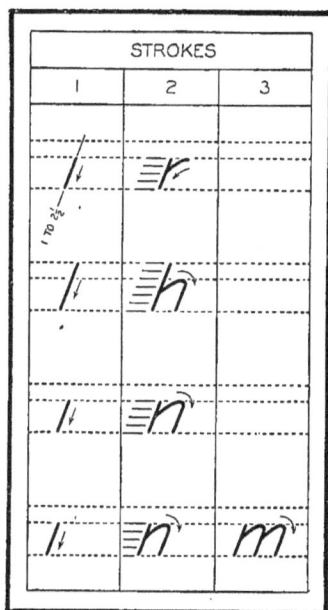

FIG. 230. LETTERING PLATE. r, h, n, m

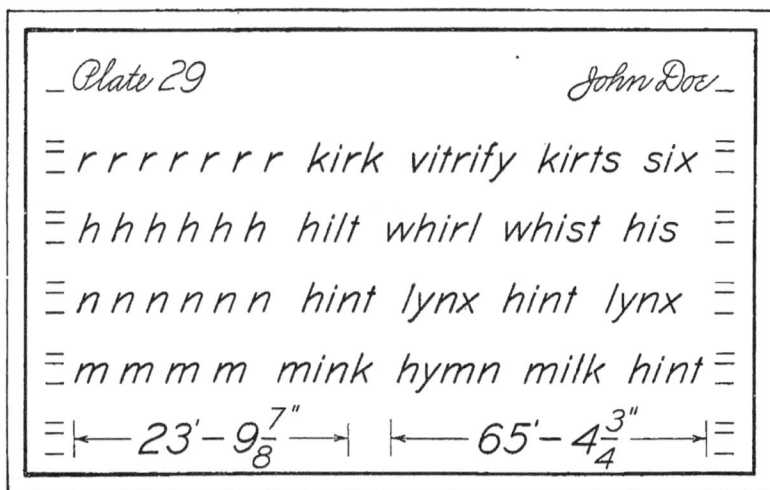

FIG. 231. LETTERING PLATE 29. r, h, n, m

If a pyramid were rolled on a flat surface its faces would come in contact with triangular areas such as shown for the square pyramid in Fig. 232. In this case there are four triangles. For a hexagonal pyramid there would be six triangles, etc. The length of the edges of the pyramid from the vertex to the corner of the base are all equal. Therefore if an arc of radius equal to their length were drawn these lines would all end in the arc. The sides of the base of the pyramid will appear in the pattern as

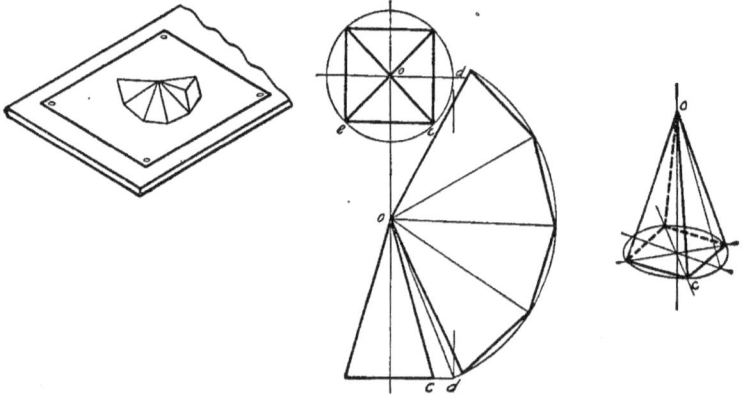

FIG. 232. ROLLING A PYRAMID TO OBTAIN THE DEVELOPMENT OF ITS LATERAL SURFACE

FIG. 233. TYPE PROBLEM. DEVELOPMENT OF A SQUARE PYRAMID

chords of this arc. In the case of the square pyramid as shown in Fig. 233, none of the edges from the vertex to the base are shown in their true length. A construction such as that described for finding the lengths of the elements of the cone must be made for finding the lengths of these edges. Referring to Fig. 233, the base of the pyramid is inscribed in a circle which corresponds to the base circle of the cone. If the edge o c, for example, is imagined turned as was the element of the cone, into a position corresponding to the contour element of the cone, it will show in its true length in the front view. o d, Fig. 233, therefore represents the true length of the edge o c. With this length as a radius, an arc may be drawn and the points representing the lower corners of the pyramid may be located on it by stepping off lengths equal to the side of the base, such as e c.

DATA FOR DRAWING PLATE 30

Given: The orthographic views of the object shown in Fig. 234.

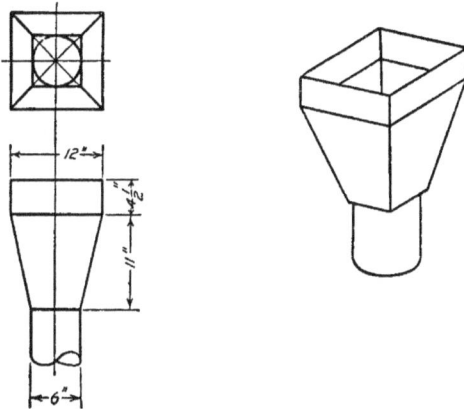

FIG. 234. END FOR GRAIN CONVEYOR

FIG. 235. LETTERING PLATE. u, o, c, e

Required: To draw a pattern for the object shown in Fig. 234 or any similar object as assigned by the instructor.

Instructions: Draw the orthographic views and make a construction for the pattern similar to that shown in Fig. 233.

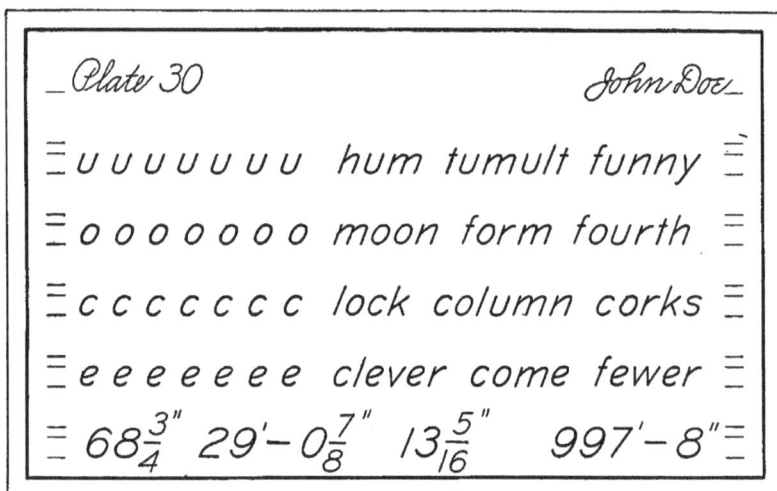

Plate 30   John Doe

Ξ u u u u u u u hum tumult funny Ξ

Ξ o o o o o o o moon form fourth Ξ

Ξ c c c c c c c lock column corks Ξ

Ξ e e e e e e e clever come fewer Ξ

Ξ $68\frac{3}{4}''$ $29'-0\frac{7}{8}''$ $13\frac{5}{16}''$ $997'-8''$ Ξ

FIG. 236. LETTERING PLATE 30. u, o, c, e

PREPARATORY INSTRUCTIONS AND DATA FOR LETTERING PLATE 30

Curved Strokes. The major axes of the oval letters of this plate are in the direction of the slope.

Given: Plate 30 to reduced size. Fig. 236.

Required: To make the plate to an enlarged scale.

PREPARATORY INSTRUCTIONS FOR DRAWING PLATE 31

The true length of the edge of a pyramid cut at a slant is found as previously described. Finding the true length of the edges from the vertex to the point where the edge strikes the cut involves the same theory discussed in connection with the pattern for the cone, page 209, Fig. 227. The construction is also the same.

DATA FOR DRAWING PLATE 31

A problem for this plate may be supplied by the instructor if desired.

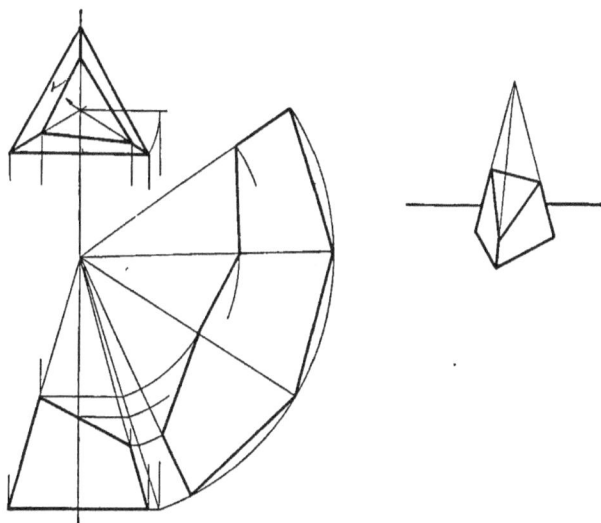

FIG. 237. DEVELOPMENT OF PYRAMID CUT AT AN ANGLE TO ITS AXIS

PREPARATORY INSTRUCTIONS FOR LETTERING PLATE 31

Curved Strokes. The major axes of the ovals of this plate make 45° with the horizontal.

DATA FOR LETTERING PLATE 31

Given: Plate 31 to reduced size. Fig. 239.
Required: To make the plate to an enlarged scale.

PREPARATORY INSTRUCTIONS FOR DRAWING PLATE 32

Intersection of Surfaces. The objects thus far considered have been of the form of geometrical solids. There is another class of patterns which involves the laying out of the line where

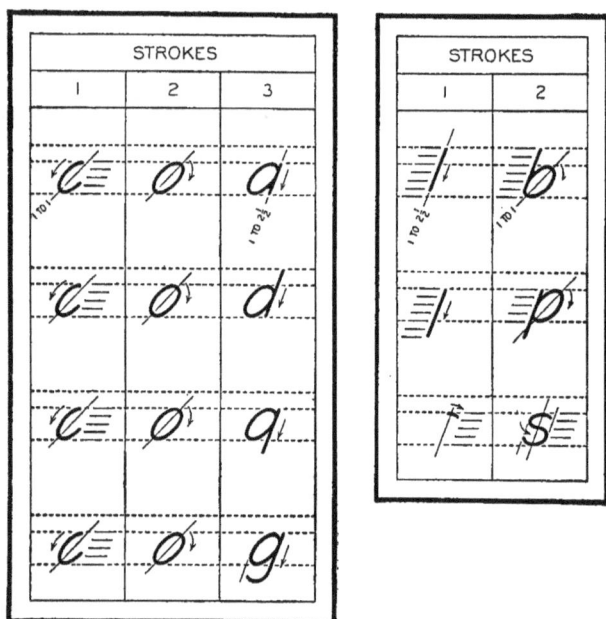

FIG. 238. LETTERING PLATE. a, d, q, g, b, p, s

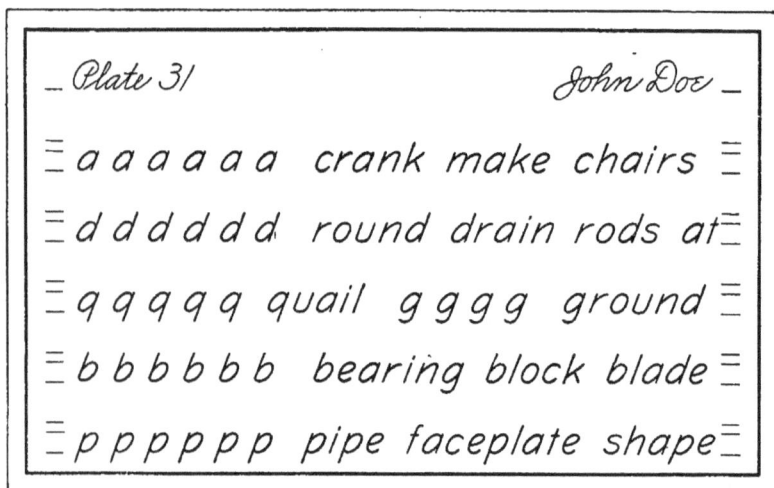

FIG. 239. LETTERING PLATE 31. a, d, q, g, b, p, s

the surfaces of two solids meet or intersect. An example of the intersection of two prisms is afforded in the case of the roof of

a house, as shown in Fig. 240. The line of intersection is the line
a b c where the roofs meet. It is quite evident that this broken
line a b c lies in the surface of the main roof and also in the surface of the dormer roof, or, in other words, it is a broken line
which is common to both roofs. This illustrates the general definition of a line of intersection, which is as follows: *The line of*

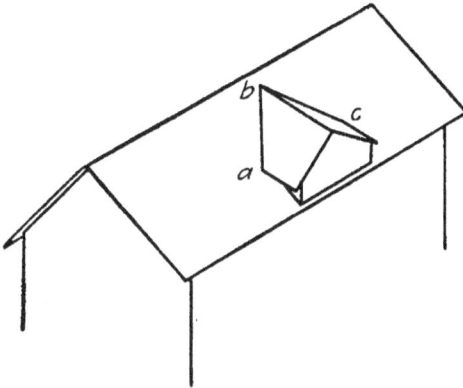

FIG. 240. ILLUSTRATION OF AN INTERSECTION

*intersection between two surfaces is the line which lies in both
surfaces.* In the following discussion of the laying out of patterns of objects containing intersecting surfaces, this definition,
if kept clearly in mind, will help in grasping the principles on
which the methods are based.

In Fig. 241 is shown the orthographic view of a cylinder intersecting a square prism. In this case the right side view is unnecessary for the purpose of laying out a pattern. It is drawn to
show the method of constructing the line of intersection in this
view, as in some cases it will be necessary to draw a view corresponding to this one. It also gives a better idea of the method
by which the points on the line of intersection are located for
transferring to the pattern.

Development of the Cylindrical Surface. As in Plate 26, the
edge of the upper base of the cylinder which is at right angles
to the axis of the cylinder will develop into a straight line. The
length of this line may be determined, as before, by dividing the

base circle into a number of equal small divisions and stepping
them off with the dividers. To obtain the development of the line
of intersection, elements are drawn in the surface of the cylinder,
preferably from the points already located in the base circle.
Lines are then drawn in the patterns to represent them. Their
lengths may be transferred from the orthographic views by means

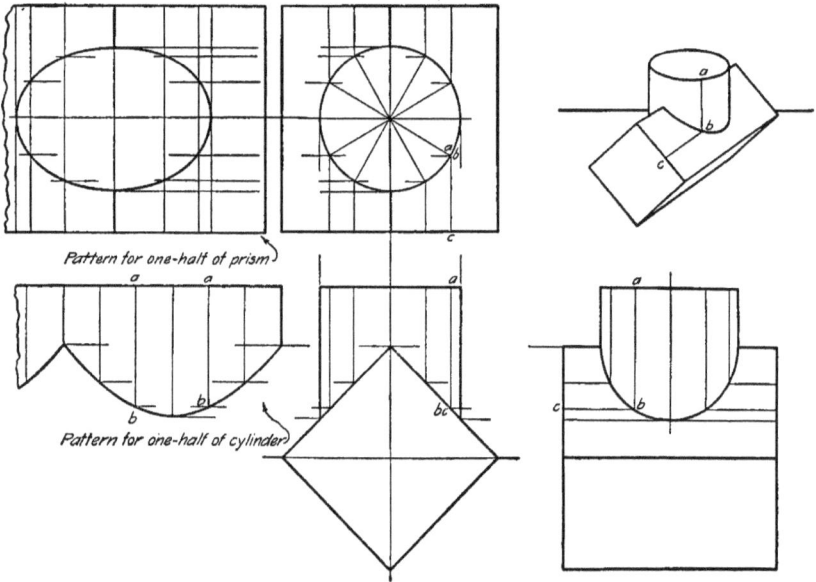

FIG. 241. TYPE PROBLEM. INTERSECTION OF A SQUARE PRISM AND A
CYLINDER

of the dividers or projected by horizontal lines as shown. in
Fig. 241.

The pattern for the entire surface of the prism is laid out as
in Plate 24. The hole opening into the cylinder or the line of
intersection is determined in the pattern as follows: In the front
view the lateral surfaces of the prism are seen edgewise, and consequently the points in which the cylinder strikes the surfaces
of the prism are seen in the points where the lines representing
these elements cross the lines representing the surfaces of the
prism. Example: b is the point in which the element a b of the

cylinder strikes the surface of the prism. If a line is drawn in
the surface of the prism through point b and parallel to the
lateral edges of the prism, the distance of this line from the edge
of the prism may be located on the pattern. The true

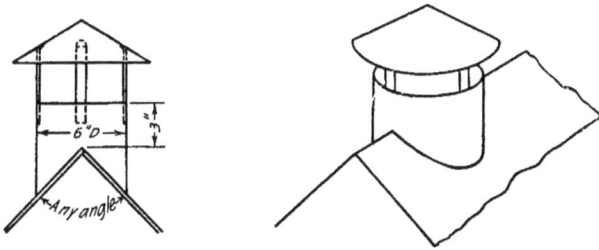

FIG. 242. ROOF CAP AND VENTILATOR

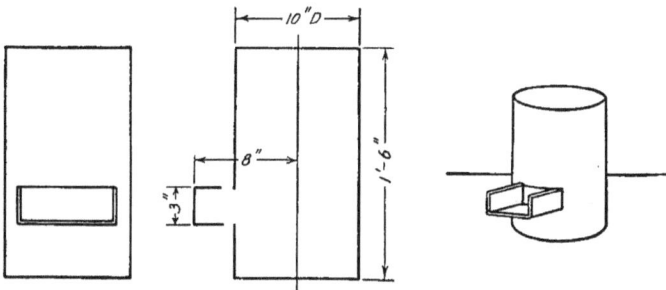

FIG. 243. SOLDERING STOVE

length of this line, which is shown in the top view, may
be transferred to the pattern with the dividers or projected from
the orthographic view as indicated in the drawing. A similar
construction should be made for the other points on the line of
intersection.

DATA FOR DRAWING PLATE 32

Given: The orthographic views of the objects shown in
Figs. 242 and 243.

Required: To draw the orthographic views and construct patterns for the object shown in Fig. 242 or 243, or any similar object as assigned by the instructor.

PREPARATORY INSTRUCTIONS FOR LETTERING PLATE 32

Composition. In the following composition plates the spacing of letters and words should be given as much consideration as the forms of the letters. The student should strive to produce a good general effect in the plate

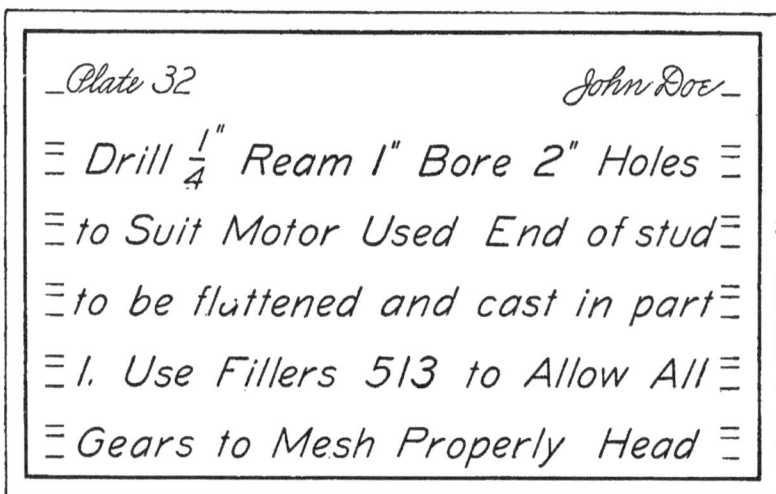

FIG. 244. LETTERING PLATE 32

DATA FOR LETTERING PLATE 32

Given: Plate 32 to reduced size. Fig. 244.
Required: To make the plate to an enlarged scale.

PREPARATORY INSTRUCTIONS FOR DRAWING PLATE 33

The laying out of patterns for the surfaces of two intersecting cylinders involves the same general principles as described for the intersection of the prism and cylinder in Plate 32. The base of the smaller of the two cylinders should be divided into a number of equal parts and its surface developed as before.

The entire surface of the larger cylinder is laid out and the points in which the elements of the smaller cylinder strike the

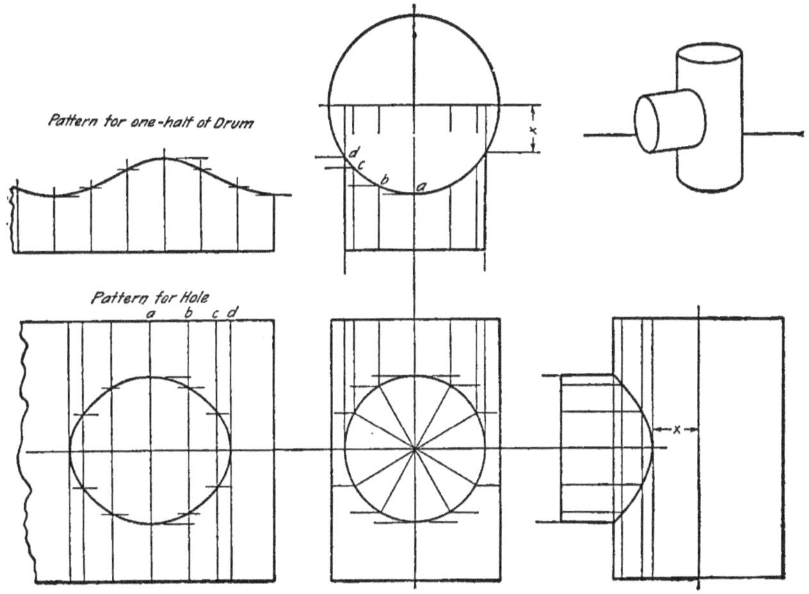

Pattern for one-half of Drum

Pattern for Hole
a b c d

FIG. 245. TYPE PROBLEM. INTERSECTION OF TWO CYLINDERS

surface are located by drawing elements of the larger cylinder through these points. The spacing of these elements is obtained by stepping off the arcs a b, b c, etc., Fig. 245, between the points representing these elements in the end view (top view in this case) of the cylinder.

DATA FOR DRAWING PLATE 33

Given: The orthographic views of the objects shown in Figs. 246 and 247.

Required: To draw the orthographic views and construct patterns for the objects shown in Fig. 246 or 247, or any similar object assigned by the instructor

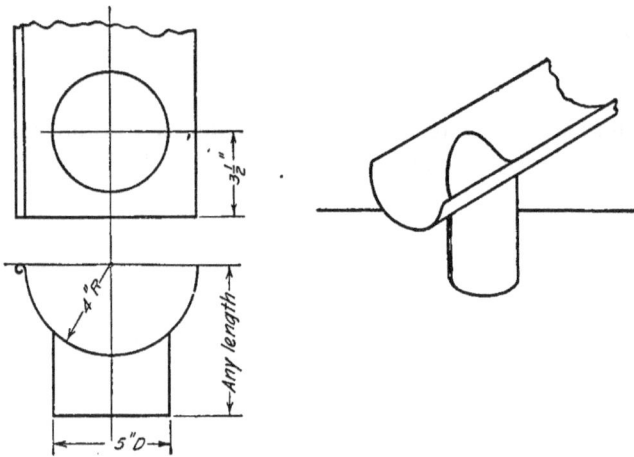

FIG. 246.   EAVE TROUGH AND DOWN SPOUT

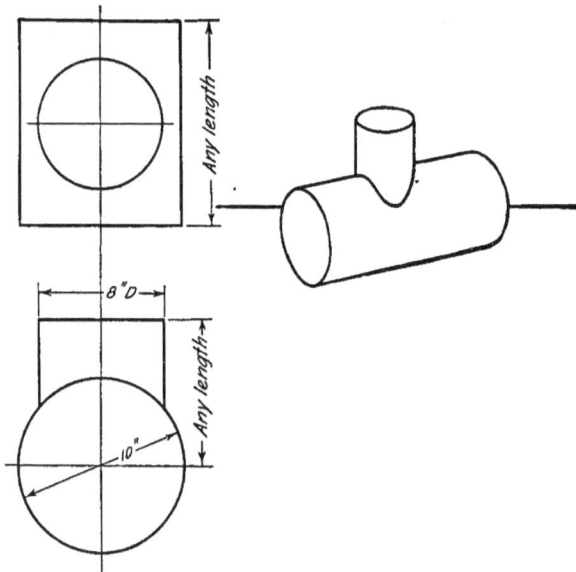

FIG. 247.   FURNACE SMOKE PIPE

DATA FOR LETTERING PLATE 33

Given: Plate 33 to reduced size. Fig. 248.

Required: To make the plate to an enlarged scale.

PREPARATORY INSTRUCTIONS FOR DRAWING PLATE 34

In the preceding plates, elements were first drawn in the cylinder to strike the surface of the other solid. In the construction of the intersection line between a cylinder and cone, example, Fig. 249, elements of the cone are first drawn striking the surface of the cylinder. This is made necessary by the fact that the cone has a slanting surface which is not seen edgewise in any view, consequently it is impossible to tell where elements drawn in the surface of the cylinder strike the surface of the cone.

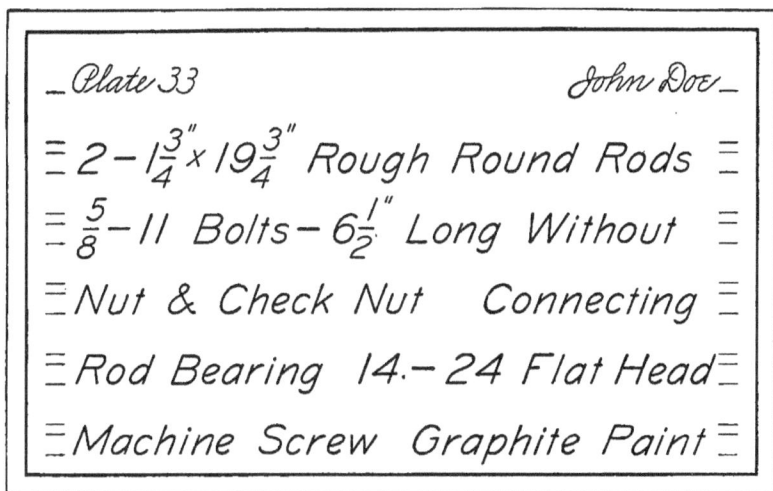

Plate 33   John Doe

= 2 - 1¾" x 19¾" Rough Round Rods =

= ⅝" - 11 Bolts - 6½" Long Without =

= Nut & Check Nut   Connecting =

= Rod Bearing 14. - 24 Flat Head =

= Machine Screw Graphite Paint =

FIG. 248. LETTERING PLATE 33

The elements of the cone should first be drawn in the front view and in such a manner as to divide the circular base of the cylinder into a number of small parts. It is evident that these parts cannot all be equal as in the preceding problems.

Draw the top views of these same elements of the cone. This is done for element a b by projecting from a to the base circle of the cone in the top view to determine the foot of the element a b in the top view, and connecting this point with the apex.

Elements should now be drawn in the surface of the cylinder to pass through the points in which the elements of the cone strike the surface of the cylinder. Example: c in the top view is projected from the point c in the front view.

The other points on the intersection are located in the top view in the same manner.

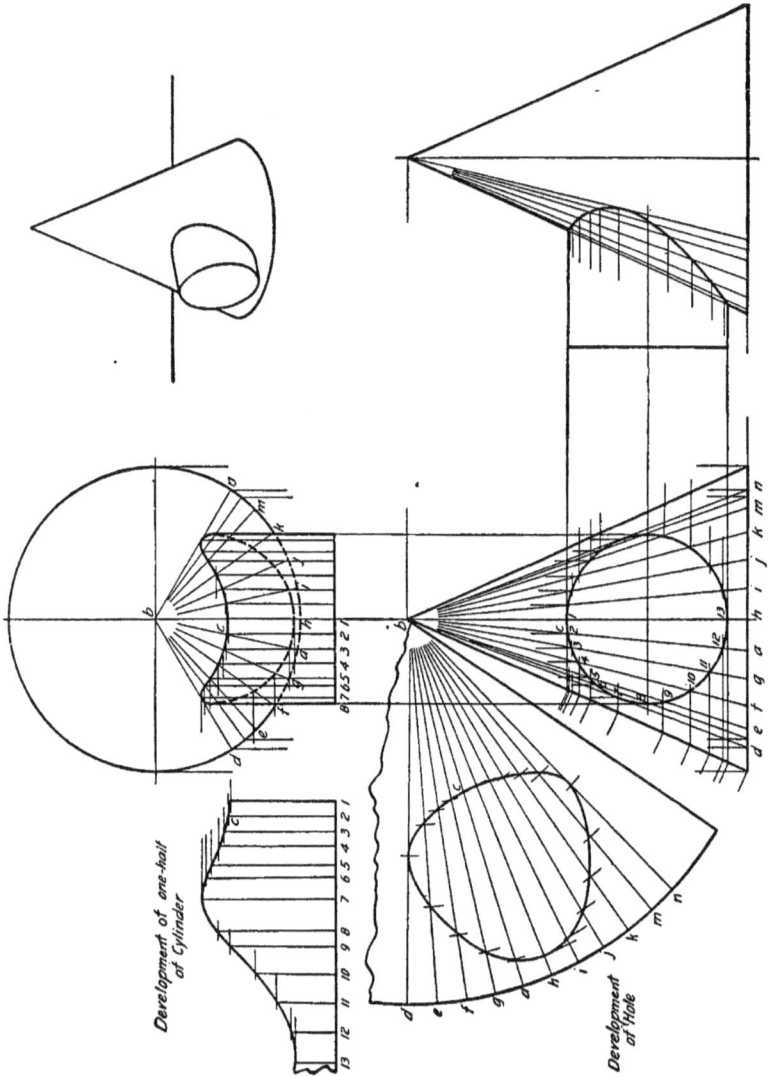

Development of one-half of Cylinder

Development of Hole

FIG. 249. TYPE PROBLEM. INTERSECTION OF CONE AND CYLINDER

The surface of the cylinder may now be developed following the usual method. Attention is again called to the fact that the elements of the cylindrical surface are not equally spaced. It will therefore be necessary to step off each division separately, taking care to place them in consecutive order in the pattern. The true lengths of the elements from the base to the line of intersection are shown in the top view and may be either transferred to the pattern with the dividers or projected from the top view as shown in Fig. 249.

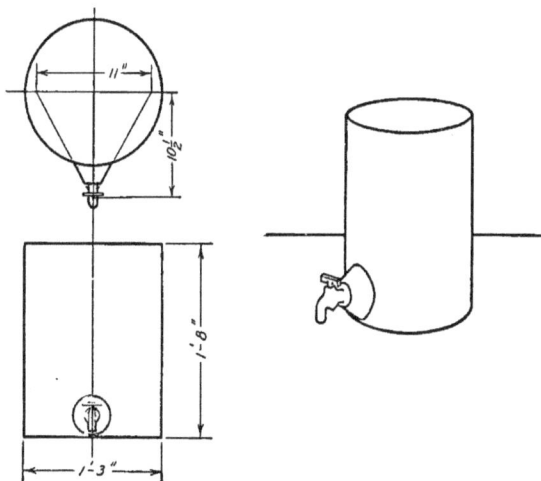

FIG. 250. OIL RECEPTACLE

The entire surface of the arc is laid out in the pattern by striking an arc of radius equal to the length of the elements of the cone and stepping off on this arc a distance equal to the circumference of the base circle. On this arc are located the feet of the elements on which points on the intersection were found, by transferring the distances between the feet of these elements from the top view of the base circle. The points on the line of intersection are located on these elements by finding the true lengths from the vertex of the cone to the line of intersection for each element. The construction for this is described in detail on page 209.

DATA FOR DRAWING PLATE 34

Given: The orthographic views of the objects shown in Figs. 250 and 251.

Required: To draw the orthographic views and make a construction for the pattern of the objects shown in Fig. 250, 251, or any similar object as assigned by the instructor.

Instructions: The fact that the cylinder in Fig. 250 is larger in proportion to the cone than in the type problem makes no difference in the principle of the problem or the method used in its solution. It is only necessary to assume a base for the cone and a vertex as indicated by the dotted lines.

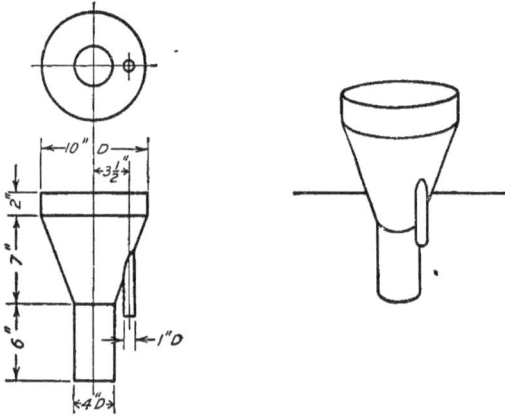

Fig. 251. Exhaust Head

In the object shown in Fig. 251, the axis of the cylinder is parallel to the axis of the cone. The elements of the cone should always be drawn first in the view in which the cylindrical surface shows as a circle, which in this case is the top view.

DATA FOR LETTERING PLATE 34

Given: Plate 34 to reduced size. Fig. 252.

Required: To make the plate to an enlarged scale.

A Bill of Stock is a tabulated form, such as the bill of material, but which gives the rough and sometimes the finished sizes for each different piece of timber and the number of each size required, together with a list of all other materials to be used in the project. Such a tabulated summary makes it possible to cut all stock and to calculate the cost of all materials for any project in wood. Example: Fig. 267 shows a bill of stock for a table.

Sectional Views. Very often a drawing is not clear because the interior of the object is complex or because a part of it is obscured by other lines. In such cases the object may be repre-

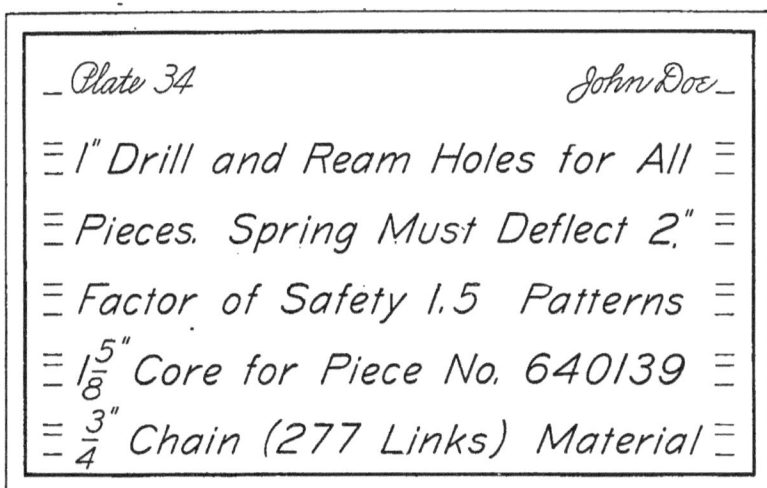

Plate 34   John Doe

= 1" Drill and Ream Holes for All =

= Pieces. Spring Must Deflect 2" =

= Factor of Safety 1.5 Patterns =

= $1\frac{5}{8}$" Core for Piece No. 640139 =

= $\frac{3}{4}$" Chain (277 Links) Material =

FIG. 252.   LETTERING PLATE 34

sented more clearly if a portion of it is imagined cut away to expose the hidden part. The most common examples of this method of representation are: (1) half-section in which the object is cut into two similar parts through an axis of symmetry, and (2) quarter-section in which the object is cut in to the center on two planes at right angles. These sections are described in detail on pages 98 and 99 and illustrated on page 94.

Other methods of sectioning may be used, depending upon the form of the object or part which it is desired to make clear. Fig. 253 illustrates a case where the section is taken on a broken line, A O B. In drawing the section view, the cut surface O A is considered revolved into the same plane with O B. Fig. 254 illustrates what is called a partial section. The ragged line indicates that a part of the shaft has been broken away.

The cross-section of an object is often given by showing a

revolved section in one of the views, Fig. 255 or 265. Where the
section cannot well be revolved a line may be drawn across the
view of the part at the place where the section is taken and the

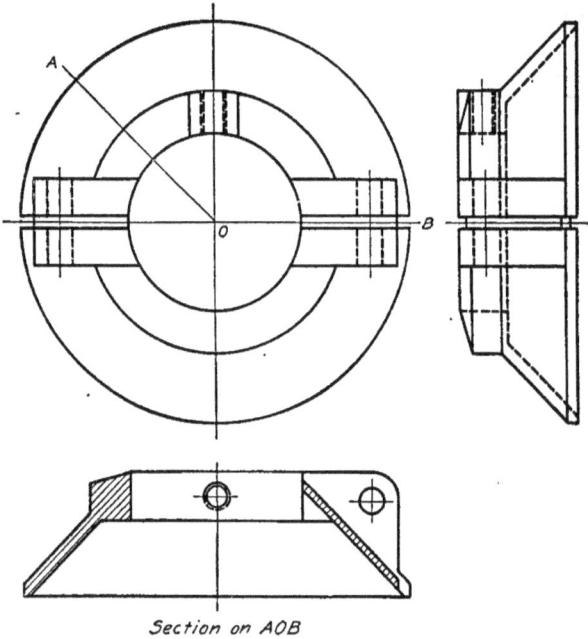

Section on AOB

FIG. 253.   BROKEN LINE SECTION

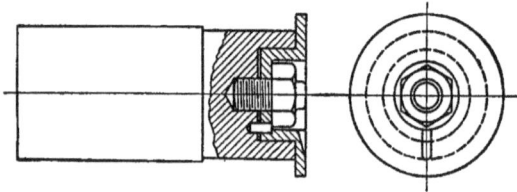

FIG. 254.   PARTIAL SECTION

section drawn in an open space near the view. Reference should
be made to the line on which the section is taken. Fig. 256. Such
parts as spokes or arms of wheels, solid shafts or rods, screws,
bolts, studs, and nuts are not represented as cut when the section
plane passes through their axes. Fig. 257. Ribs and webs are

not sectioned when the section plane is parallel to their lateral faces.

When a section is taken through an assembly, adjacent parts are crosshatched in different directions to aid in distinguishing one from another.

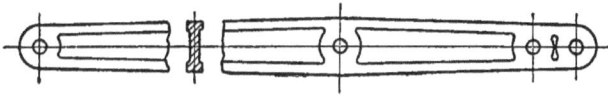

FIG. 255. REVOLVED SECTION

Various combinations of lines are used to represent sections of different materials. No standard section notation has ever been universally adopted. It is customary to add a note giving the

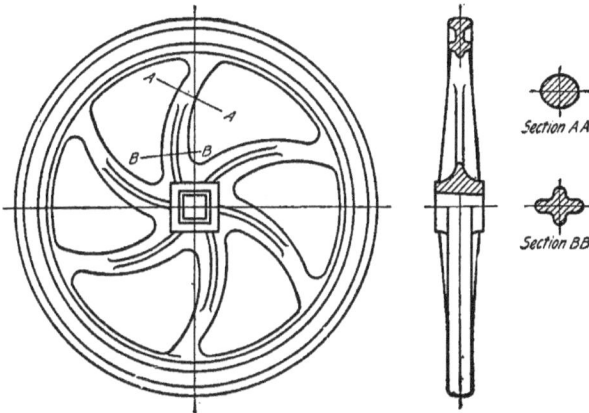

FIG. 256. REMOVED SECTIONS

name of the material unless a local section notation is in use. Except for a few cases where it is desirable to distinguish between the metals in adjacent parts, such as the babbit and the casting of a bearing, nothing is gained by using characteristic section lines since, in general, a note must be added to insure proper interpretation. Fig. 258 shows a few sections in common use.

Breaks. Where it is desirable to omit part of a shaft or rod, either may be broken and the break indicated as shown in Fig.

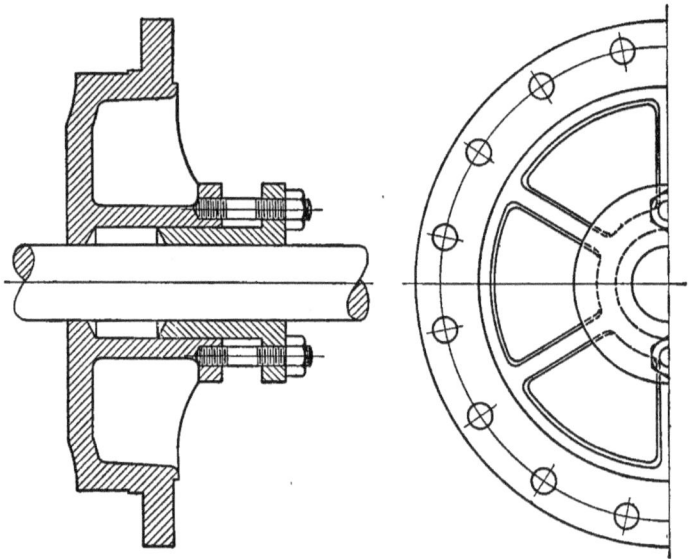

FIG. 257. SECTION THROUGH RIBS, SHAFTS, BOLTS, ETC.

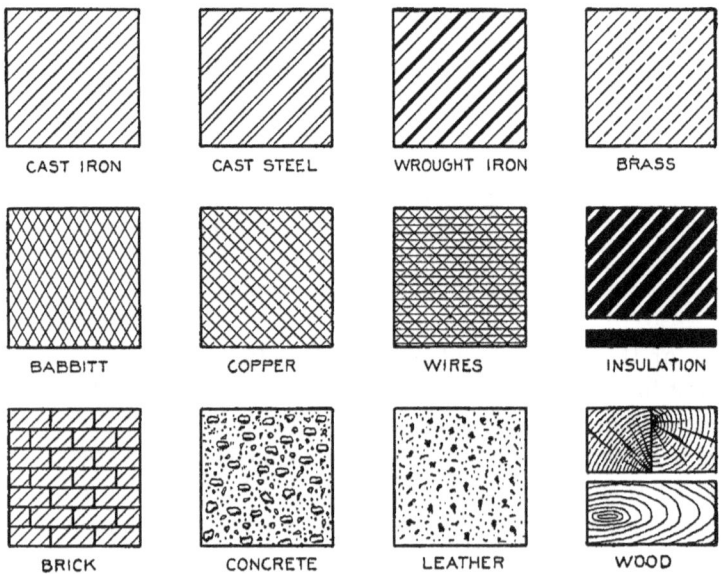

CAST IRON　　　CAST STEEL　　　WROUGHT IRON　　　BRASS

BABBITT　　　COPPER　　　WIRES　　　INSULATION

BRICK　　　CONCRETE　　　LEATHER　　　WOOD

FIG. 258. CONVENTIONAL CROSS-SECTIONING

259. The ragged line representing the break is drawn freehand in both the pencil and the ink drawing.

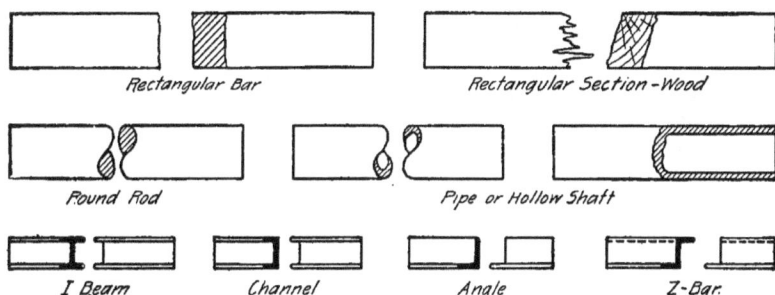

FIG. 259. CONVENTIONAL BREAKS

Wood Screws are made of steel or brass. They have heads of various shapes as shown in Fig. 260. The size of screws is given in terms of their number and their length, which is indicated by giving the gage number. The threads of wood screws are represented conventionally as shown in Fig. 260.

Furniture and Cabinet Details. Various joints and a few other common constructions used in furniture and cabinet construction are shown in Figs. 261 and 262.

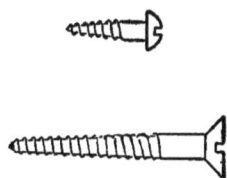

FIG. 260. CONVENTIONAL REPRESENTATION OF WOOD SCREWS

FURNITURE AND CABINET PROBLEMS

PREPARATORY INSTRUCTIONS FOR DRAWING PLATE 35

The problems for this plate were selected with the idea of giving practice in drawing and dimensioning projects in furniture making and also to set before the student typical examples

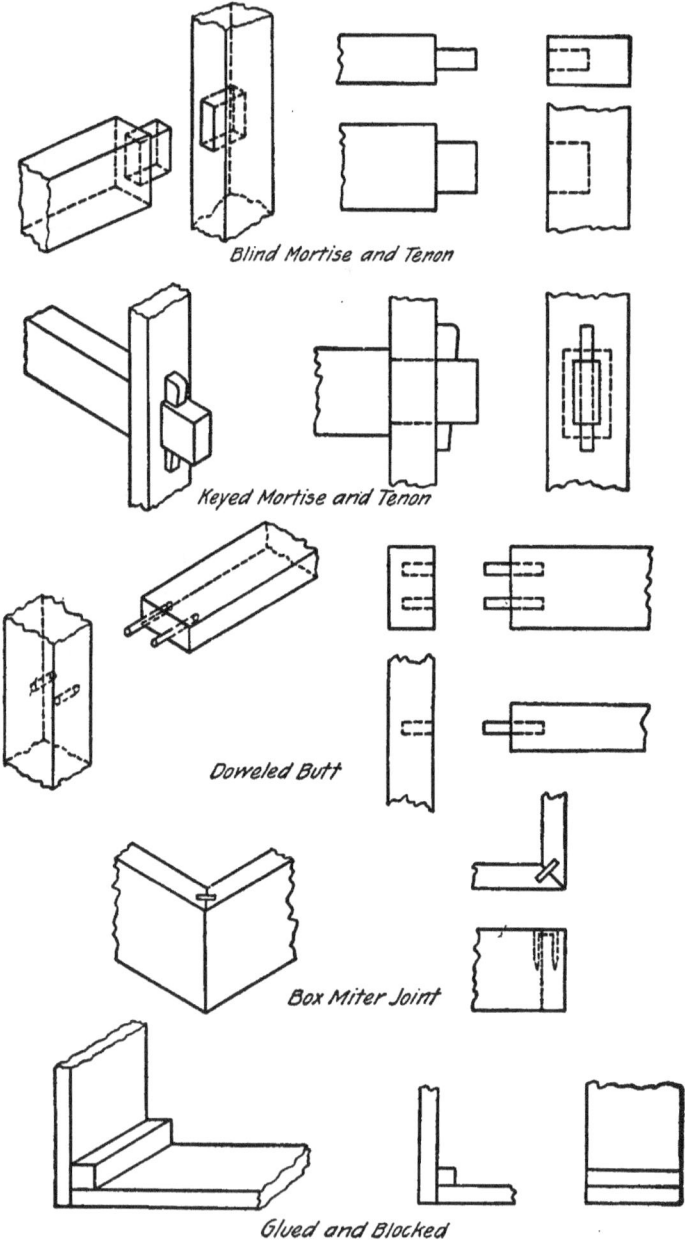

Blind Mortise and Tenon

Keyed Mortise and Tenon

Doweled Butt

Box Miter Joint

Glued and Blocked

FIG. 261. JOINTS

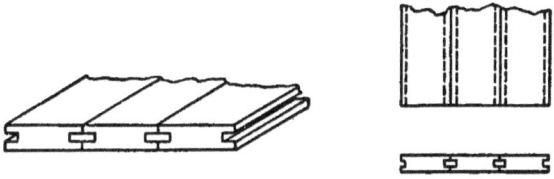

Spline Joint

Dado and Rabbet

Dado, Tongue, and Rabbet

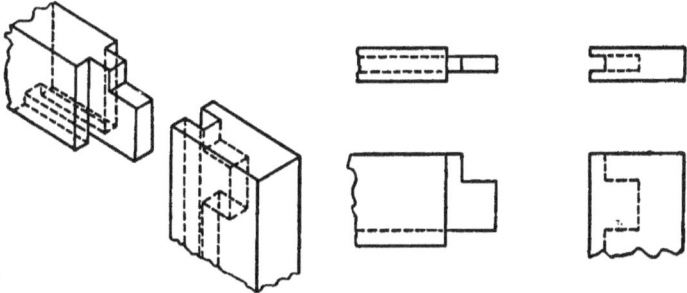

Haunched Mortise and Tenon

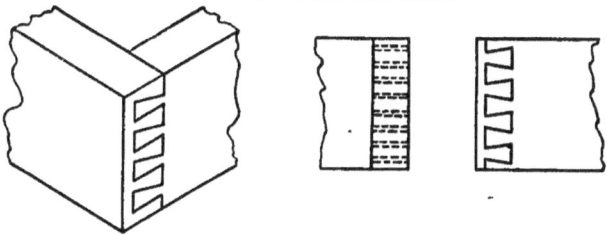

Lap Dovetail

Fig. 262. Joints

(233)

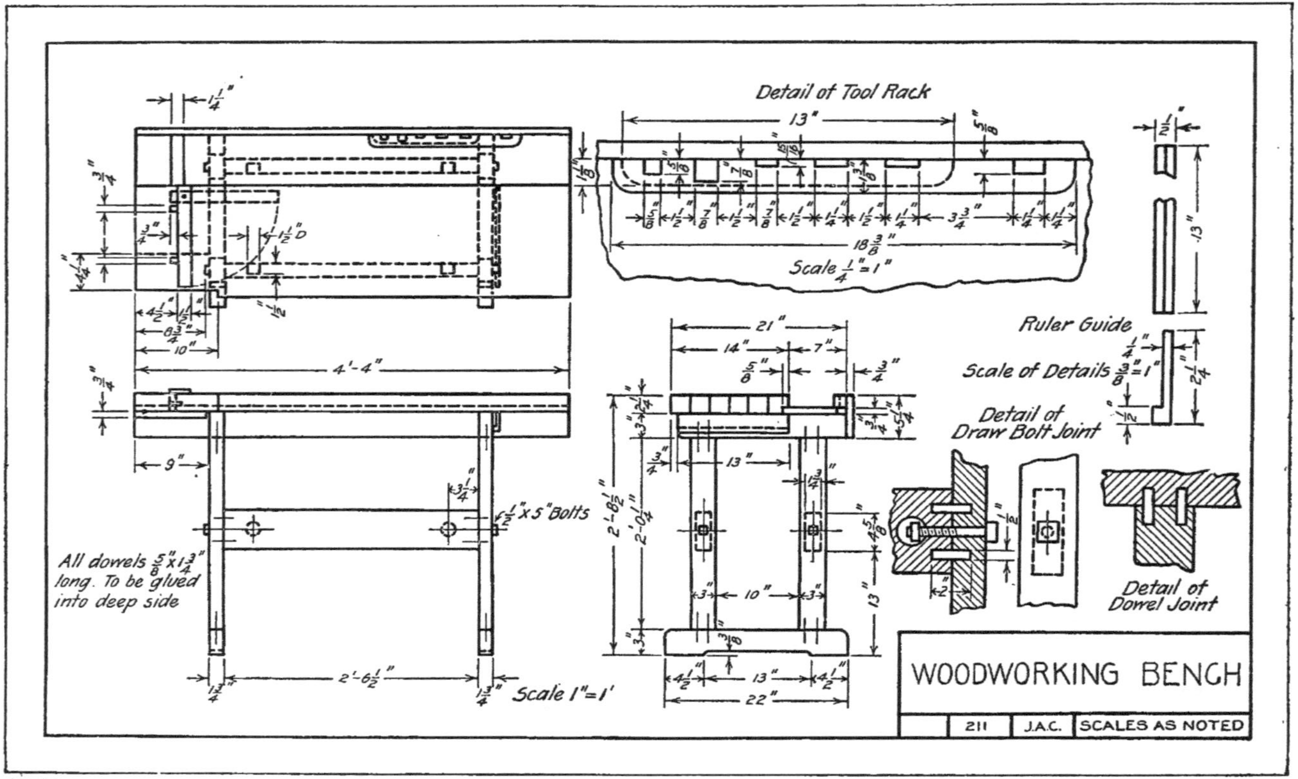

FIG. 263. TYPE PROBLEM. WORKING DRAWING OF WOOD WORKING BENCH

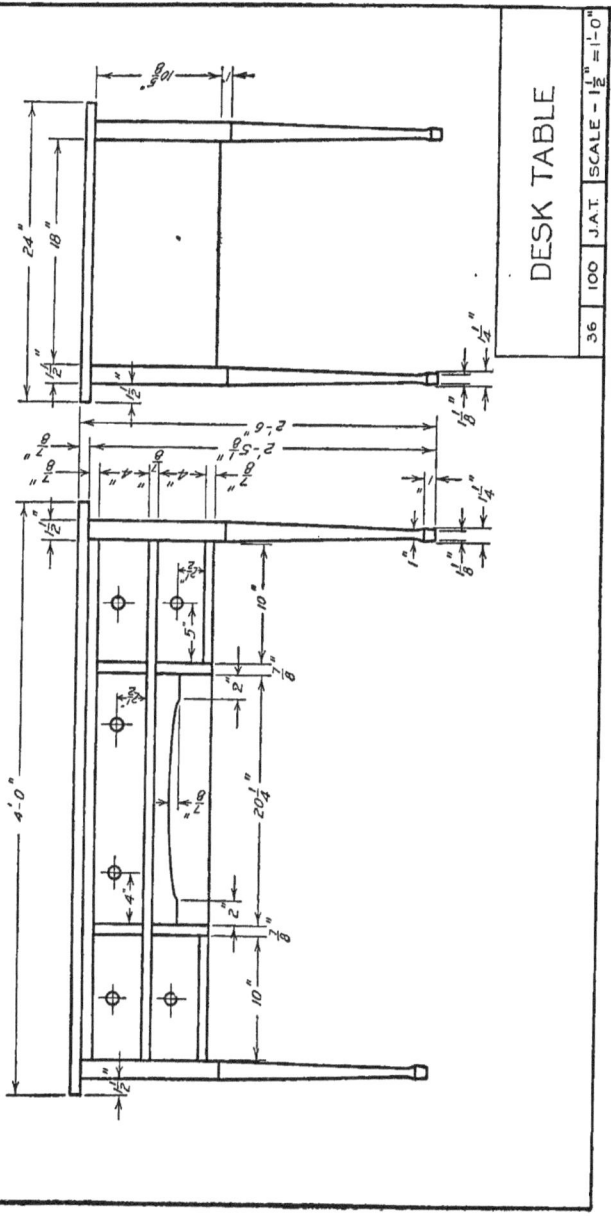

Fig. 264. Type Problem. Desk Table

DESK TABLE

| 36 | 100 | J.A.T. | SCALE - $1\frac{1}{2}$" = 1'-0" |

(235)

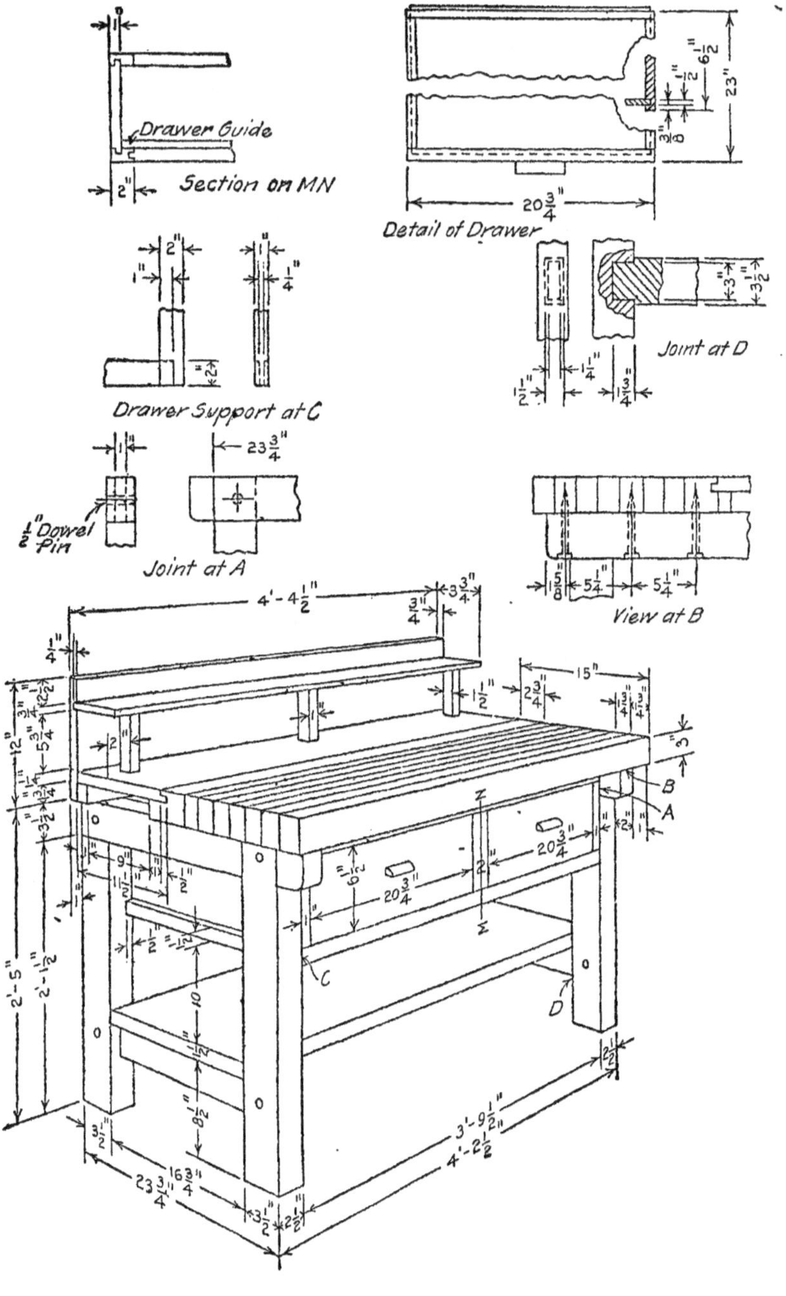

Drawer Guide

Section on MN

Detail of Drawer

Drawer Support at C

Joint at D

Joint at A

½" Dowel Pin

View at B

FIG. 265. MACHINE SHOP BENCH

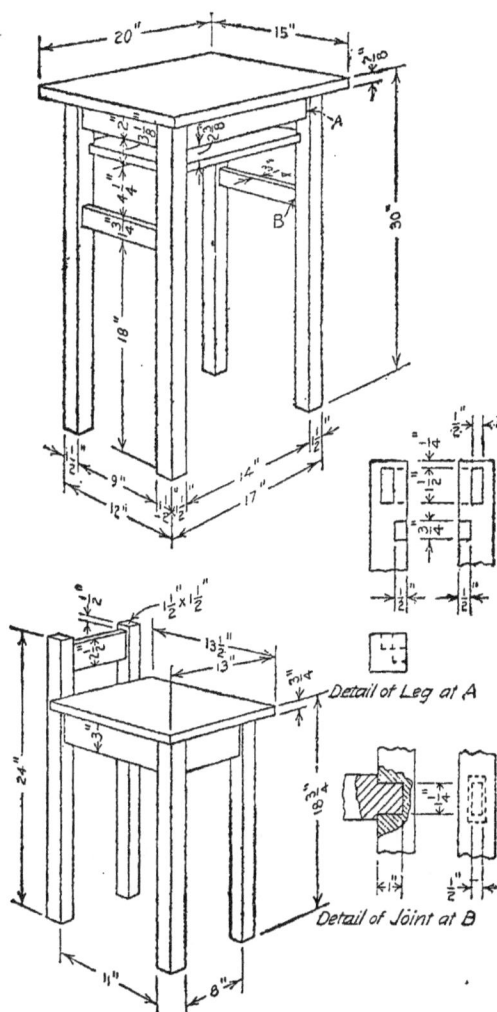

FIG. 266. PHONE TABLE AND CHAIR

BILL OF STOCK						
PCS	SIZE	NAME	MATERIAL	NO. FT.	PER. FT.	TOTAL
1	1"X 30"X48"	TOP	OAK	10	$.12	$1.20
2	3/4"X5 1/2"X41 1/2"	RAILS	"	3.5	"	.42
2	3/4"X5 1/2"X25 1/2"	"	"	2.16	"	.26
2	3/4"X 3"X 25 1/2"	"	"	1.08	"	.13
4	1/2"X 2"X 16 3/4"	SLATS	"	.94	"	.11
2	1/2"X4"X16 3/4"	"	"	.94	"	.11
1	3/4"X12"X41 1/4"	BOTTOM	"	3.5	"	.42
4	2 1/2"X2 1/2"X30"	LEGS	"	5.2	"	.62
3	1/2"X4"X25"	DRAWER	"	2.08	"	.25
1	1"X 4"X 24"	"	"	.67	"	.08
1	3/8"X23 1/2"X25"	DRAW. BOT.	PINE	4.80	.045	.18
2	1"X2"X25"	DRAW. GUIDE	OAK	.69	.12	.08
				TOTAL COST		3.86

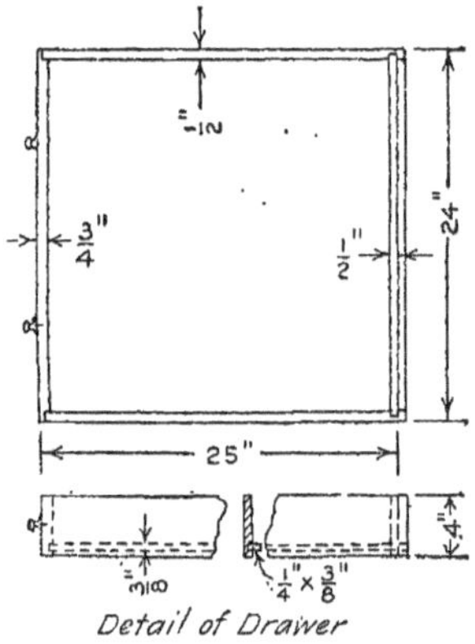
25"
24"
Detail of Drawer

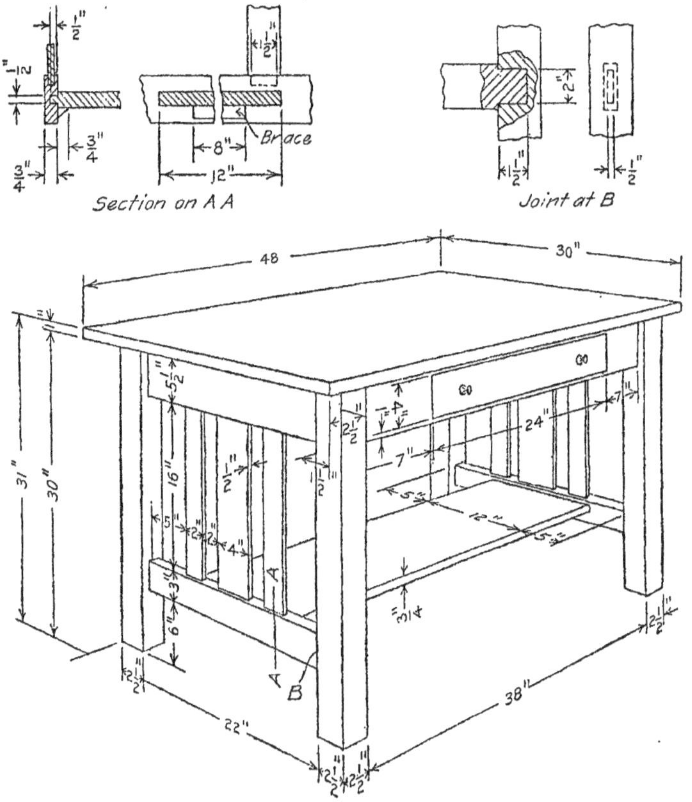
Brace
12"
Section on A A
Joint at B
48
30"
31"
30"
24"
22"
38"
A
B

FIG. 267. LIBRARY TABLE

of furniture construction. In order to cover this field as thoroughly as possible, two type problems of widely different character are presented. Figs. 263 and 264. The student should fix

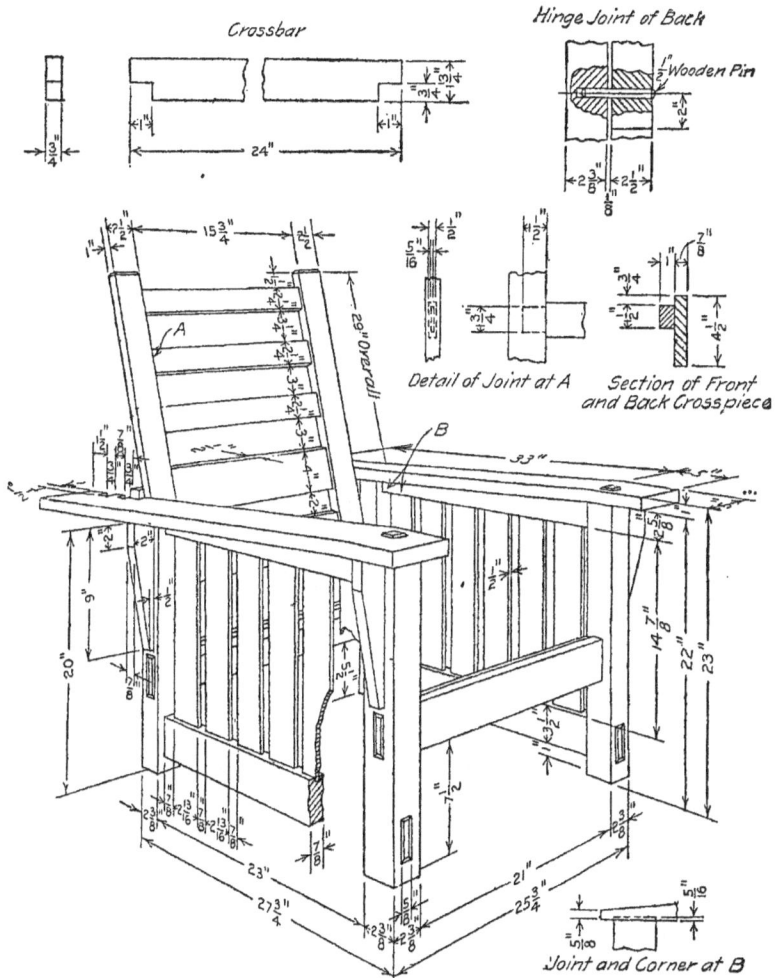

FIG. 268. MORRIS CHAIR

in mind the correct proportion of the joints used in these problems as well as their names and uses by referring to the discussion and figures on pages 231, 232, and 233.

DATA FOR DRAWING PLATE 35

Given: The perspective sketch of the object shown in Fig. 265, 266, 267, or 268.

Required: To draw the orthographic views of the object shown in Fig. 265, 266, 267, or 268, or any similar problem as assigned by the instructor.

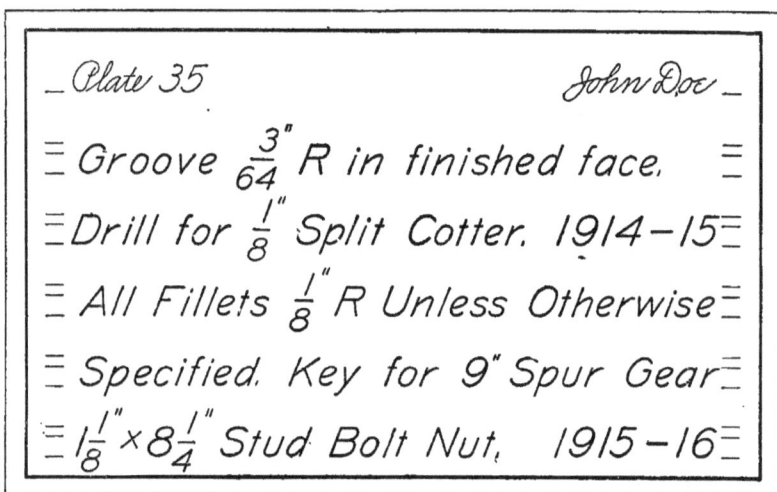

$= Groove \frac{3}{64}'' R$ in finished face.

$= Drill$ for $\frac{1}{8}''$ Split Cotter. 1914 – 15

$= All$ Fillets $\frac{1}{8}''$ R Unless Otherwise

$= Specified.$ Key for $9''$ Spur Gear

$= 1\frac{1}{8}'' \times 8\frac{1}{4}''$ Stud Bolt Nut, 1915 – 16

Plate 35   John Doe

FIG. 269. LETTERING PLATE 35

The problems for this plate are designed to give practice in making working drawings for various kinds of cabinet work and to familiarize the student with typical cabinet construction.

DATA FOR LETTERING PLATE 35

Given: Plate 35 to reduced size. Fig. 269.
Required: To make the plate to an enlarged scale.

DATA FOR DRAWING PLATE 36

Given: The pencil mechanical drawing, Plate 35.
Required: To make a tracing of Plate 35.

DATA FOR LETTERING PLATE 36

Given: Plate 36 to reduced size. Fig. 270.

Required: To make the plate to an enlarged scale.

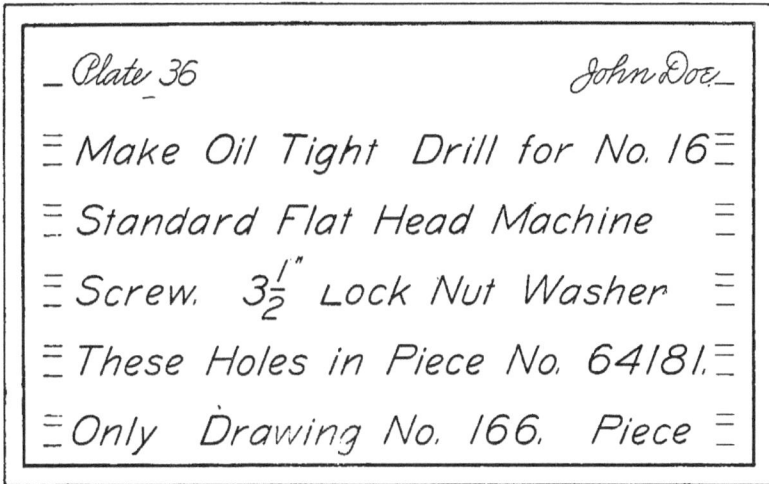

_ *Plate 36*   *John Doe_*

≡ *Make Oil Tight Drill for No. 16*≡

≡ *Standard Flat Head Machine*   ≡

≡ *Screw. 3½″ Lock Nut Washer*   ≡

≡ *These Holes in Piece No. 64181.*≡

≡ *Only Drawing No. 166. Piece*   ≡

FIG. 270. LETTERING PLATE 36

DATA FOR DRAWING PLATE 37

Given: The orthographic views of the objects shown in Fig. 273, 274, 275, or 276.

Required: To draw to large scale two views of objects shown in Fig. 273, 274, 275, or 276, with sections and details of joints and paneling; or any similar problem assigned by the instructor.

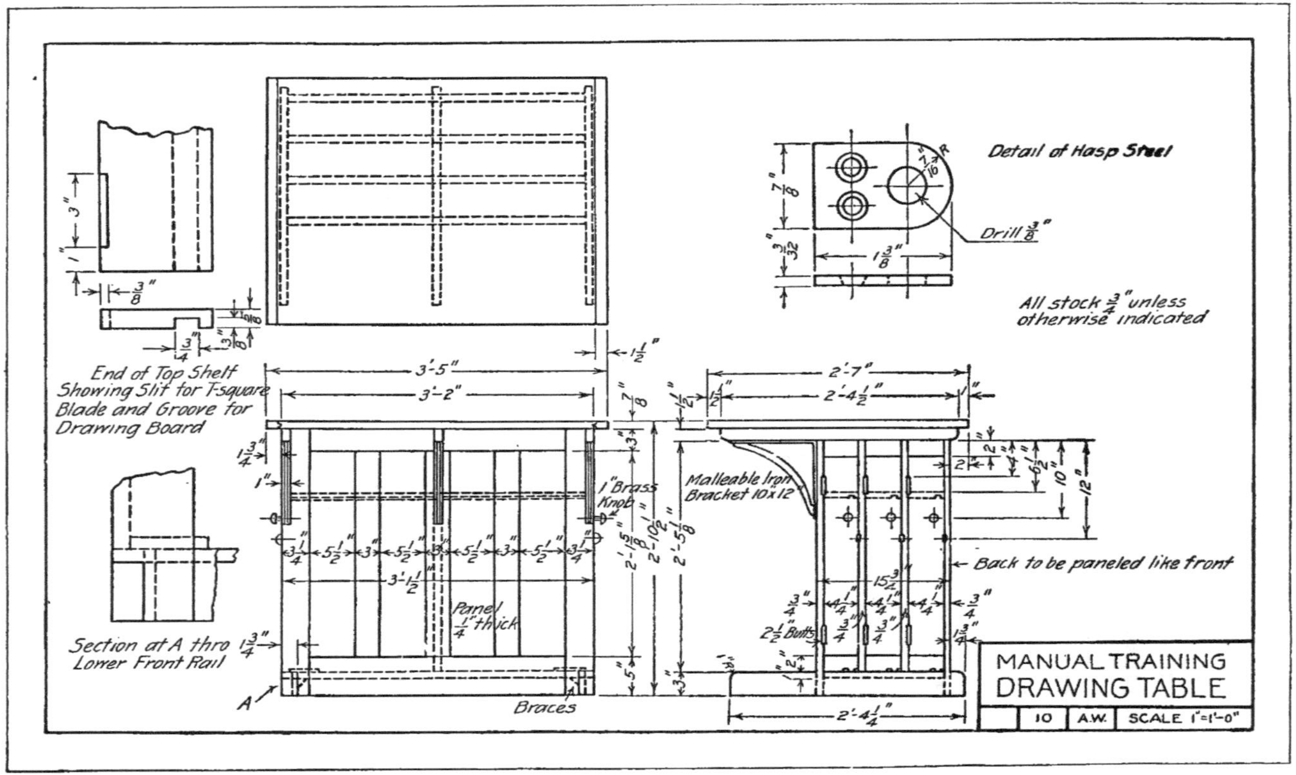

FIG. 271. TYPE PROBLEM. DRAWING TABLE

(242)

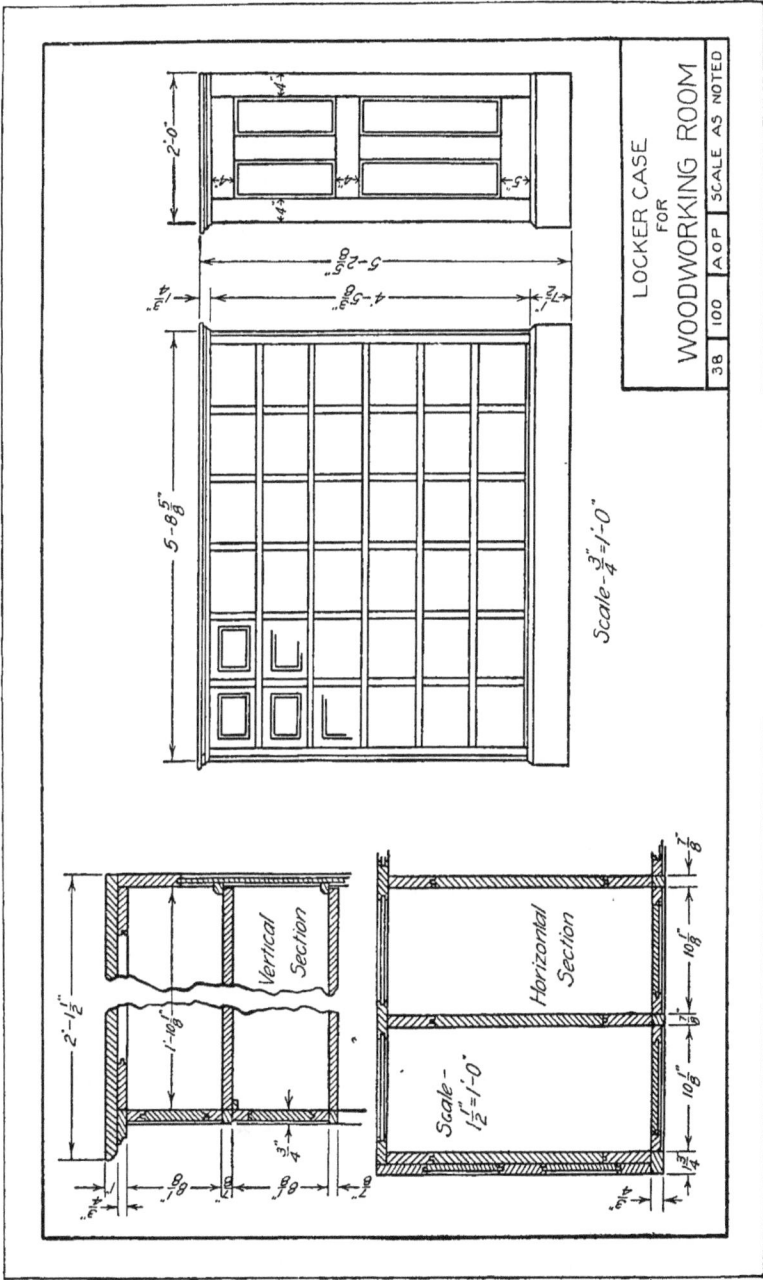

LOCKER CASE
FOR
WOODWORKING ROOM

| 38 | 100 | A O P | SCALE AS NOTED |

(243)

FIG. 272. TYPE PROBLEM. LOCKER CASE

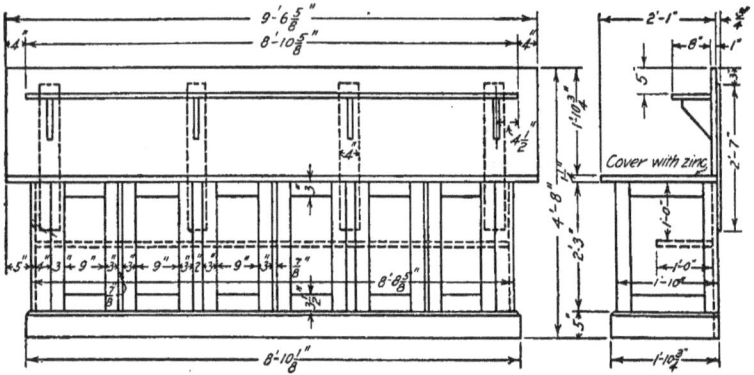

FIG. 273. GLUE BENCH

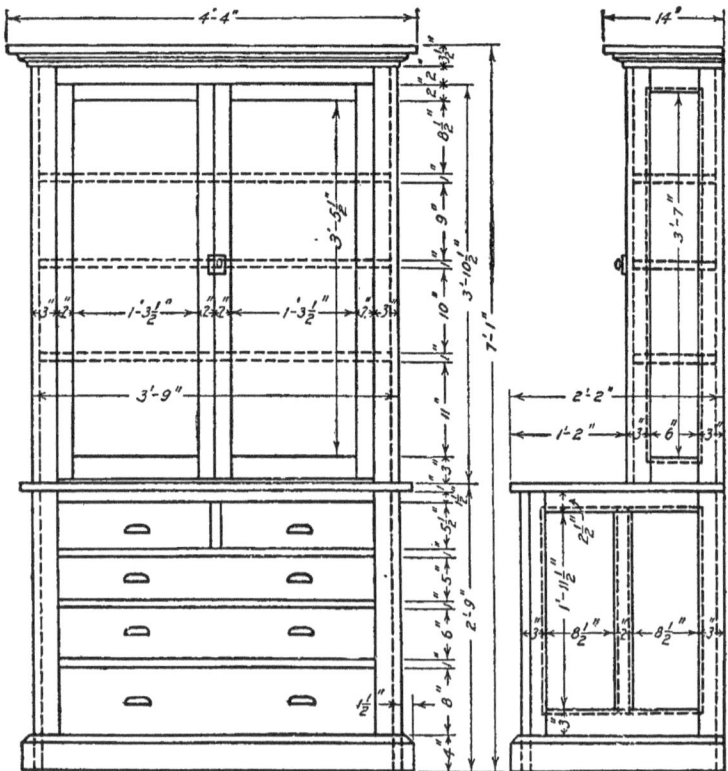

FIG. 274. CABINET FOR DRAWING ROOM

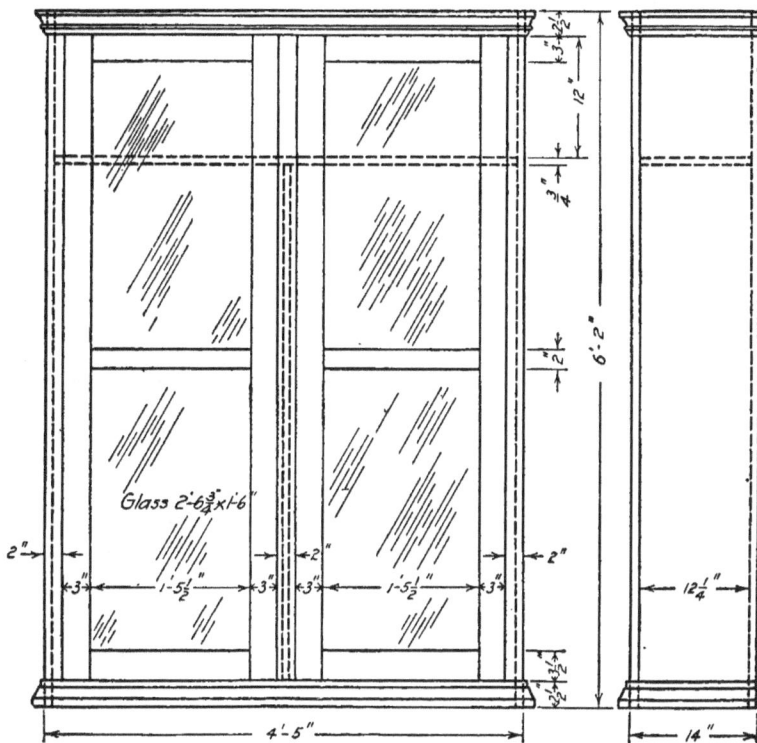

FIG. 275. CABINET FOR CLOTHES PRESS

Glass 2'-6¾"x1'-6"

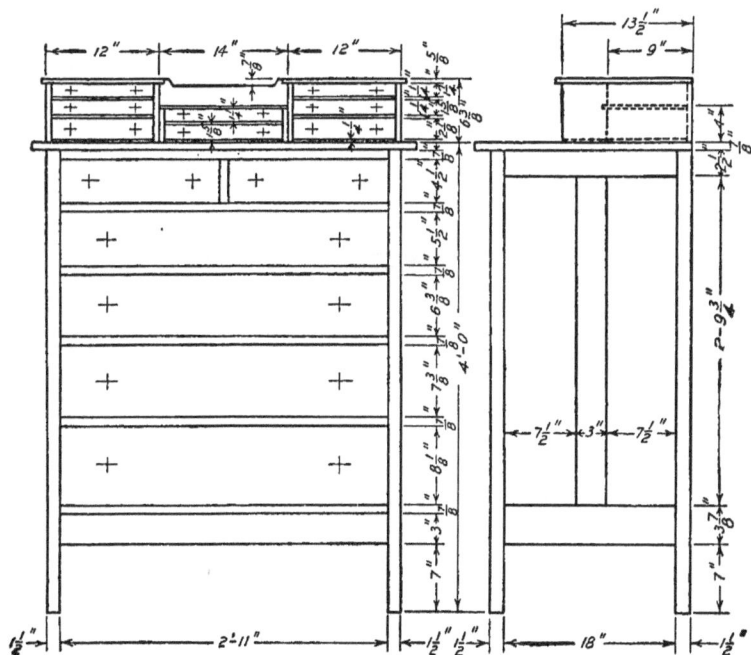

FIG. 276. CHIFFONIER

(245)

DATA FOR LETTERING PLATE 37

Given: Plate 37 to reduced size. Fig. 277.
Required: To make the plate to an enlarged scale.

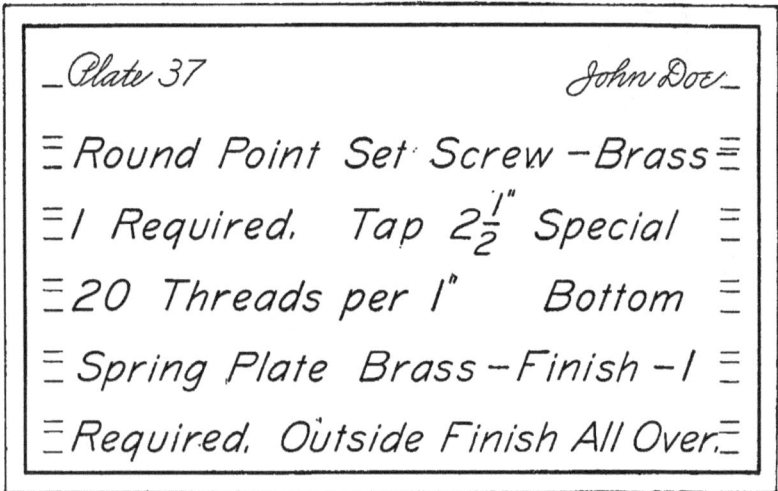

Plate 37   John Doe

≡ Round Point Set Screw —Brass≡

≡/ Required.   Tap $2\frac{1}{2}''$ Special   ≡

≡20 Threads per $1''$ Bottom   ≡

≡ Spring Plate Brass —Finish —/ ≡

≡Required. Outside Finish All Over.≡

FIG. 277. LETTERING PLATE 37

DATA FOR DRAWING PLATE 38

Given: The pencil mechanical drawing for Plate 37.
Required: To make a tracing of Plate 37.

DATA FOR LETTERING PLATE 38

Given: Plate 38 to reduced size. Fig. 278.
Required: To make the plate to an enlarged scale.

MACHINE DRAWING

A *Bill of Material* may be given on the drawing or on a separate sheet. It is a tabulated form in which such information as the following is given:

1. Number of each part required on one complete machine or structure.

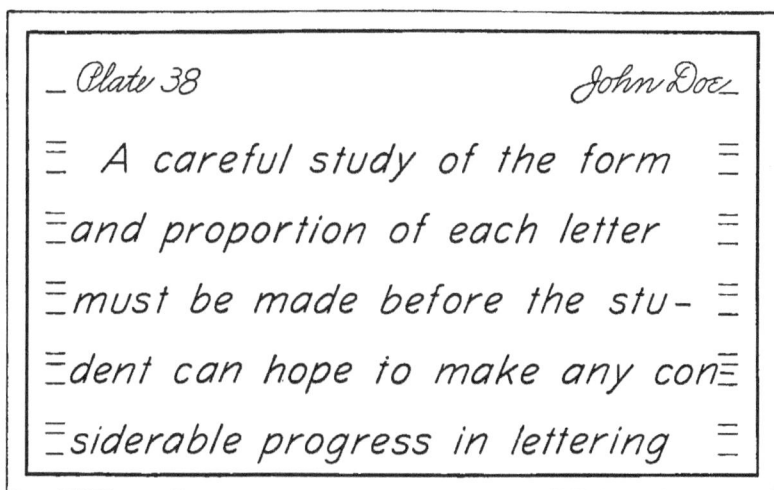

_ *Plate 38*   *John Doe*

≡ A careful study of the form ≡

≡and proportion of each letter ≡

≡must be made before the stu- ≡

≡dent can hope to make any con≡

≡siderable progress in lettering ≡

FIG. 278.   LETTERING PLATE 38

2. Description or name of piece.
3. Mark or number by which a piece is designated on the drawing.
4. General drawing number.
5. Shop drawing number.
6. Erection drawing number.
7. Material from which each piece is made.
8. Pattern number if cast.
9. Where used.
10. Estimated weight.
11. Order number.

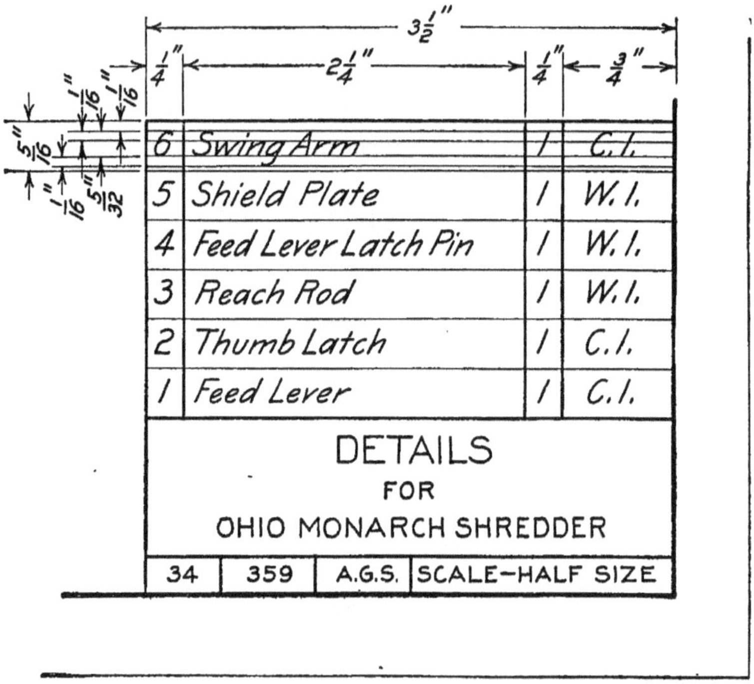
3 1/2"
1/4"
2 1/4"
1/4"
3/4"
1/16"
1/16"
5/16"
1/16"
5/32"
6 Swing Arm 1 C.I.
5 Shield Plate 1 W.I.
4 Feed Lever Latch Pin 1 W.I.
3 Reach Rod 1 W.I.
2 Thumb Latch 1 C.I.
1 Feed Lever 1 C.I.
DETAILS
FOR
OHIO MONARCH SHREDDER
34 359 A.G.S. SCALE–HALF SIZE

FIG. 279. BILL OF MATERIAL

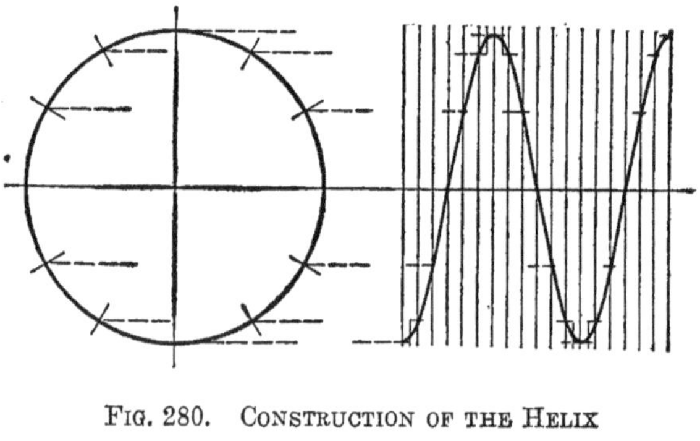

FIG. 280. CONSTRUCTION OF THE HELIX

The bill of material includes standard parts such as bolts and screws which are not detailed on the drawings. A simple bill of material is shown in Fig. 279.

Screw Threads. The curve of the screw thread is the helix. It is generated by a point which moves on the surface of a cylinder and which advances uniformly in the direction of the axis of the cylinder and at the same time has a uniform motion around

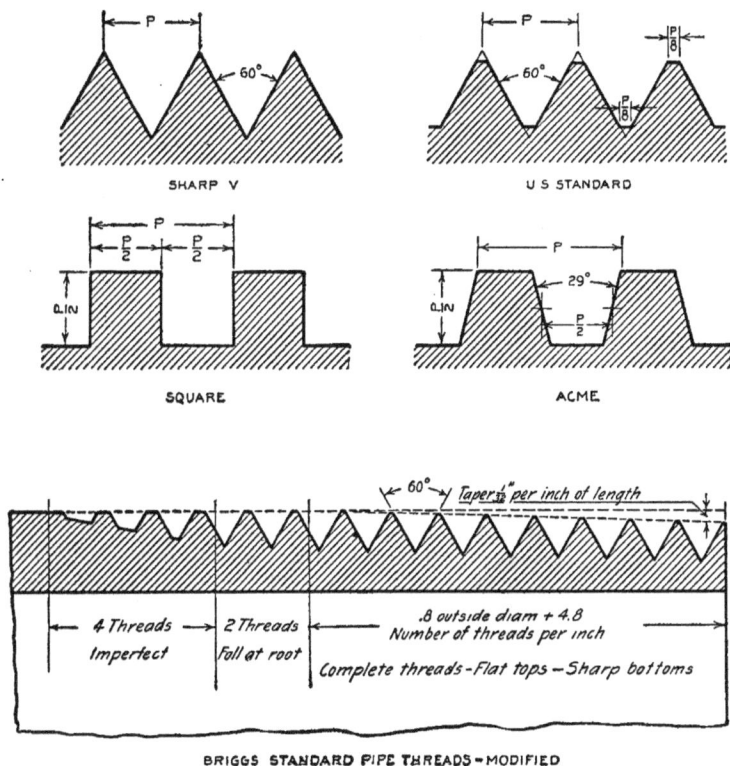

FIG. 281. PROPORTION OF COMMON THREAD FORMS

its axis. Fig. 280 shows the construction for the helix. The distance in the direction of the axis traversed by a point in one revolution is called the *pitch*. Pitch in the case of a thread is its advance in the direction of the axis in one revolution.

In Fig. 281 the proportions of the several common thread forms are shown to a large scale.

The V-thread is shown in Fig. 282 as it would actually appear with the edges drawn as helices. On account of the difficulty of constructing and drawing these curves they are usually conven-

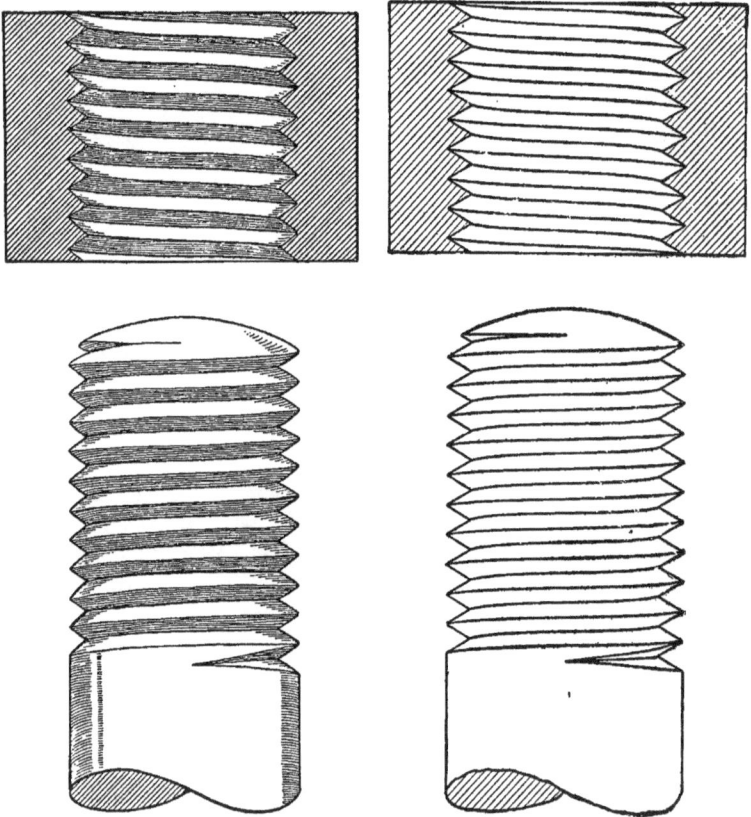

FIG. 282. V-THREAD, SHOWING HELICES

tionalized into straight lines as shown in Fig. 283. The method commonly used for representing screws up to about one inch in diameter, as measured on the drawing, is still further simplified by omitting the short inclined lines forming the "saw teeth."

Fig. 284. On the pencil drawing no distinction is made in the weight of the two sets of parallel lines drawn across the screw, but on the tracing it is customary to make a striking contrast between the longer and shorter lines as shown.

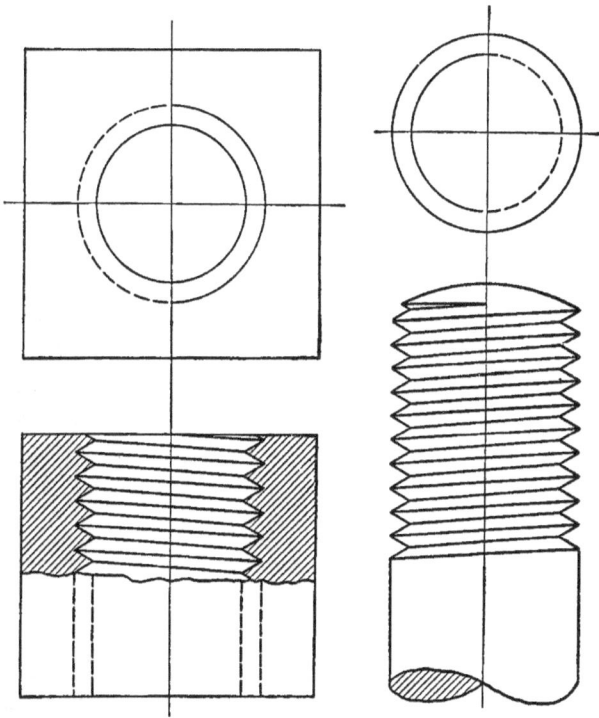

FIG. 283. V-THREAD. CONVENTIONAL REPRESENTATION FOR LARGE SIZES

In this course the shorter lines will be made object-line width and the longer lines center-line width. The angle at which these lines are drawn is estimated. It remains practically constant for all sizes of standard screws as the pitch of the thread increases with the diameter of the screw. It will be noted that the lines in the section view of the nut make the opposite angle to the horizontal that those on the screw make because of the fact that the

part of the nut shown matches the *invisible* half of the screw. The
lines are usually spaced by eye. Guide lines should be drawn to
limit the length of the shorter lines.

In the conventional end view of the bolt, the circle representing the outer edges of the thread is a full line, while one-half of
the circle representing the inner edges of the thread is a dotted
line and the other half is a full line.

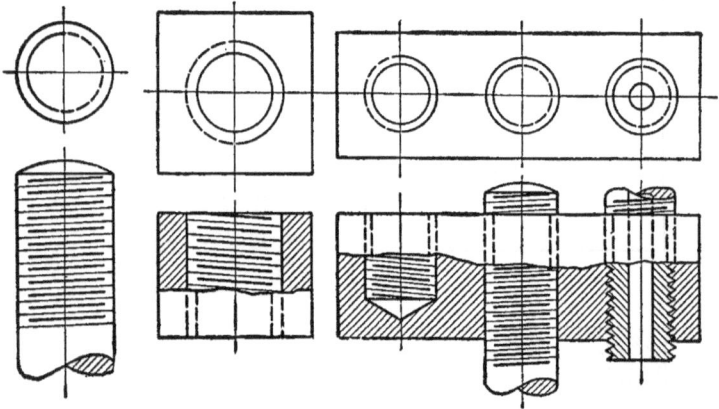

FIG. 284. V-THREAD. CONVENTIONAL REPRESENTATION FOR SMALL SIZES

In the conventional end view of the nut, the circle representing the inner edges of thread is a full line, while one-half of the
circle representing the outer edges of the thread is a dotted line
and the other half is a full line.

A study of the relation of these conventions to the form of
the object should enable the student to fix in mind the principles
on which they are based. With this relation in mind it will be
unnecessary for him to refer to the figures in rendering the
convention.

The United States Standard (U. S. S.) or Sellers thread,
Fig. 281, differs from the sharp V-thread in that the outer and
inner edges of the thread are flattened. The same convention
is used for representing it that is used for the sharp V-thread.

The Square Thread is shown in Fig. 285 with the edges drawn as helices. Fig. 286 is a conventional representation of the screw and nut in which the helices have been replaced by straight lines.

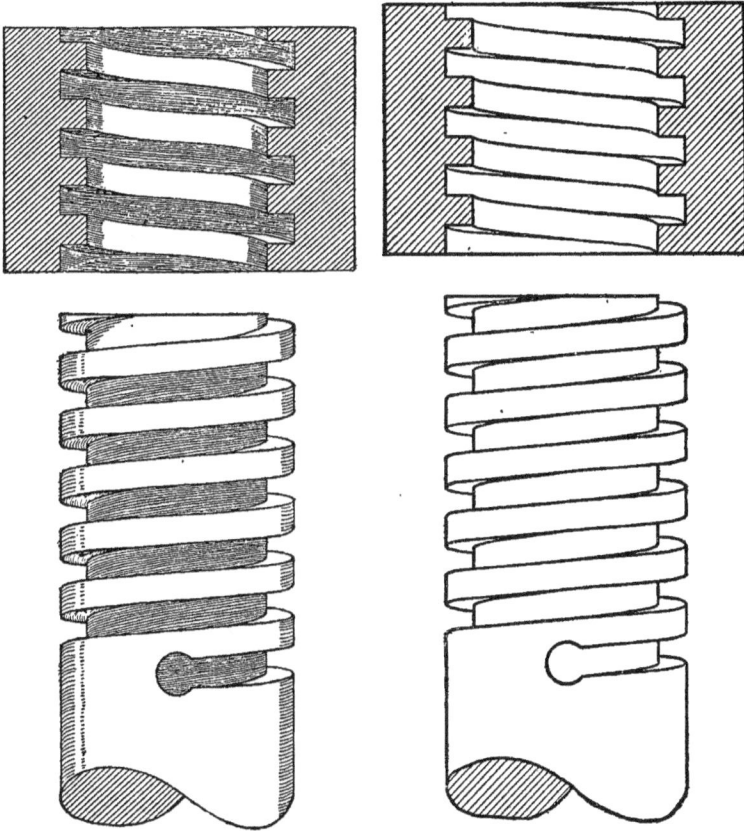

FIG. 285. SQUARE THREAD. EDGES DRAWN AS HELICES

For small sizes, the method shown in Fig. 287 is generally used because of its simplicity. The Acme screw thread is represented conventionally as shown in Fig. 288. It is convenient in drawing to make the angle between the faces of the thread 30° instead of 29°.

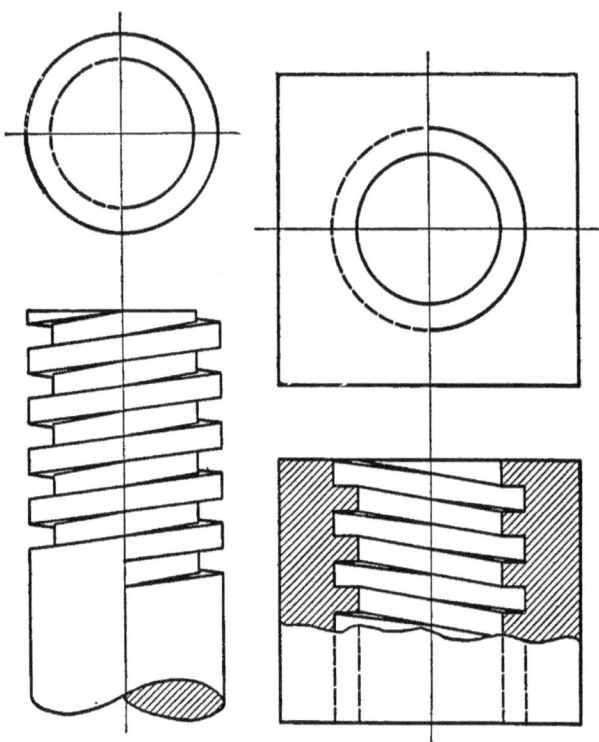

FIG. 286. SQUARE THREAD. CONVENTIONAL REPRESENTATION FOR LARGE SIZES

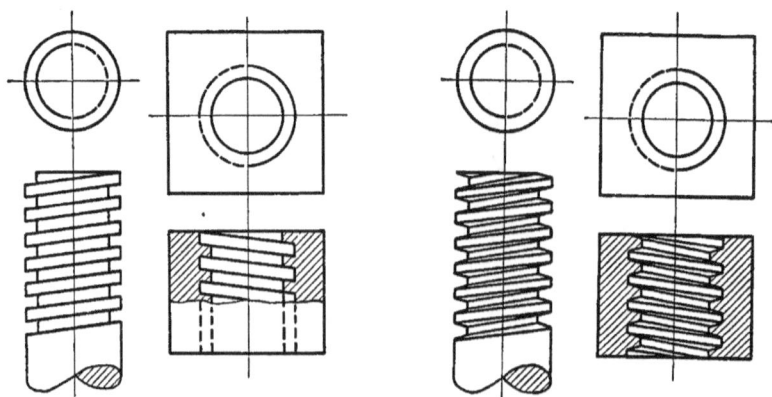

FIG. 287. SQUARE THREAD. CONVENTIONAL REPRESENTATION FOR SMALL SIZES

FIG. 288. ACME THREAD. CONVENTIONAL REPRESENTATION

Pipe Thread. The basic form of the Briggs standard pipe thread is that of the V-thread. This thread is rounded slightly .at the outer and inner edges. A modified form in which the threads have flat outer edges and sharp inner edges is shown in Fig. 289. This form is used by manufacturers because of the comparative ease with which taps and dies are made for cutting the threads.

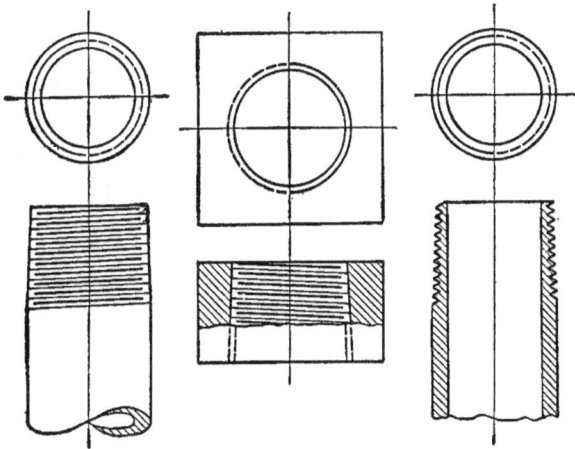

FIG. 289. PIPE THREADS. CONVENTIONAL REPRESENTATION

The threaded portion of the pipe tapers one thirty-second of an inch in radius for each inch of length.

Pipe threads are represented conventionally as shown in Fig. 289.

Springs. The curve of the coil spring is the helix. Fig. 290 shows a spring in which the curves are drawn and also the conventional representation which shows the curves replaced by straight lines.

Bolts and Nuts. A bolt consists of a rod with a head on one end and a screw on the other to receive a nut. Fig. 291. What are known as United States Standard bolts and nuts are shown in Figs. 292 and 294. The proportions given by the formulae are those adopted for rough bolts and nuts. The finished nuts are $\frac{1}{16}''$ less in width and thickness than the rough nuts. The fin-

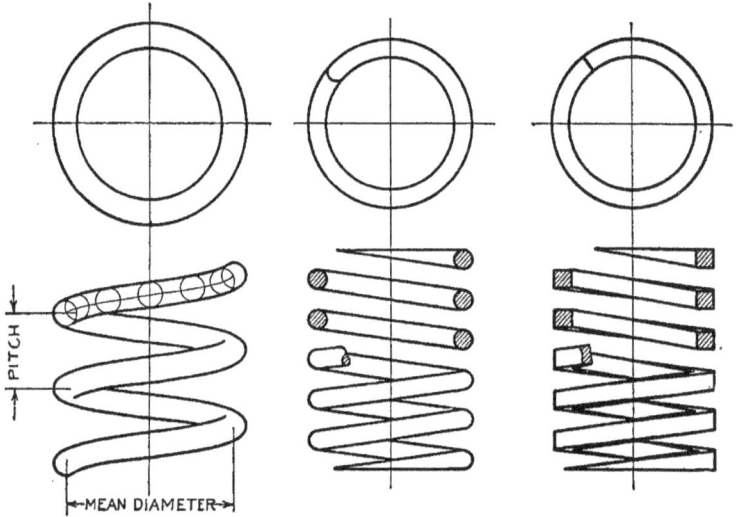

FIG. 290.   COIL SPRING SHOWING ACTUAL AND CONVENTIONAL
REPRESENTATION

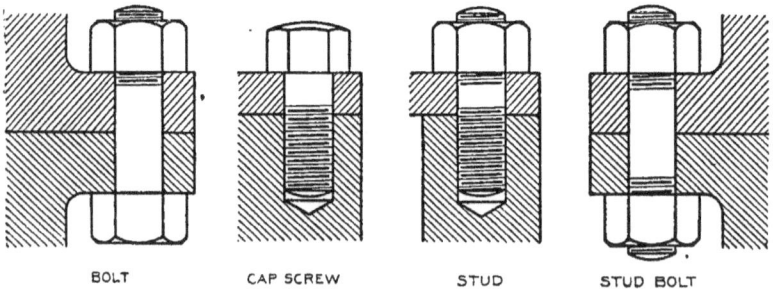

BOLT   CAP SCREW   STUD   STUD BOLT

FIG. 291.   COMMON SCREW FASTENINGS

ished heads have the same sizes as the finished nuts. A table of standard sizes may be found in an engineering handbook. United States Standard threads are used on these bolts.

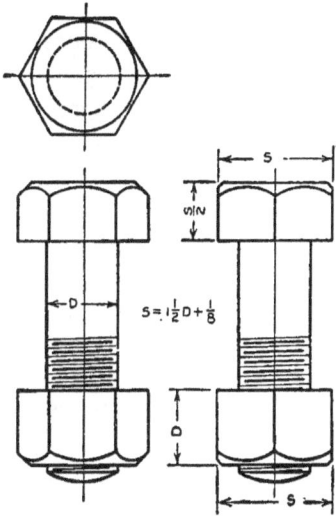

FIG. 292. ACTUAL PROPORTIONS.
HEXAGONAL HEAD—U. S. STANDARD
BOLTS AND NUTS

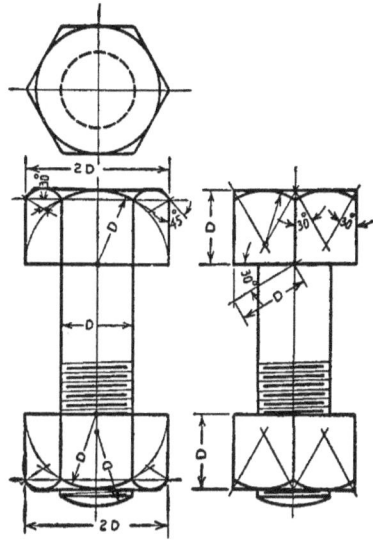

FIG. 293. CONVENTIONAL REPRE-
SENTATION. HEXAGONAL HEAD—U.
S. STANDARD BOLTS AND NUTS

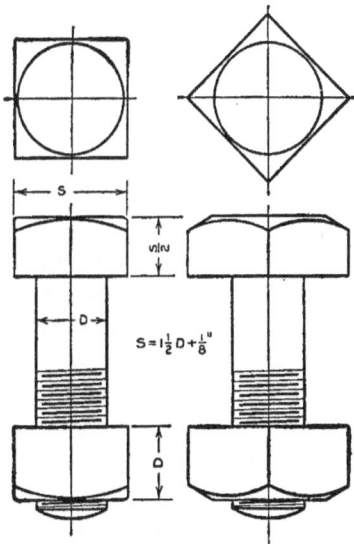

FIG. 294. ACTUAL PROPORTIONS.
SQUARE HEAD—U. S. STANDARD
BOLTS AND NUTS

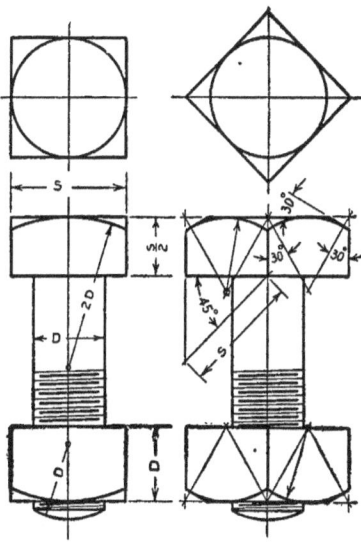

FIG. 295. CONVENTIONAL REPRE-
SENTATION. SQUARE HEAD—U. S.
STANDARD BOLTS AND NUTS

Figs. 293 and 295 show the conventional methods of representing hexagonal and square bolt heads and nuts. Hexagonal heads and nuts are usually drawn to show three faces, whereas square heads and nuts are drawn to show two faces. When this is done the hexagonal forms are easily distinguished from the square forms.

Since the proportions of the head and nut of standard bolts are fixed, it is only necessary to give three dimensions, viz., the length of the bolt under the head, the length of the threaded portion, and the diameter.

A Stud is a rod threaded at both ends. One end is screwed into a threaded hole. The other end receives a nut. In Fig. 291 a standard nut is used.

A stud placed through two unthreaded holes with a nut at each end is called a *stud bolt*. Fig. 291.

Cap Screws are similar in form to bolts. They hold two parts together by passing through an unthreaded hole in one and a threaded hole in the other. Fig. 291. Heads of various forms are used as shown in Fig. 296.

Machine Screws are similar to cap screws in form. They differ from them by being measured in decimals instead of even fractions of an inch.

Tap Bolts have the same form as cap screws except that they are not finished before threading, are threaded for their full length, and are used for rough work.

Set Screws are used ordinarily to prevent relative motion of two parts such as a pulley and shaft. The screw is passed through a threaded hole in one part and the point is forced against another part. The proportions of the set screws and the shapes of the different points are shown in Fig. 297.

Multiple Threads. It is sometimes necessary to increase the distance traversed by a nut in one revolution. If a coarse enough single thread is used to give the advance required, the strength of the bolt may be considerably diminished. To obviate this difficulty, more than one thread may be cut side by side. The advance for one revolution of a multiple thread is commonly called the ''lead,'' and the pitch is the distance between corresponding points on two successive threads. Fig. 298. The con-

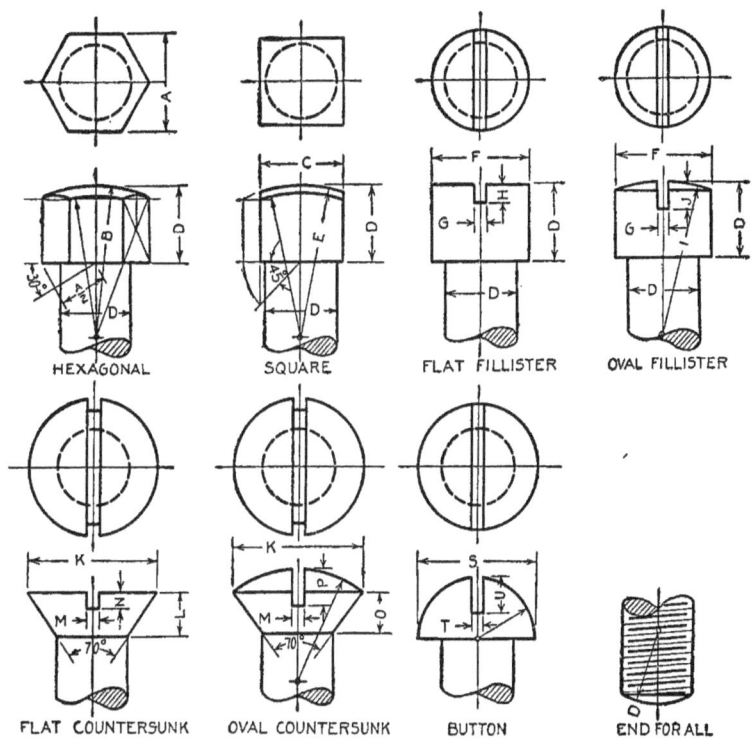

HEXAGONAL SQUARE FLAT FILLISTER OVAL FILLISTER

FLAT COUNTERSUNK OVAL COUNTERSUNK BUTTON END FOR ALL

D	A	B	C	E	F	G	H	I	J	K	L	M	N	P	O	R	S	T	U
1/8						3/16	.032	1/16	1/4	5/64	1/4	3/32	.040	3/64	5/64	9/64	3/16	.035	1/16
3/16						1/4	.040	1/16	5/16	3/32	3/8	9/64	.064	3/64	3/32	13/64	9/32	.051	3/32
1/4	7/16	21/32	3/8	9/16	3/8	5/16	.064	1/16	1/2	3/32	15/32	5/32	.072	1/16	1/8	17/64	3/8	.072	1/8
5/16	1/2	3/4	7/16	21/32	7/16	5/16	.072	5/64	5/8	1/4	5/8	7/32	.102	5/64	5/32	11/64	15/64	.091	5/32
3/8	9/16	27/32	1/2	3/4	9/16	3/32	.091	3/32	3/4	9/64	3/4	17/64	.114	3/32	3/16	13/64	9/16	.102	3/16
7/16	5/8	15/16	9/16	27/32	5/8	7/64	.102	7/64	7/8	11/64	13/64	17/64	.114	3/32	7/32	13/64	21/32	.114	7/32
1/2	3/4	1 1/8	5/8	15/16	3/4	1/8	.114	1/8	1 1/16	3/16	7/8	17/64	.128	3/32	1/4	13/32	3/4	.114	1/4
9/16	13/16	1 7/32	11/16	1 1/32	13/16	13/64	.114	9/64	1 1/8	7/32	1	5/16	.133	7/64	9/32	15/32	27/32	.114	9/32
5/8	7/8	1 5/16	3/4	1 1/8	7/8	1/8	.128	5/32	1 1/4	15/64	1 1/8	23/64	133	1/8	5/16	35/64	15/16	133	5/16
3/4	1	1 1/2	7/8	1 5/16	1	3/16	.133	3/16	1 1/2	9/32	1 3/8	7/16	.133	5/32	3/8	21/32	1 1/4	133	3/8
7/8	1 1/8	1 11/16	1	1 11/16	1 1/8	7/32	.133	7/32	15/32	21/64									
1	1 1/4	1 7/8	1 1/8	1 7/8	1 1/4	1/4	.165	1/4	1 3/4	3/8									
1 1/8	1 3/8	2 1/16	1 3/8	2 1/16															
1 1/4	1 1/2	2 1/4	1 1/2	2 1/4															

FIG. 296. VARIOUS FORMS OF CAP SCREW HEADS

ventions for multiple threads are distinguished from those for single threads by increasing the angle of the cross lines

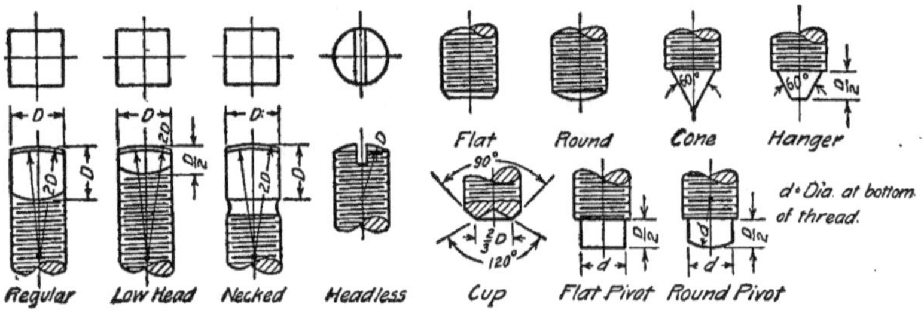

FIG. 297. SET SCREW HEADS AND POINTS

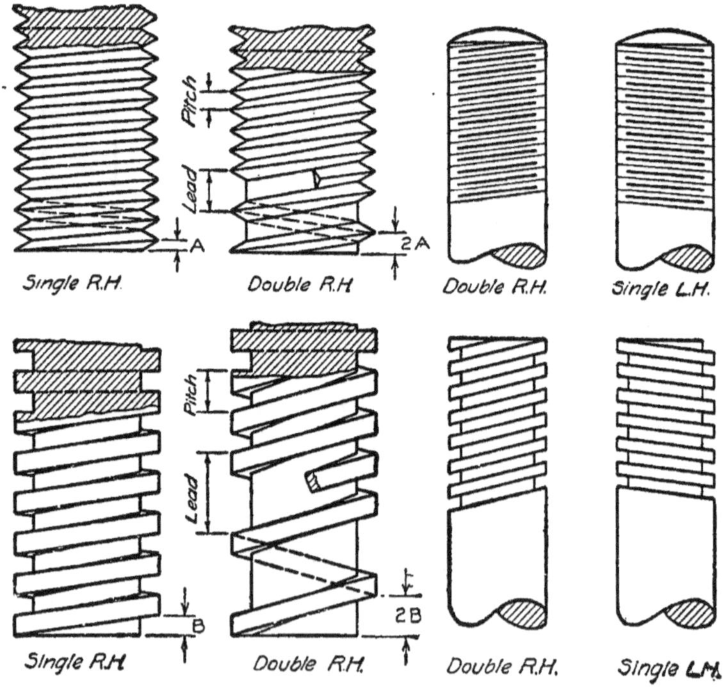

FIG. 298. CONVENTIONAL REPRESENTATION OF MULTIPLE THREADS

and by a note indicating the kind of threads as double, triple, quadruple, etc.

Methods of Indicating Finish. Where and how a part is to be finished may be shown by symbols or notes, or both. In case a hole is to be bored, drilled, reamed, cored, etc., a note is usually made in connection with the dimension figure. Fig. 299. A cylindrical surface to be turned, ground, polished, rough finished, etc., may have the method of finishing indicated in the same way. In case all surfaces of the object are to be finished and the method can be left to the workman's judgment, a note may be made: FINISH ALL OVER. Where only certain surfaces are to be finished,

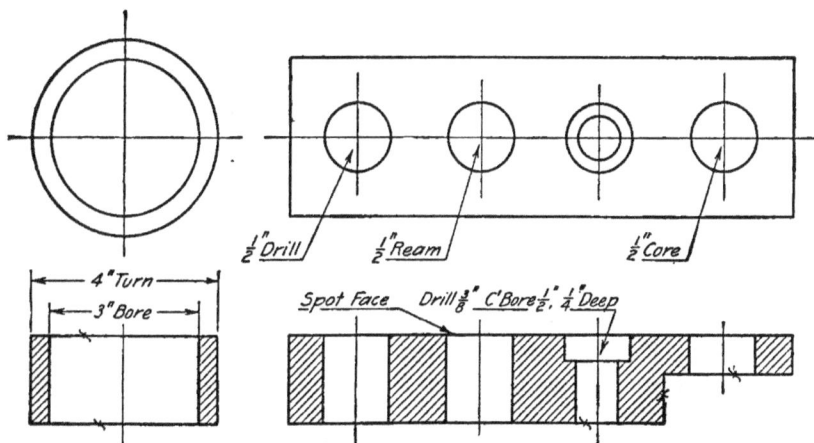

FIG. 299. METHODS OF INDICATING FINISH

the character **f** may be placed across the lines which represent these surfaces viewed edgewise. Fig. 299.

While the indication of finish is a very small part of a drawing, it is nevertheless a very important detail. The omission of a finish mark may mean the making of a large number of castings from a pattern on which no stock has been allowed for finish.

SKETCHING FROM THE OBJECT

PREPARATORY INSTRUCTIONS FOR DRAWING PLATE 39

Freehand sketches may be made by a designer to get an idea of the form of certain parts in working out his design. A designer or chief draftsman may use them as a means of conveying his ideas to a junior draftsman.

In case a machine is broken, time may often be saved by sketching the broken parts in the shop and having parts made to replace them instead of sending to the manufacturer of the machine for repairs. When a change of design is contemplated and the original drawings are not to be had, sketches of the parts affected may be made from the existing machine and the desired changes incorporated in the mechanical drawing made from sketches. When time permits and it is desirable to have a permanent record of the drawing a mechanical drawing should be made from the sketch, but in an emergency the sketch, if carefully drawn and checked, may be used as a shop drawing.

In making the orthographic sketches of Chapter II, the fact that certain views of the object were shown in correct proportion and were dimensioned made the task of drawing the other views of the object to larger scale a simple process.

The drawing of orthographic sketches from dimensioned perspective sketches, Chapter III, increased the difficulty of selecting and arranging the views, and to some extent, the dimensions.

Compared to sketching from orthographic and perspective views the average beginner will find the making of an orthographic sketch from the object a rather infangible problem. He will find it difficult to represent in outline an object which to the eye stands out in relief in light and shadow. At the same time he must keep in mind the fact that only two dimensions can be represented in each view. He is, also, confronted with the necessity of establishing center lines, datum lines, etc., which are not edges of the object but are of prime importance in the drawing.

He must select dimensions to show the proper relation between the details of the object. These dimensions must also be selected to show similar distances on parts which are fitted to the object. He must use his judgment as to the accuracy with which each measurement should be made, as to the allowance for inaccuracies of workmanship, inaccuracies inherent in the process of manufacture, etc.

Selecting Views. In selecting the views of an object to be drawn, the principles developed in previous chapters should be used. In general only necessary views are drawn, but in the sketch additional views, partial views, sections, etc., may be drawn in preference to complicating the necessary views with lines.

Methods Used in Drawing. After an inspection of the object and after a decision has been reached as to what views are to be drawn, the student should place the object, if it is removable, so that he can obtain the required views without changing its position. Very often the shifting of the object leads to errors in the relative position of the views, such as placing the left side view to the right instead of to the left of the front view. With the object always in the same position and the principles as to relation of views, developed in former chapters, well in mind, such errors are not likely to occur.' The views should show the object in as good proportion as can be obtained without scaling it. Time should not be wasted in taking dimensions at this stage and attempting to lay them out to scale.

The first step in the construction of the drawing is to locate center or other reference lines. Circles should be constructed by first drawing two center lines at right angles. The radii should then be estimated from the intersection on these lines, and the circle drawn through the four points located.

If the object is of cylindrical form, it will usually be found advantageous to draw the circular view first because of the ease with which the other views may be drawn by projecting the diameters from the circular view. In some cases where the views of the details of the object are interdependent, it will be necessary to construct two or more views simultaneously.

The use of the coördinate paper greatly facilitates the alignment and proportioning of the details in different views. The student should learn to use the ruled lines merely as a guide in locating and proportioning the views. The use of the squares as units of measurement for the purpose of drawing the object to scale is not to be considered; for while it is admitted that their use will aid in proportioning the drawing, it is not one of the functions of a freehand sketch to show the object in accurate proportion, and the counting of the squares entails a serious waste of time.

Selection and Arrangement of Dimensions. When the views of the object are complete and have been checked carefully to make sure that they, together with necessary supplementary notes, fully represent the object, the question of dimensioning should next be considered.

To dimension an object properly the draftsman must have some knowledge of the process through which it must go in the shop to become a finished product. If it be a casting he must know what dimensions the patternmakers will use in making the pattern; if it has finished surfaces he must know with what machines each is finished and give the dimensions in such a way that the machinist may use them directly. Example: The diameter of a part to be turned in the lathe should be given rather than the radius, since the most convenient and accurate method of measuring a cylindrical surface is by means of the caliper or micrometer.

Enough dimensions should be given to determine completely the sizes and relation of the details of the object. When a sketch is made at some distance from the place at which it is to be used either to furnish information for a mechanical drawing or as a shop drawing, the draftsman must be sure that all necessary dimensions are given. However, he should guard against giving unnecessary or useless dimensions in an attempt to avoid omitting necessary dimensions. All finished surfaces, special fits, etc., should be marked in such a way that they cannot be misunderstood. The nature of the sketch admits of a freer use of explanatory notes than would be tolerated on the mechanical drawing.

Details which are required to be accurately located on the object should be referred by dimensions to center lines or finished surfaces. As the dimensions to be given are planned, the extension and dimension lines should be drawn, but the dimension figures should not be inserted until all such lines are drawn.

When the extension and dimension lines are drawn the arrowheads should be made.

MEASUREMENTS

MEASURING INSTRUMENTS

The following paragraphs contain a short description of the
more common tools used in taking measurements from the object
for the purpose of dimensioning a sketch.

The Folding Rule. Rules are made of various lengths which
may be folded and carried in the pocket. The smallest divisions
are usually $\frac{1}{16}''$ and $\frac{1}{10}''$. Their construction makes the division
into smaller fractions of an inch unwarrantable as these rules
cannot be depended upon to read accurately to smaller units.

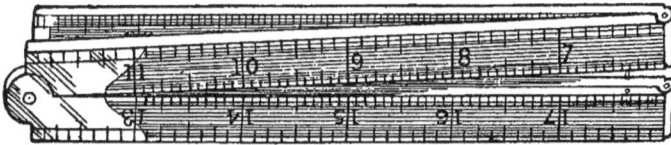

FIG. 300. FOLDING RULE

A two-foot rule will be found very serviceable where accuracy is
not required. While convenient in measuring long distances,
they are in general suitable only for rough work. Fig. 300.

The Steel Tape. Steel tape may be had in lengths of 3 feet to
200 feet or more. As in the case of the rules, their divisions are
coarse and cannot be used for accurate measurements. Fig. 301.

The Steel Scale. For accurate measurements steel scales are
used. These scales may be had in lengths of 1″ to 72″, and with
various combinations of graduations on the two edges of each
side. The most common graduations are $\frac{1}{8}''$, $\frac{1}{16}''$, $\frac{1}{32}''$, $\frac{1}{64}''$, and
$\frac{1}{100}''$. Fig. 302.

The Adjustable Square. Fig. 302 shows a square in which
the blade is adjustable in the stock. The blade is an ordinary

steel scale with a groove made to receive a hook which serves to clamp the blade in the stock. The stock is furnished with a level. This instrument will be found useful in many ways.

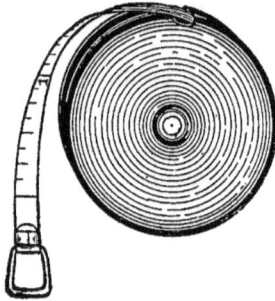

FIG. 301. STEEL TAPE

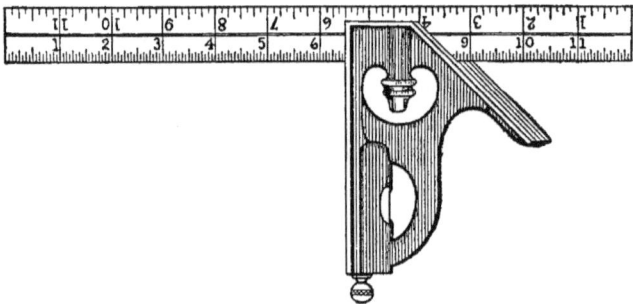

FIG. 302. ADJUSTABLE SQUARE

Calipers. Calipers are used for obtaining measurements of length or diameter where the scale cannot be applied directly. After they are set to the distance which is to be measured they are placed upon a scale and the distance read. Fig. 303 shows two forms of calipers, one adapted to outside measurements, such as diameters of shafts, etc., while the other is best suited to inside measurements, such as the diameter of holes.

Other Devices, such as the plumb bob, straightedge, and surface gauge, may be of occasional use in taking measurements from the object.

Taking Measurements

Having drawn the dimension lines, extension lines, and arrow-heads, there remain the taking of dimensions from the object and inserting them on the drawing. In doing this, judgment must be exercised in determining with what degree of accuracy each measurement should be taken. Examples: Dimensions between rough surfaces usually need not be given closer than the nearest $\frac{1}{16}''$ or $\frac{1}{32}''$, while the inside diameter of the bushing in which a

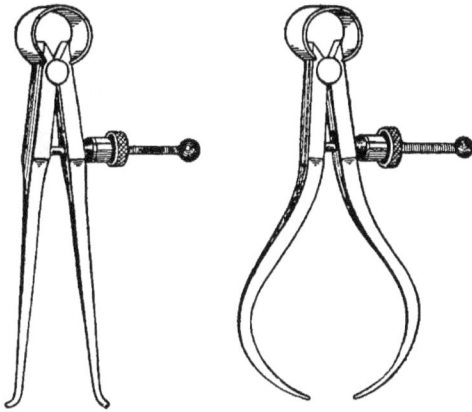

Fig. 303. Inside Caliper.   Outside Caliper

shaft is to run would probably be given .003″ or .004″ larger than the diameter of the shaft.

Judgment must also be exercised in determining whether irregularities such as the uneven thickness of castings, lack of symmetry, apparent discrepancies in spacing of holes, etc., are intentional and essential to the design and construction of the object, or whether they are non-essentials which have come about through natural causes in the process of manufacture or poor workmanship, and should be eliminated from the drawing.

The problems arising in the taking of measurements from the object are so varied that no attempt will be made here to discuss the subject fully. However, a few examples may be given which will illustrate the use of the measuring instruments and also the general principles involved in securing dimensions.

The distance between points on the same plane surface such as the distance between two parallel edges of the surface and the

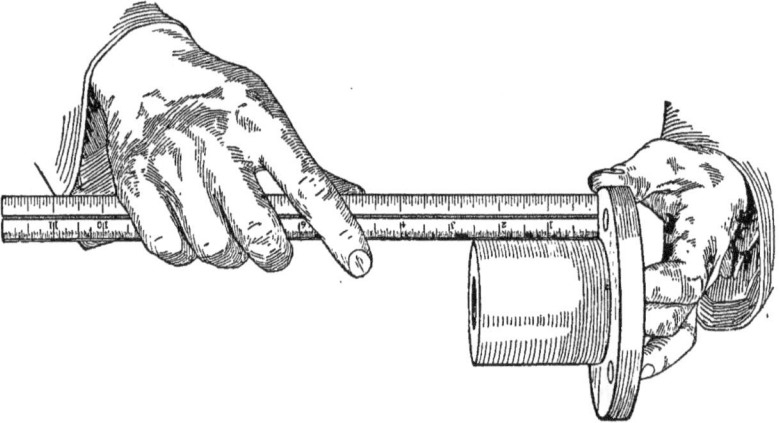

FIG. 304. MEASURING A LINEAR DISTANCE WITH THE SCALE

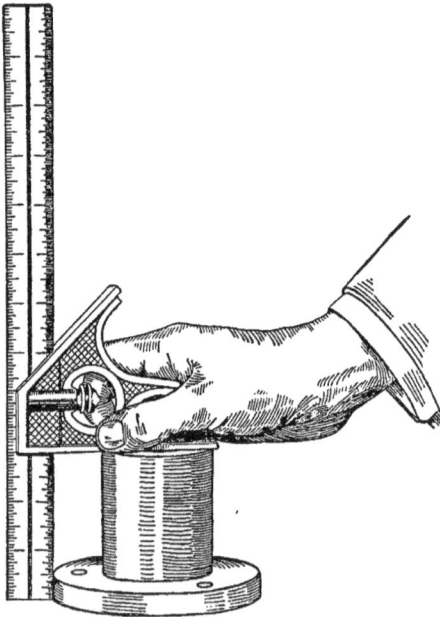

FIG. 305. MEASURING A LINEAR DISTANCE WITH THE SQUARE

length of cylinders may be measured directly with the rule or steel scale, as shown in Fig. 304. This method is only applicable for accurate measurement when the corners are sharp. When the corners are rounded, the same dimension may be obtained by using the square or caliper, as shown in Fig. 305 or 306.

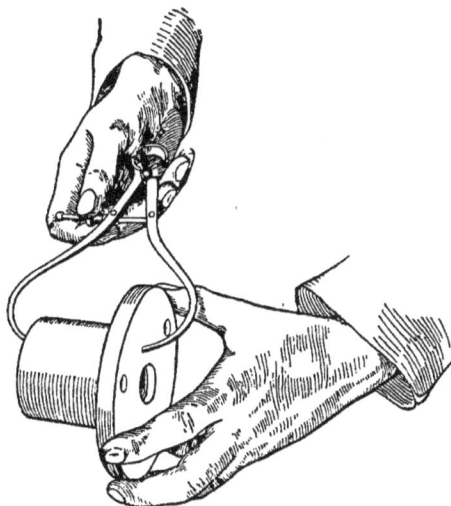

FIG. 306. MEASURING A LINEAR DISTANCE WITH THE CALIPER

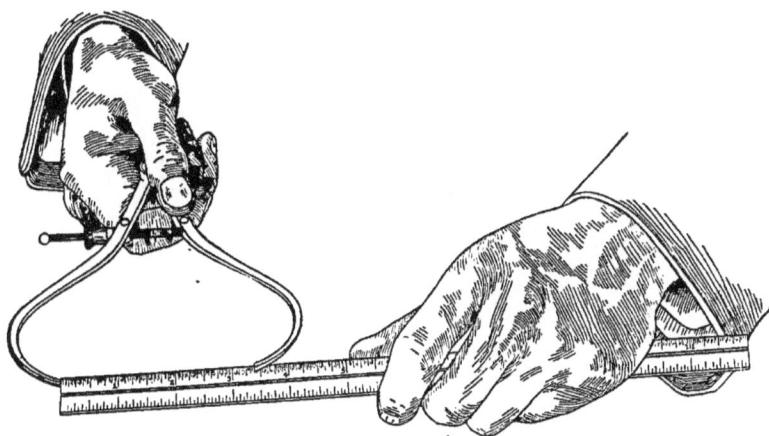

FIG. 307. READING THE CALIPER MEASUREMENT FROM THE SCALE

The use of the square here needs no explanation. The caliper must be set very carefully so that its points touch both surfaces between which the distance is to be measured, but not with enough pressure to spring the caliper. The proper adjustment is obtained by means of the thumb screw on the adjustable caliper or by tapping the leg against a solid object in the case of the plain caliper. The distance between the points of the caliper is

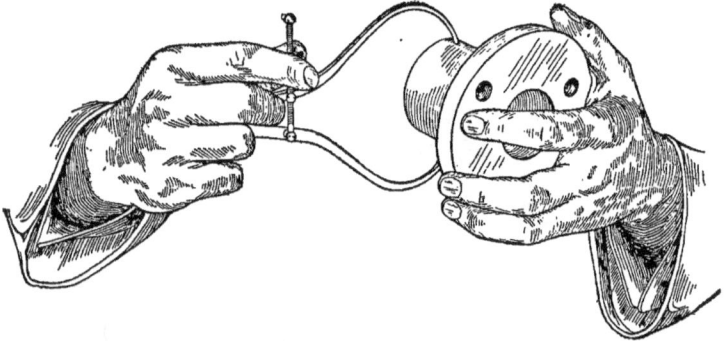

FIG. 308. MEASURING THE DIAMETER OF A CYLINDER WITH THE CALIPER

measured with the steel scale, as shown in Fig. 307. Note that one point of the caliper rests against the end of the scale so that the operator's attention may be given entirely to reading the scale division at the other point.

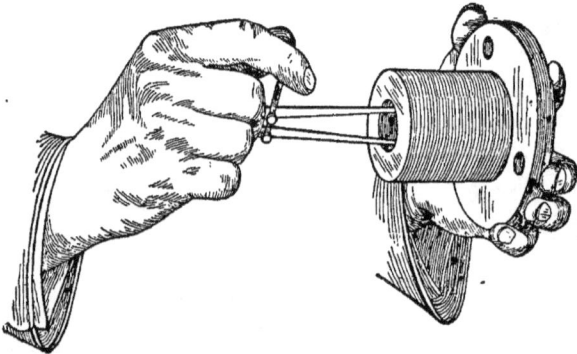

FIG. 309. MEASURING THE DIAMETER OF A HOLE WITH THE CALIPER

The outside caliper is used in obtaining dimensions of curved surfaces. See Fig. 308. It is adjusted and the measurement taken from the scale as previously described.

The inside caliper is used in measuring the diameters of holes and the openings between surfaces where the scale cannot be applied. Fig. 309. Measurements are obtained from the inside

caliper by placing it over the scale, as shown in Fig. 310. Note
that the scale is placed against a smooth surface and at right

FIG. 310. READING MEASUREMENTS FROM THE INSIDE CALIPER

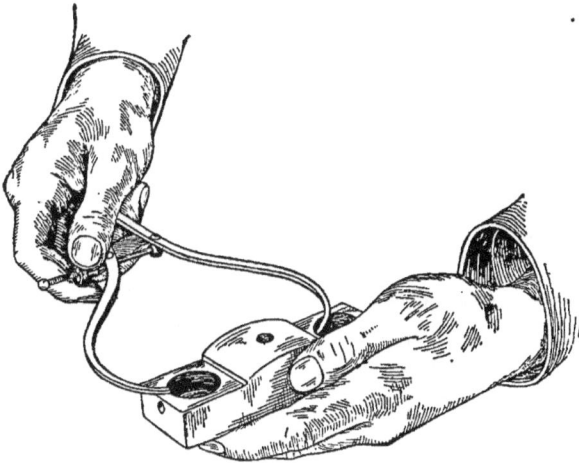

FIG. 311. MEASURING THE CENTER TO CENTER DISTANCE OF EQUAL HOLES

angles to it. One point of the inside caliper is placed against the
smooth surface. By this method the scale division opposite the
other point may be easily and accurately read.

When, as is very often the case, it is necessary to locate centers of holes with reference to each other or with reference to some finished surface or datum line, a difficulty arises from the

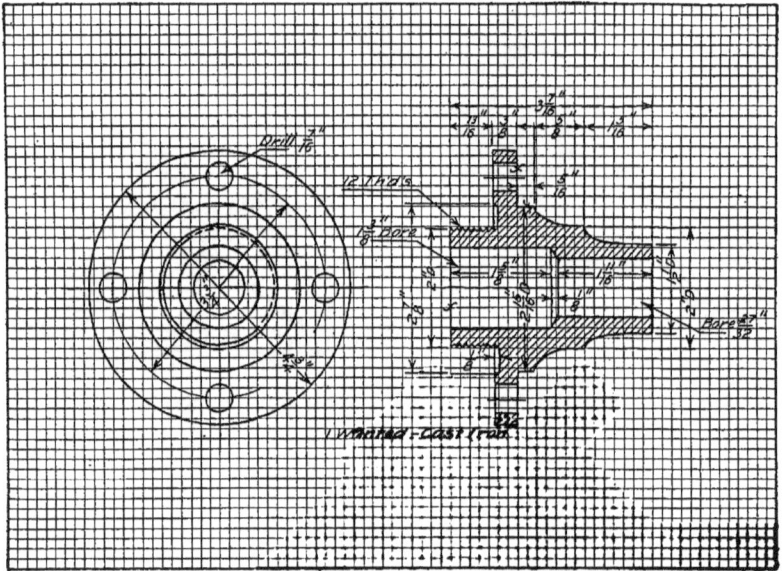

FIG. 312. TYPE PROBLEM. CYLINDER HEAD. FREEHAND SKETCH

fact that a center line does not exist on the object and must be established or the dimension obtained in a roundabout way.

In the case of two holes of equal diameter, the center-to-center distance may be obtained by measuring from the near edge of one to the far edge of the other. Fig. 311. The center-to-center distance of holes of unequal diameter may be obtained by measuring from the near edge of one to the near edge of the other and adding one-half the diameter of each. The distance from an edge or surface to the center of a hole may be had by adding one-half the diameter of the hole to the distance from the edge or surface to the near edge of the hole.

Fig. 311 shows an object the form of which makes it necessary to use the caliper in measuring the distance between the centers

of the two holes. The corners of cast parts are usually rounded
or filleted. The radii of these curves are not easily measured, but
usually it is unnecessary to measure them accurately. The radii
of small fillets may often be estimated entirely by eye or the ·
scale held against the object at one point of tangency and the
radius estimated by placing the thumb nail at the division on the

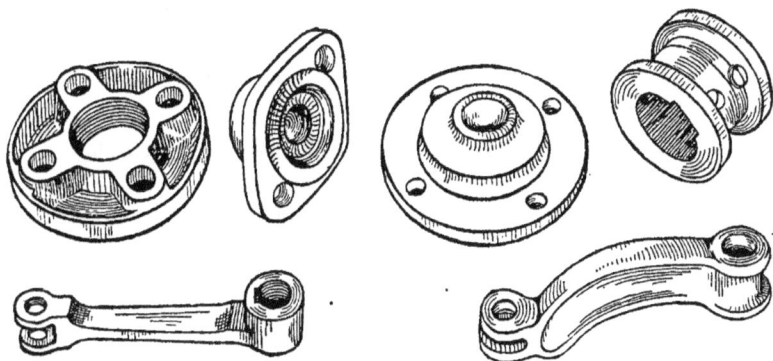

FIG. 313. TYPICAL OBJECTS FOR FIRST DRAWING FROM MODEL

scale opposite the other tangent point. A very satisfactory
method applicable in some cases is to place the object over a sheet
of paper and trace around the corner or fillet with a sharp pencil.
The center of the arc thus obtained may be found by trial with
the dividers and the radius measured.

Checking. Where a number of detail dimensions have been
· taken which make up the length of a larger detail or the whole
length of the object, this over-all dimension should be checked
by direct measurement as well as by addition of the detail
dimensions.

DATA FOR DRAWING PLATE 39

Given: A simple machine part or model preferably finished
all over. Fig. 313 shows typical objects for this plate.

Required: To make a freehand orthographic sketch.

Instructions: The following is a brief summary of the steps arranged in sequential order to be taken in making a sketch from the object. It is believed that by carefully observing the steps of this outline the draftsman will be able to make the sketch complete and accurate with a minimum amount of effort, and to do the work in the least amount of time.

1. Select views.

2. Draw views (proportioning details by eye without taking dimensions).

3. Plan dimensions—draw dimension and extension lines.

4. Draw arrowheads.

5. Take dimensions from the object and place figures.

6. Mark finished surfaces.

7. Print all notes, including the name of the part drawn, the number required, and the material from which each part is to be made.

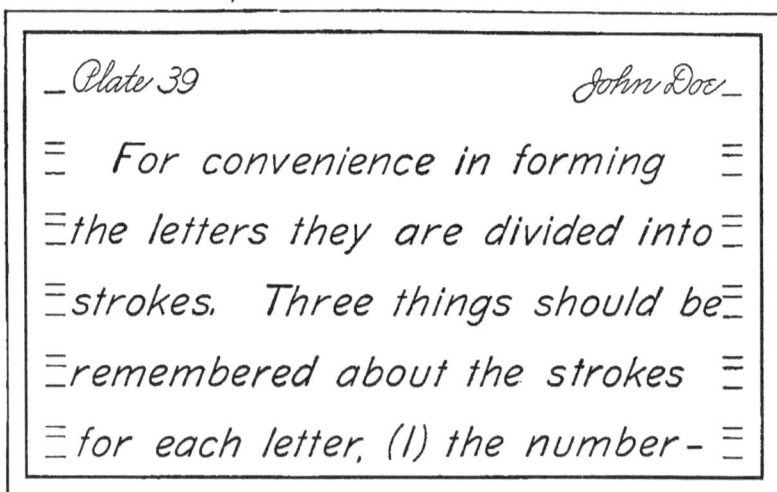

Plate 39   John Doe

≡ For convenience in forming ≡

≡the letters they are divided into≡

≡strokes. Three things should be≡

≡remembered about the strokes ≡

≡for each letter, (1) the number - ≡

FIG. 314. LETTERING PLATE 39

DATA FOR LETTERING PLATE 39

Given: Plate 39 to reduced size. Fig. 314.

Required: To make the plate to an enlarged scale.

DATA FOR DRAWING PLATE 40

Given: The orthographic sketch. Plate 39.

Required: To make a mechanical drawing from Plate 39.

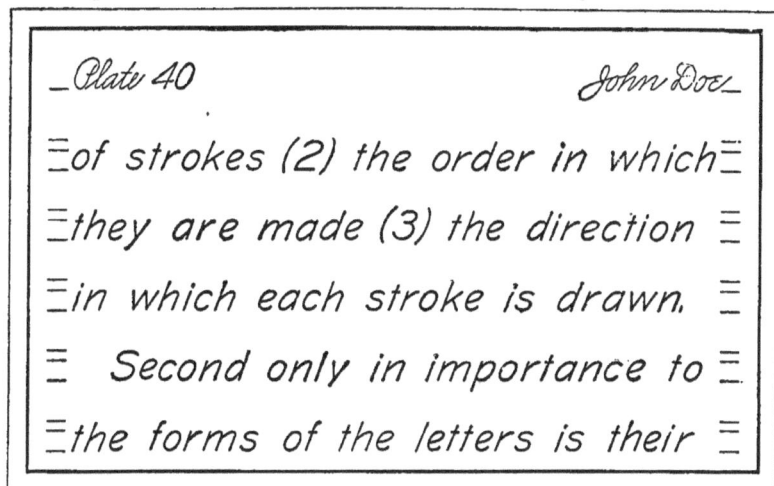

_Plate 40   John Doe

=of strokes (2) the order in which=

=they are made (3) the direction =

=in which each stroke is drawn. =

= Second only in importance to =

=the forms of the letters is their =

FIG. 315. LETTERING PLATE 40

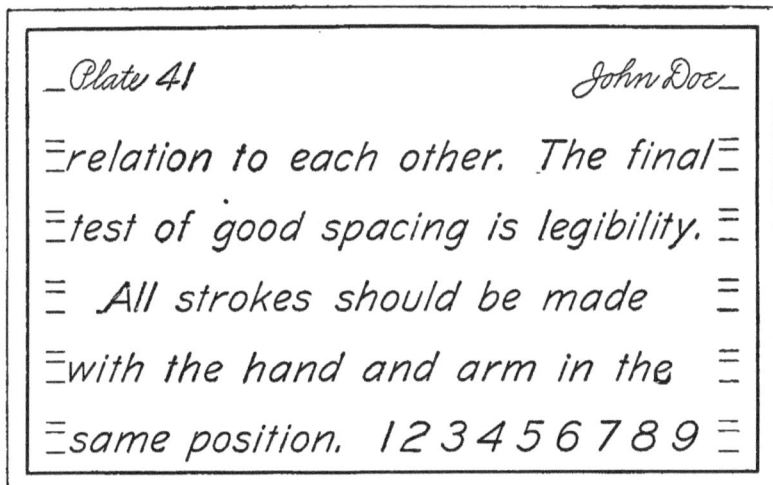

_Plate 41   John Doe

=relation to each other. The final=

=test of good spacing is legibility. =

= All strokes should be made   =

=with the hand and arm in the   =

=same position. 1 2 3 4 5 6 7 8 9 =

FIG. 316. LETTERING PLATE 41

DATA FOR LETTERING PLATE 40

Given: Plate 40 to reduced size. Fig. 315.

Required: To make the plate to an enlarged scale.

DATA FOR DRAWING PLATE 41

Given: The mechanical drawing, Plate 40.
Required: To make a tracing from Plate 40.

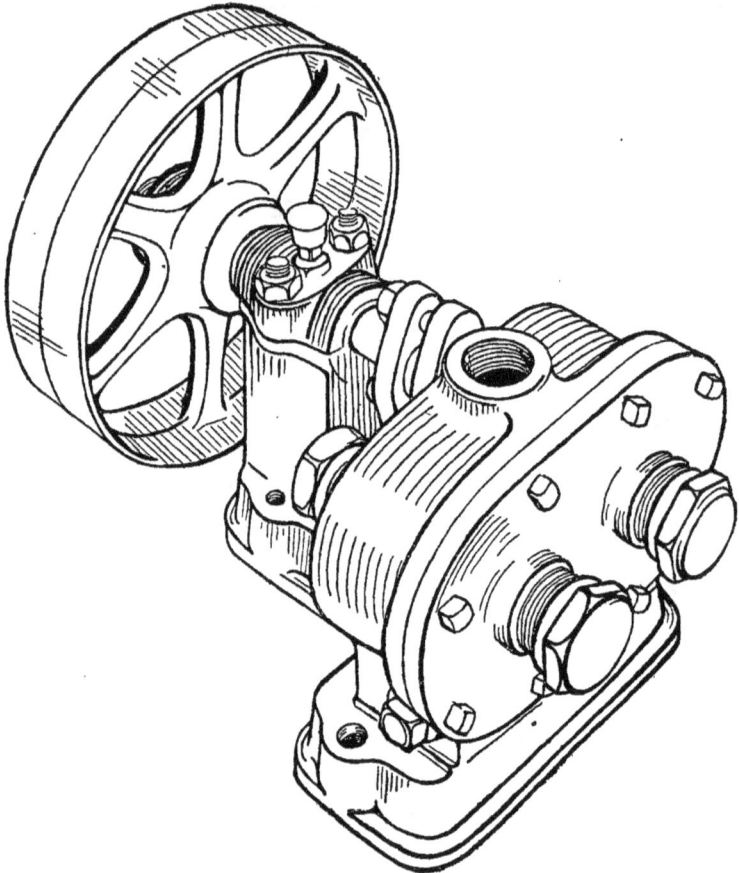

FIG. 317. TYPICAL MODEL OF COMPLETE MACHINE

DATA FOR LETTERING PLATE 41

Given: Plate 41 to reduced size. Fig. 316.
Required: To make the plate to an enlarged scale.

PREPARATORY INSTRUCTIONS FOR DRAWING PLATE 42

The model for this plate should be a complete machine or some unit of a machine which is composed of several parts. The

parts of the model then can be divided into several groups and
each group assigned to a student.* Fig. 317 shows a typical
model, the parts of which are divided into groups. Fig. 318. The
detail drawings of this model will be used later (Plate 48) in
making an assembly drawing.

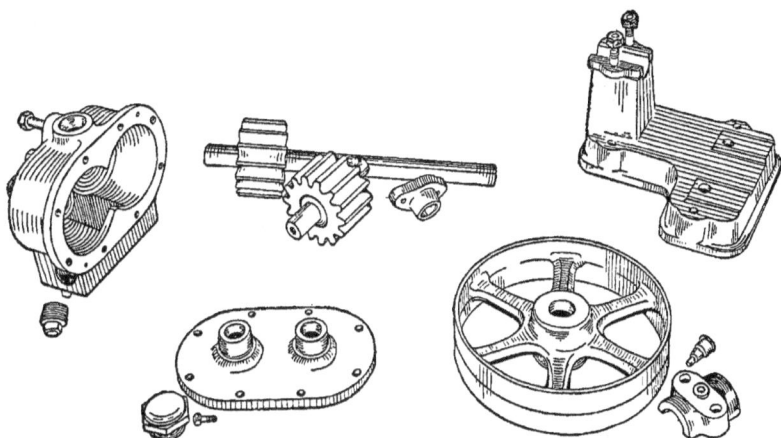

FIG. 318. SHOWING GROUPS OF PARTS OF MACHINE FOR ASSIGNMENT

DATA FOR DRAWING PLATE 42

Given: A part or group of parts of a machine.

Required: To make an orthographic sketch of each part
assigned by the instructor.

Instructions: In making the sketches proceed according to
the steps outlined for Plate 39.

More than one part may be drawn on each sheet, provided the
views are not too small or crowded too closely together.

In drawing and dimensioning these objects the student should
check each detail with the parts which are related to it or depend
upon it in any way.

Note should be made of the name of each part, the number
required, and the material from which it is made.

* This plan gives best results when there are from 3 to 6 students
working on each model.

DATA FOR LETTERING PLATE 42

Given: Plate 42 to reduced size. Fig. 319.

Required: To make the plate to an enlarged scale.

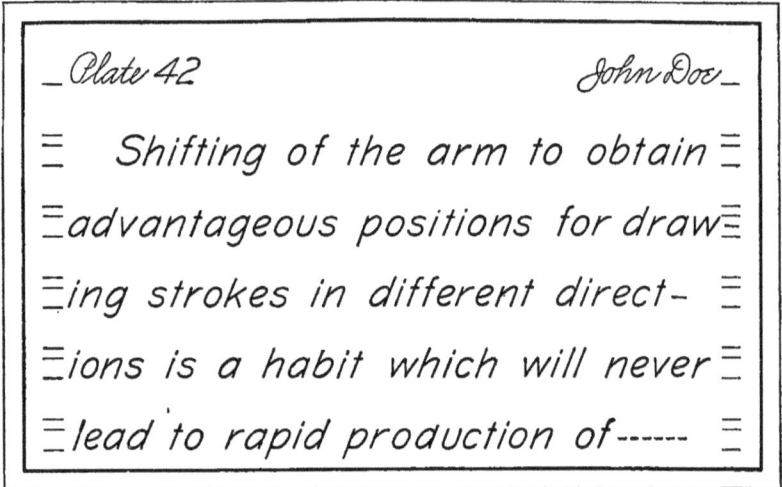

Plate 42   _John Doe_

≡ *Shifting of the arm to obtain* ≡

≡*advantageous positions for draw*≡

≡*ing strokes in different direct-* ≡

≡*ions is a habit which will never* ≡

≡*lead to rapid production of*------ ≡

FIG. 319. LETTERING PLATE 42

PREPARATORY INSTRUCTIONS FOR DRAWING PLATE 43

When making the mechanical drawing, all of the parts in each group should be drawn on one sheet if possible. The arrangement of the views should be such as to make the best use of the space available, and at the same time produce a pleasing effect for the sheet as a whole. This will require careful study. The solution will depend largely on the draftsman's judgment. In general, it may be said that the distance between views of different objects should be greater than that between views of the same object. The enclosing rectangles for each view may be drawn lightly to make sure that sufficient space has been allowed for the drawing of all parts before drawing the views, or better yet, a rectangle equal in size to the enclosing rectangle for the views of each part may be cut from paper and moved about until the best possible arrangement is secured.

Before starting to plan the arrangement of the sheet, the areas occupied by the bill of material and the title block should be laid out. The bill of material as shown in Fig. 279 contains the *reference figure* corresponding to the one placed near the views of the object, the *name of the object*, the *number required*, and the *materials from which it is made.* The width of the bill of material is equal to the width of the title block, and the height depends upon the number of parts to be listed. See Fig. 279 for detail dimensions.

In some shops the information referred to above is given for each part near the views of that part and is called a sub-title.

The title for a sheet containing the drawings of several parts must be a general one in which the word "details" usually takes the place of the name of the part drawn. See Fig. 322. It is often convenient to use different scales for the various objects, in which case the scale for each should be printed with the views of that part and the words, "Scales as noted," printed in the usual place in the title.

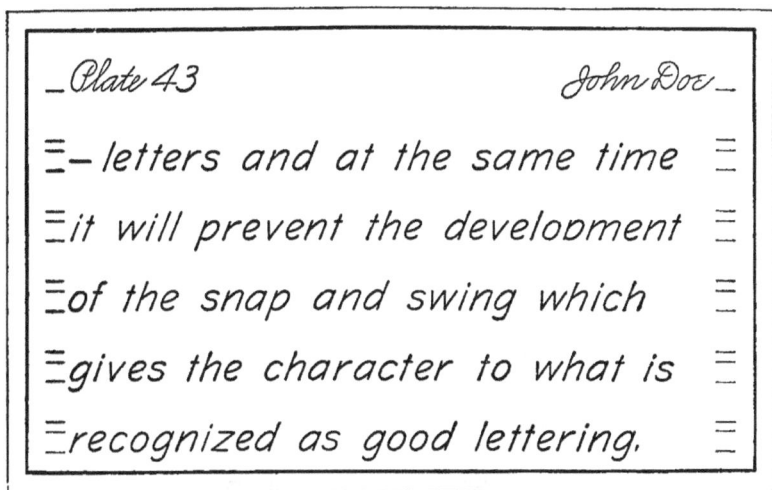

Plate 43   John Doe

=_ letters and at the same time =

=it will prevent the development =

=of the snap and swing which =

=gives the character to what is =

=recognized as good lettering. =

FIG. 320.   LETTERING PLATE 43

DATA FOR DRAWING PLATE 43

Given: The orthographic sketch, Plate 42.

Required: To make a mechanical drawing from Plate 42.

DATA FOR LETTERING PLATE 43

Given: Plate 43 to reduced size. Fig. 320.

Required: To make the plate to an enlarged scale.

DATA FOR DRAWING PLATE 44

Given: The pencil mechanical drawing. Plate 43.

Required: To make a tracing of Plate 43.

Instructions: The width of the top and left sides of the rectangle enclosing the bill of material and the vertical division lines should be object line width ($\frac{1}{64}$ ″).The horizontal lines between lines of lettering should be center line width ($\frac{1}{128}$″).

DATA FOR LETTERING PLATE 44

Given: Plate 44 to reduced size. Fig. 321.

Required: To make the plate to an enlarged scale.

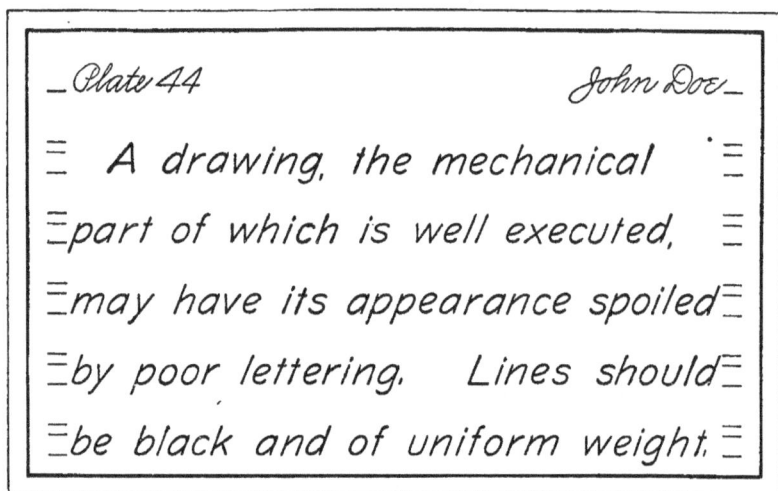

Plate 44   John Doe

A drawing, the mechanical part of which is well executed, may have its appearance spoiled by poor lettering. Lines should be black and of uniform weight.

FIG. 321. LETTERING PLATE 44

PREPARATORY INSTRUCTIONS FOR DRAWING PLATE 45

One of the problems of the draftsman is to make detail drawings from the original layout of a machine in which the parts are shown assembled. On this assembly drawing some important dimensions may be given, others may be scaled from the drawing, and the remainder must be supplied by the draftsman himself. Since this course does not presuppose a knowledge of design, all necessary dimensions will be given on the assembly drawing from which the student draws this plate.

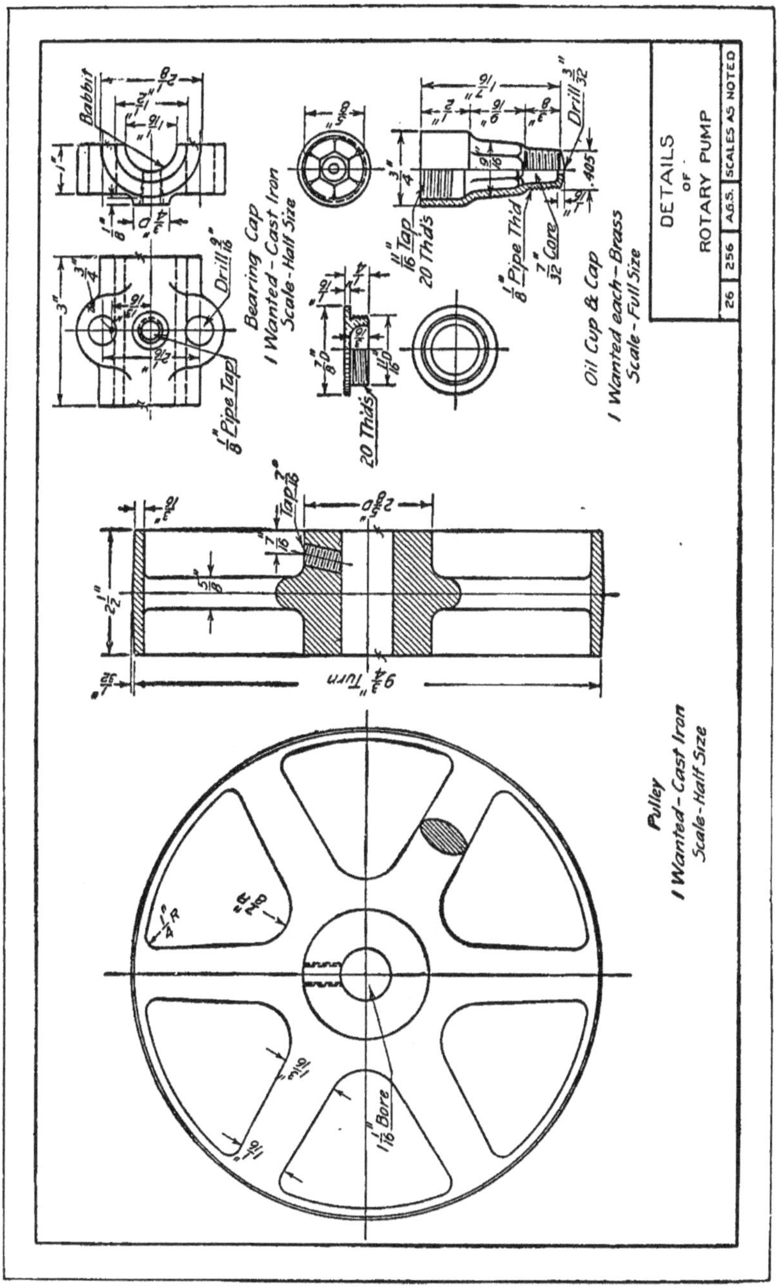

FIG. 322. TYPE PROBLEM. MECHANICAL DRAWING OF SEVERAL DETAILS ON SAME SHEET

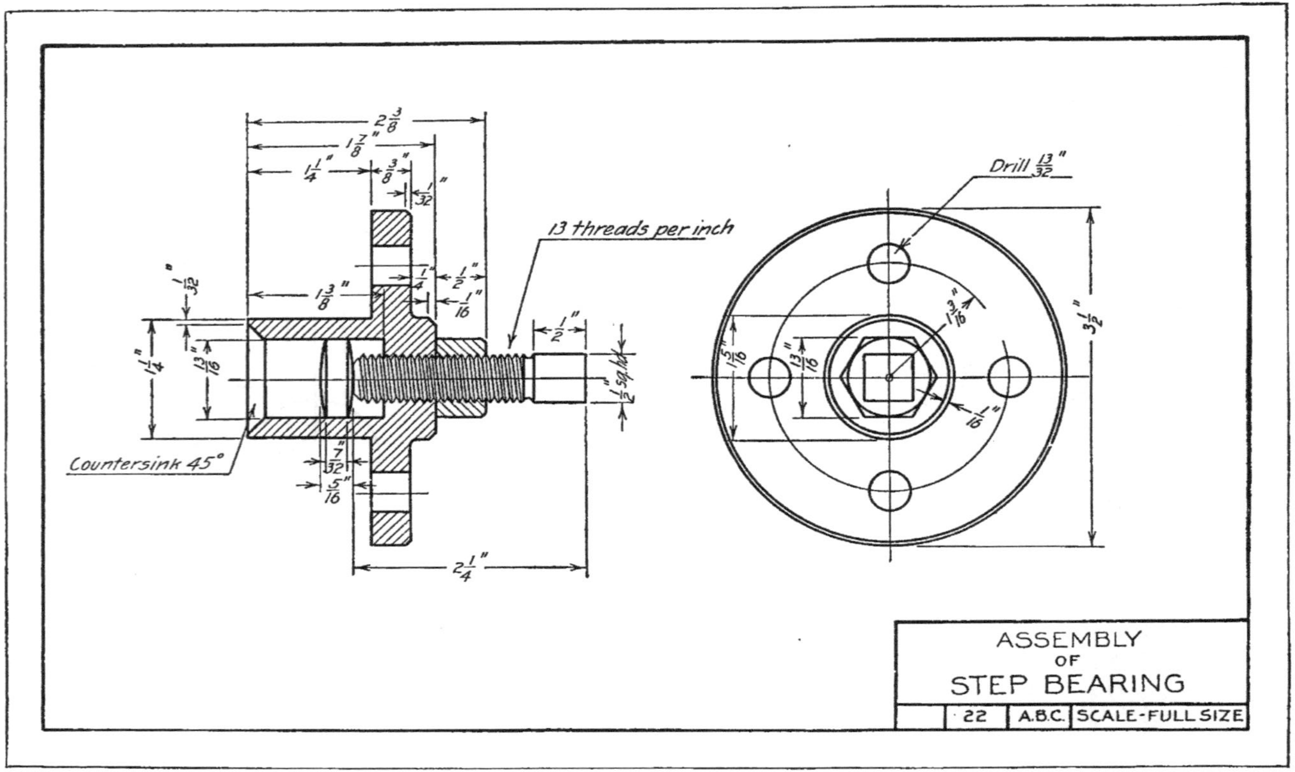

FIG. 323. TYPE PROBLEM. ASSEMBLY DRAWING OF STEP BEARING

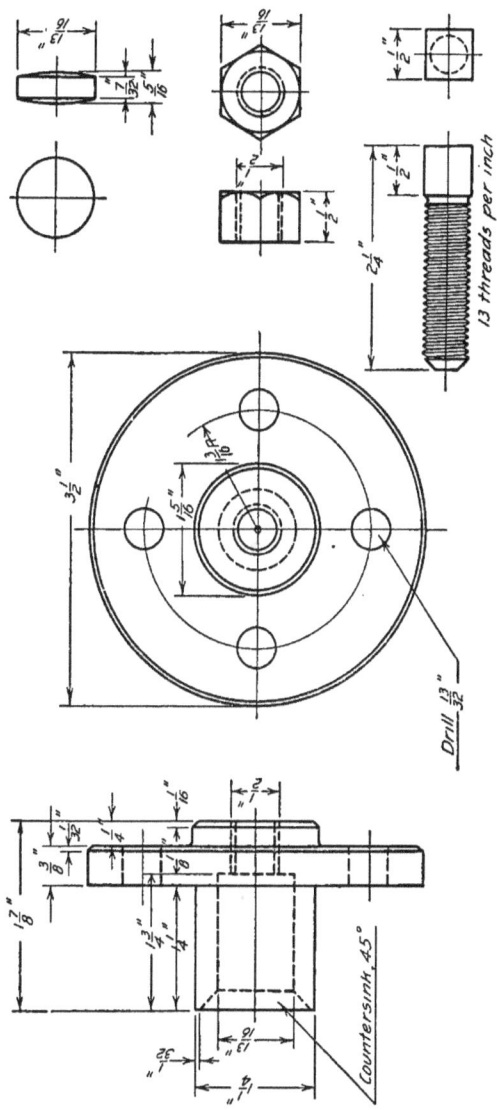

FIG. 324. TYPE PROBLEM. DETAILS OF STEP BEARING

The reading of the assembly drawing to get the correct form
for each detail will in most cases require careful study. The different parts may be distinguished when in section by various
crosshatching for different metals and by the sectioning of adjacent parts at opposite angles. But even with this aid the different views must be compared carefully to check the first impression gained of the form of each part and to make sure that no
detail has been overlooked. Each part of the object must be
dimensioned completely. It is not sufficient to give a dimension
on the views of one part and omit the same dimension on the views
of another part, even though it is evident that the dimension is
the same on both.

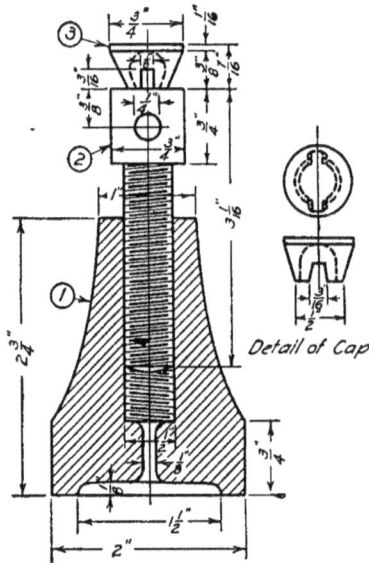

Fig. 325. Leveling Screw

DATA FOR DRAWING PLATE 45

Given: An assembly drawing of a Leveling Screw, Fig.
325; a flap valve, Fig. 326; and a letter press, Fig. 327.

Required: To make freehand orthographic detail sketch
of the object shown in Fig. 325, 326, 327, or any similar object
as assigned by the instructor.

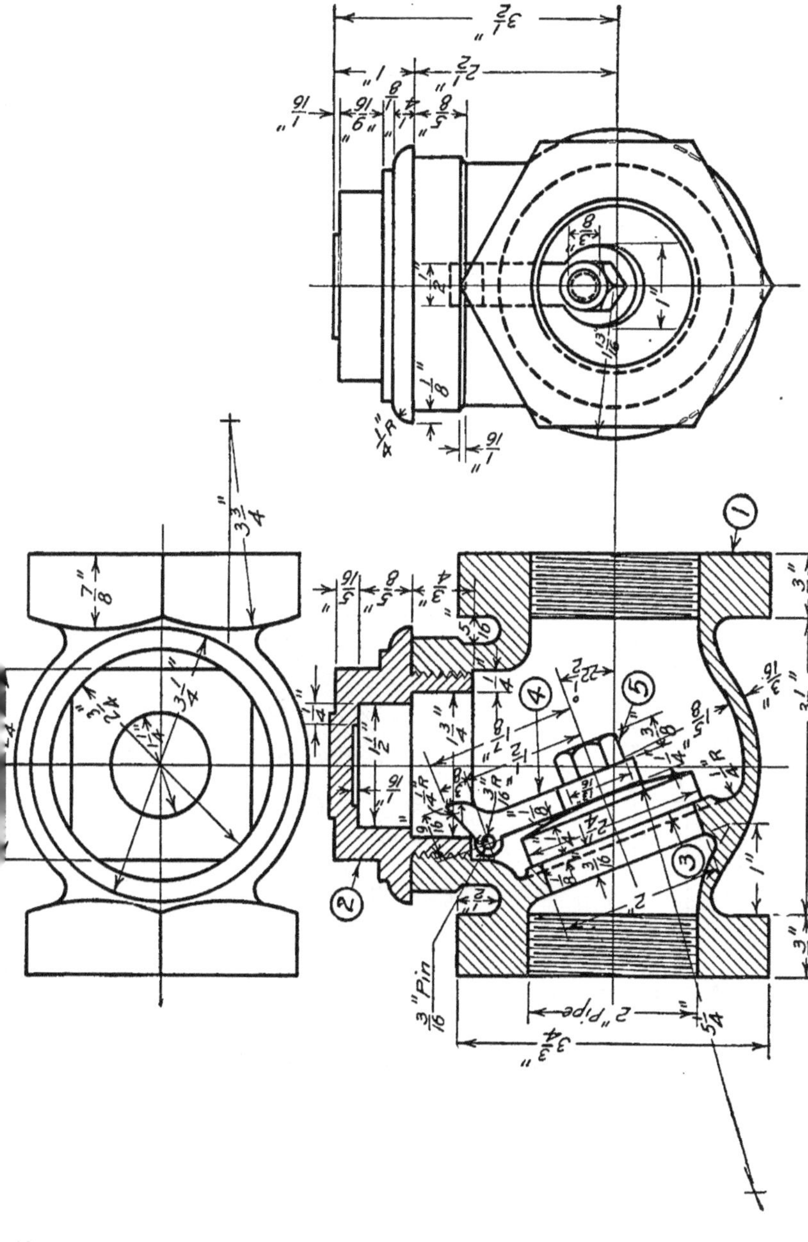

Fig. 326. ' Flap Valve

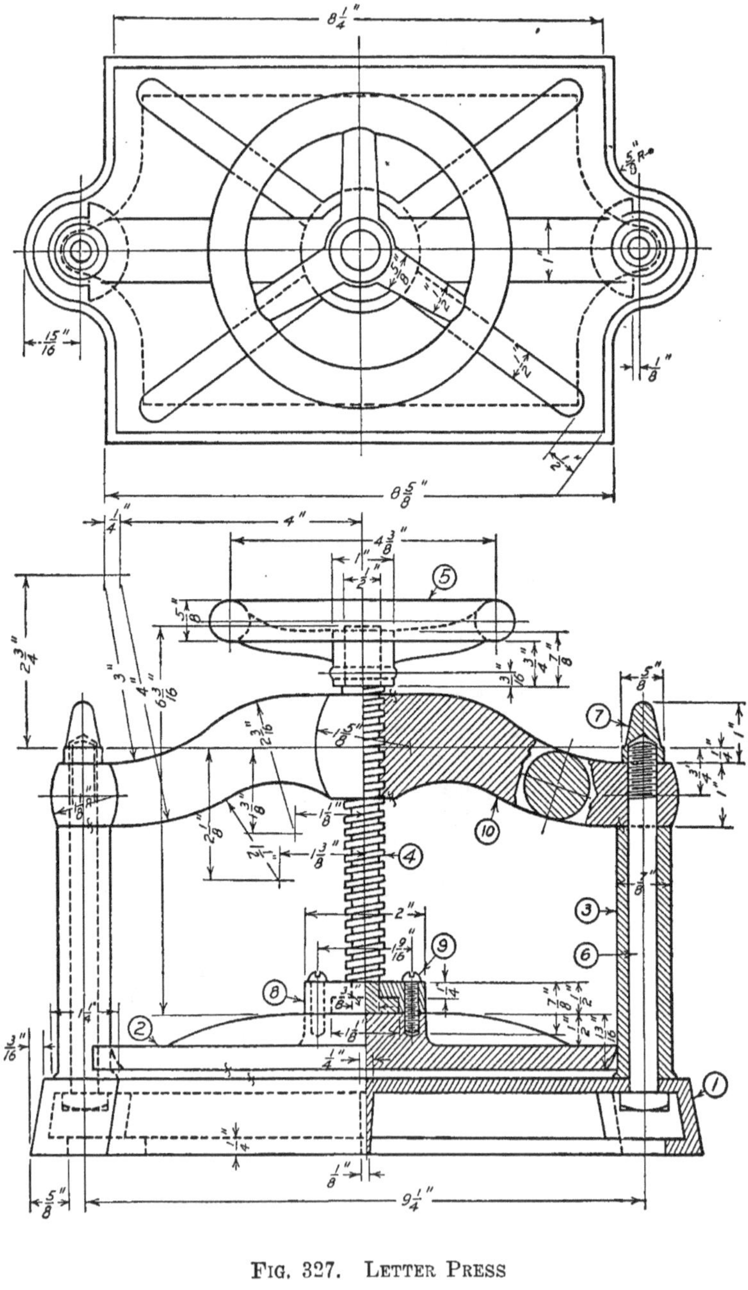

FIG. 327. LETTER PRESS

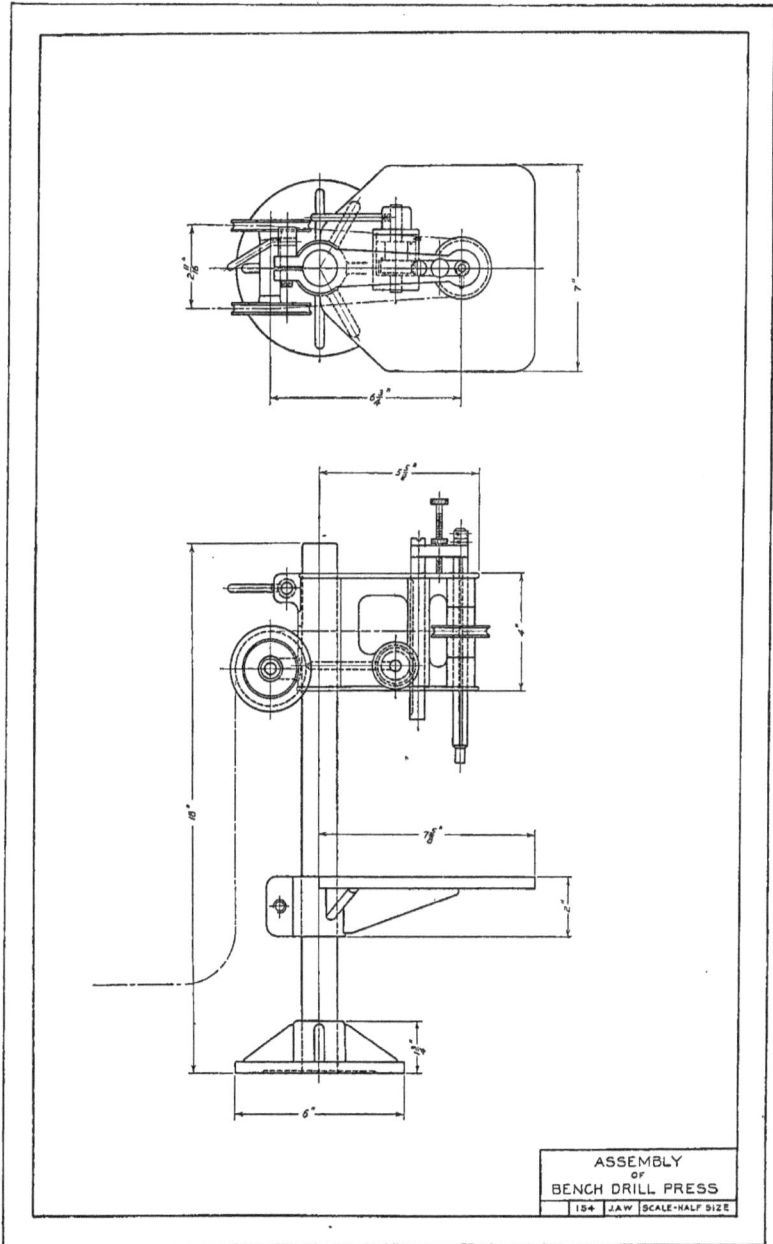

(287)

FIG. 328. TYPE PROBLEM. BENCH DRILL PRESS

A Flap Valve is used to allow a liquid or gas, such as water or steam, to flow in one direction through a pipe but not in the other. Its parts, as designated by figures in circles in Fig. 326, are named as follows:

1. Body. 2. Cap. 3. Valve. 4. Arm. 5. Cap Screw.

A *Leveling Screw* is used for leveling up work on a planer. Its parts, as designated by figures in circles in Fig. 325, are named as follows:

1. Base. 2. Screw. 3. Cap.

Fig. 327 shows a *Letter Press.* Its parts, as designated by figures in circles, are named as follows:

1. Base.
2. Plate.
3. Yoke Support.
4. Press Screw.
5. Hand Wheel.
6. Bolt.
7. Nut.
8. Clamp.
9. Button Head Screw.
10. Yoke.

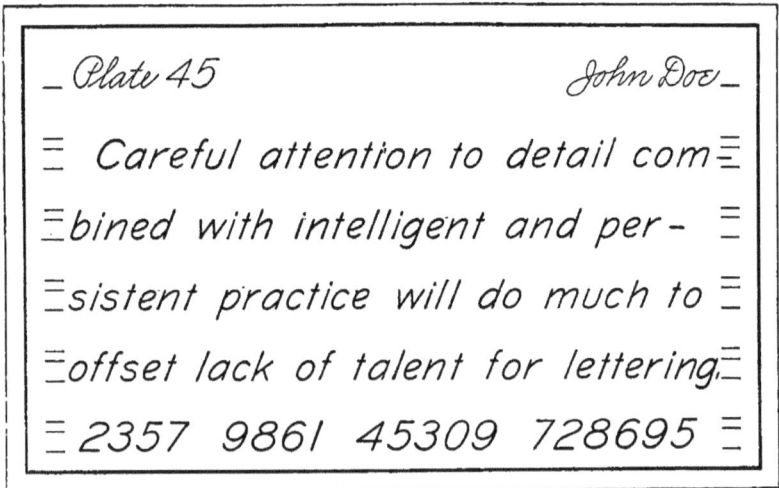

_ *Plate 45* *John Doe* _

= Careful attention to detail com=

=bined with intelligent and per - =

=sistent practice will do much to =

=offset lack of talent for lettering=

= 2357 9861 45309 728695 =

FIG. 329. LETTERING PLATE 45.

DATA FOR LETTERING PLATE 45

Given: Plate 45 to reduced size. Fig. 329

Required: To make the plate to an enlarged scale.

MEASUREMENTS

289

DATA FOR DRAWING PLATE 46

Given: The orthographic sketch. Plate 45.

Required: To make a pencil mechanical drawing from Plate 45.

DATA FOR DRAWING PLATE 47

Given: The pencil mechanical drawing. Plate 46.

Required: To make a tracing of Plate 46.

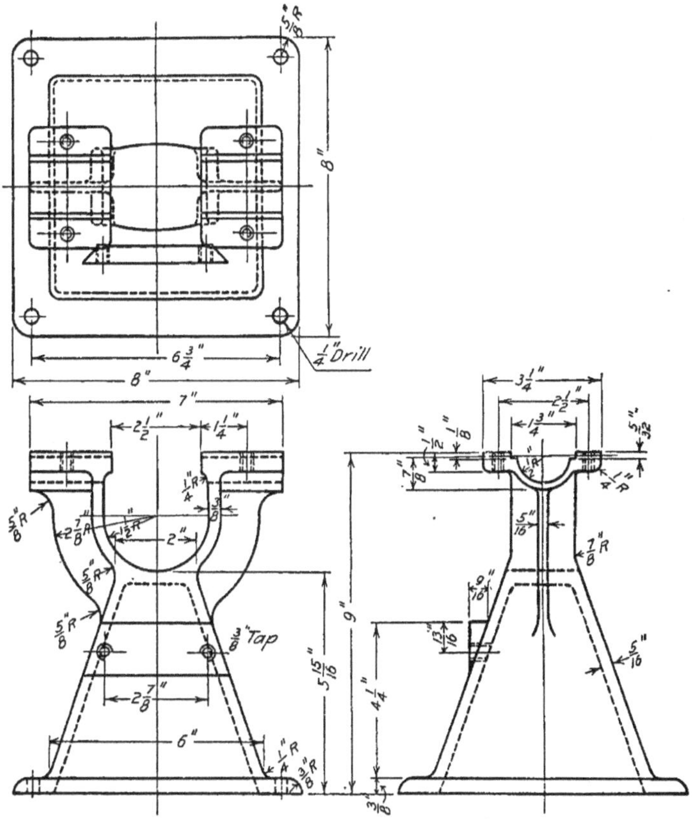

FIG. 330A. DETAILS OF BENCH GRINDER

PREPARATORY INSTRUCTIONS FOR DRAWING PLATE 48

An Assembly or General Drawing is made for showing the position and relation of parts of a machine or structure. Usually only the most important dimensions are given. Example: Fig. 328.

DATA FOR DRAWING PLATE 48

Given: The detail drawings of a bench grinder, Figs. 330A and 330B; screw punch, Fig. 331.

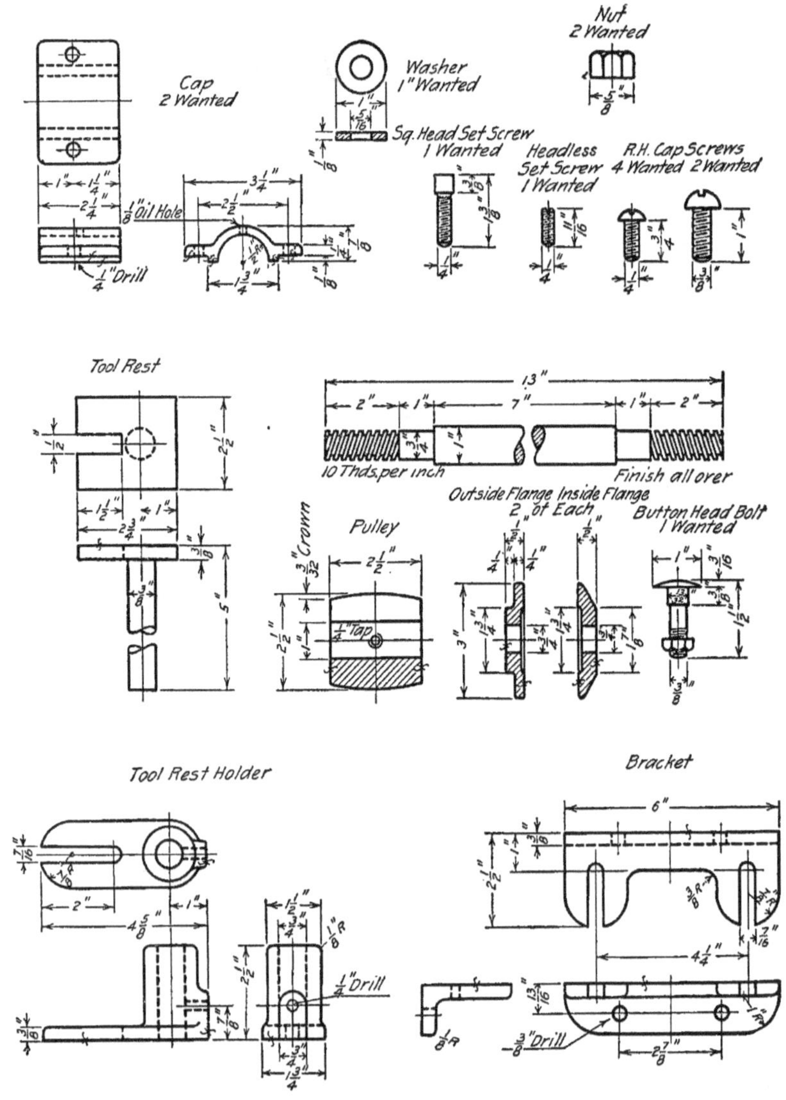

FIG. 330B. DETAILS OF BENCH GRINDER

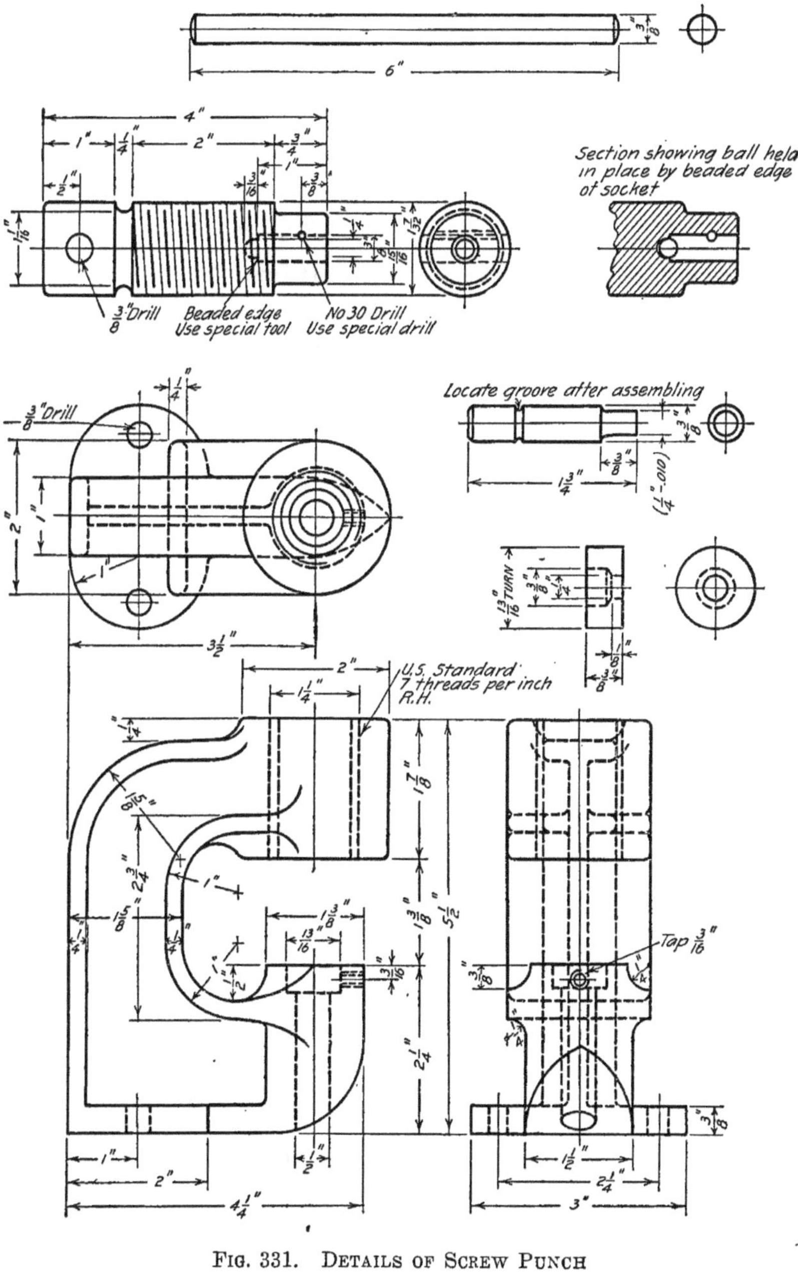

FIG. 331. DETAILS OF SCREW PUNCH

Required: To make a pencil mechanical drawing from the details of the objects shown in Figs. 330A and 330B or 331, or any similar problem assigned by the instructor. It is suggested that an assembly drawing may be made from the student detail drawings of Plate 43.

DATA FOR DRAWING PLATE 49

Given: The pencil mechanical drawing. Plate 48.

Required: To make a tracing of Plate 48.

ARCHITECTURAL DRAWING

PREPARATORY INSTRUCTIONS FOR DRAWING PLATE 50

The average man who contemplates building a house for himself finds it a great convenience to be able to make scale drawings of sufficient accuracy to test his ideas of the arrangement of rooms, dimensions, proportions, general appearance, etc., and as a means of conveying his ideas to an architect or contractor. The drawings should consist of floor plans and views of each side of the house, known as elevations.

The general arrangement of rooms on the first floor will depend on such things as the nature of the site, the owner's ideas of household conveniences, etc. As a general principle, the rooms which are used the largest percentage of time are given the most favorable location for light and ventilation. As an example, the living room is quite frequently arranged to have a south and east exposure.

In all rooms, the location of furniture, doors, windows, etc., should be carefully considered as the plan for the house is progressing. The kitchen in the average small house is planned to save steps for the housewife. The sink, stove, table, and cupboards should be arranged with a view to convenience.

In the average house, economy of space is a feature worth striving for, as the cost of the house will range from 15 cents to 22 cents per cubic foot.

The plan of the second floor will depend somewhat on that of the first floor. The partitions on the second floor should be

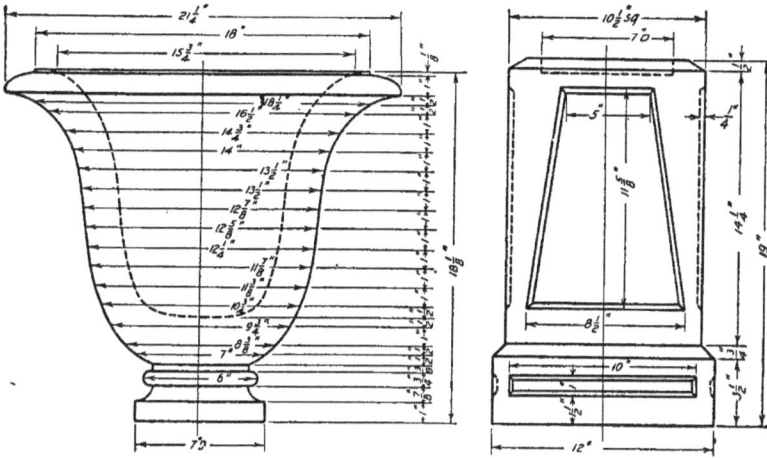

FIG. 331A. EXTRA PLATE

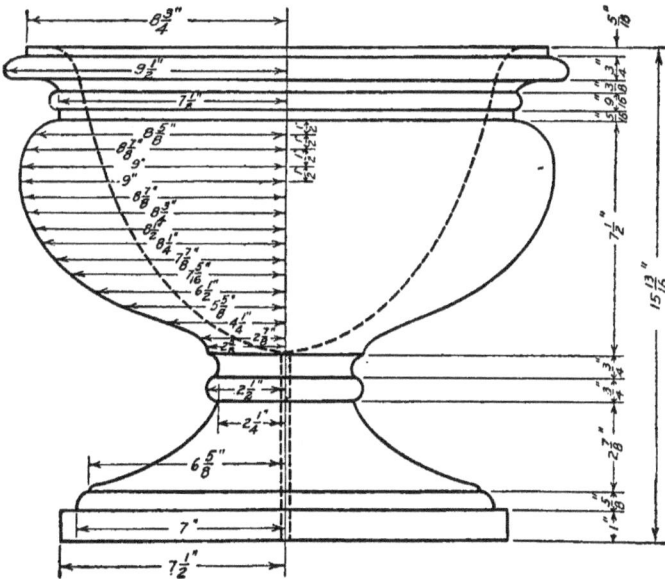

FIG. 331B. EXTRA PLATE

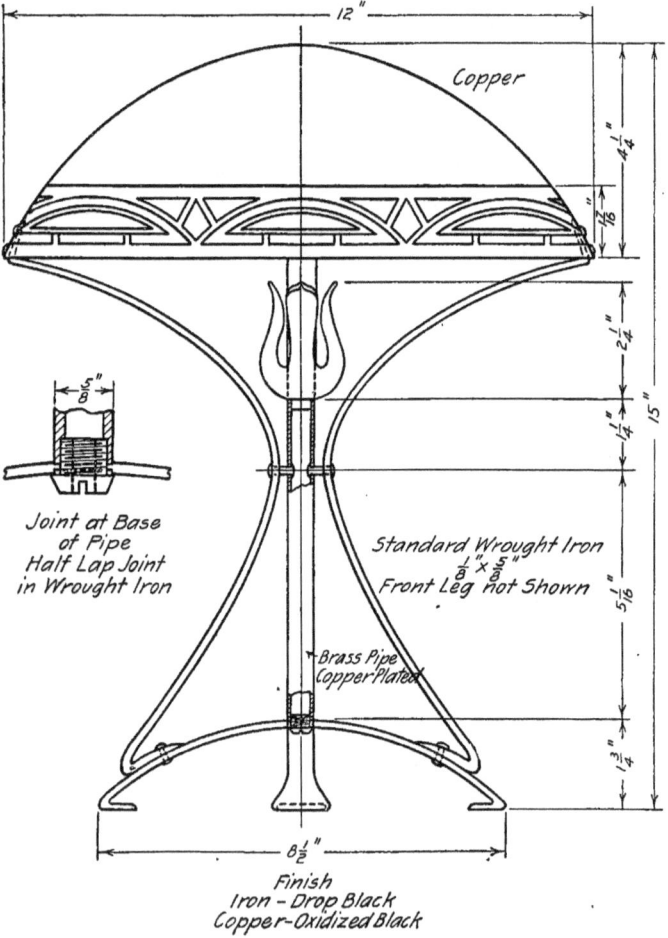

12 "

Copper

4¼"

¹⁵⁄₁₆"

2¼"

1¼"

15 "

5⁵⁄₈"

⁵⁄₈"

Joint at Base
of Pipe
Half Lap Joint
in Wrought Iron

Standard Wrought Iron
⅛" x ⅝"
Front Leg not Shown

Brass Pipe
Copper Plated

1¾"

8½"

Finish
Iron – Drop Black
Copper–Oxidized Black

FIG. 331C. EXTRA PLATE

directly over those on the first floor, as far as this arrangement
can be made. The position of the stairs must be considered and
space enough allowed for steps to reach the second floor. As a

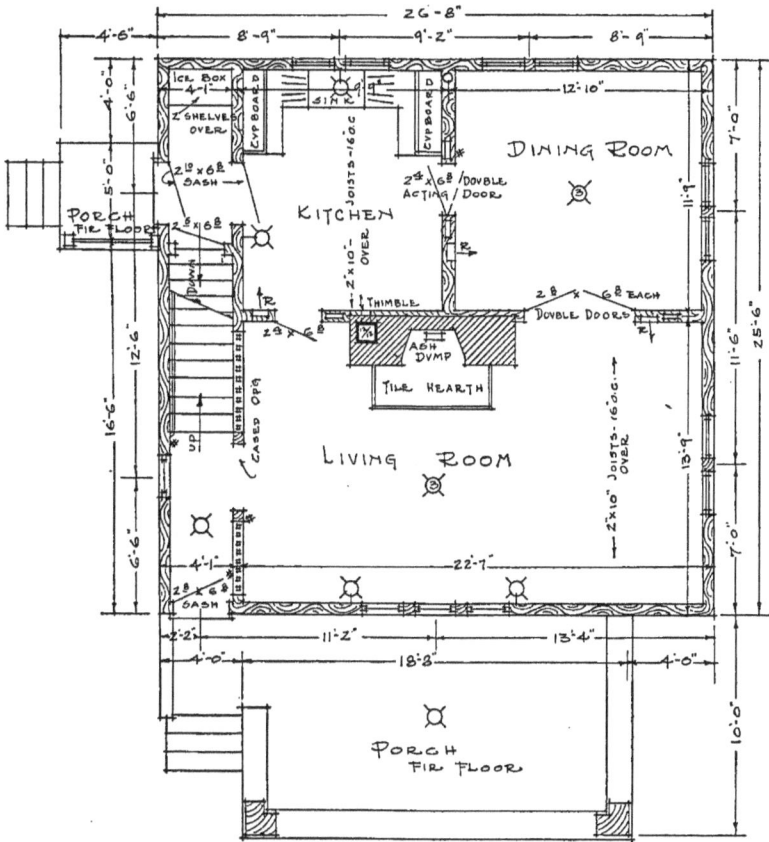

FIG. 332. FIRST FLOOR PLAN OF HOUSE

guide in laying out stairs, the sum of the tread and rise should
be about 17″ to 17½″, with a tread not less than 9″ wide for front
stairs or 8″ wide for back stairs.

In locating the bathroom, care must be taken to make it possible for the pipes and drains to lead down through the walls of the first floor. In a cold climate there is danger of freezing

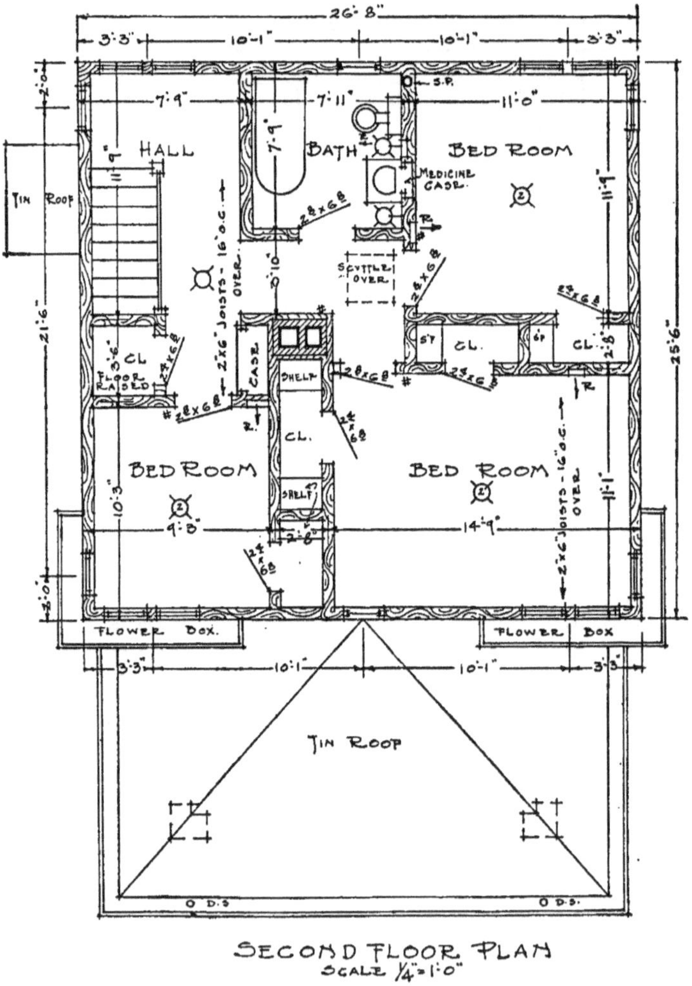

Fig. 333. Second Floor Plan of House

if they are put through an outside wall. Wherever possible, plumbing should be minimized. It is well, therefore, to have bathroom, kitchen, and basement water and sewer connections in the same vertical wall.

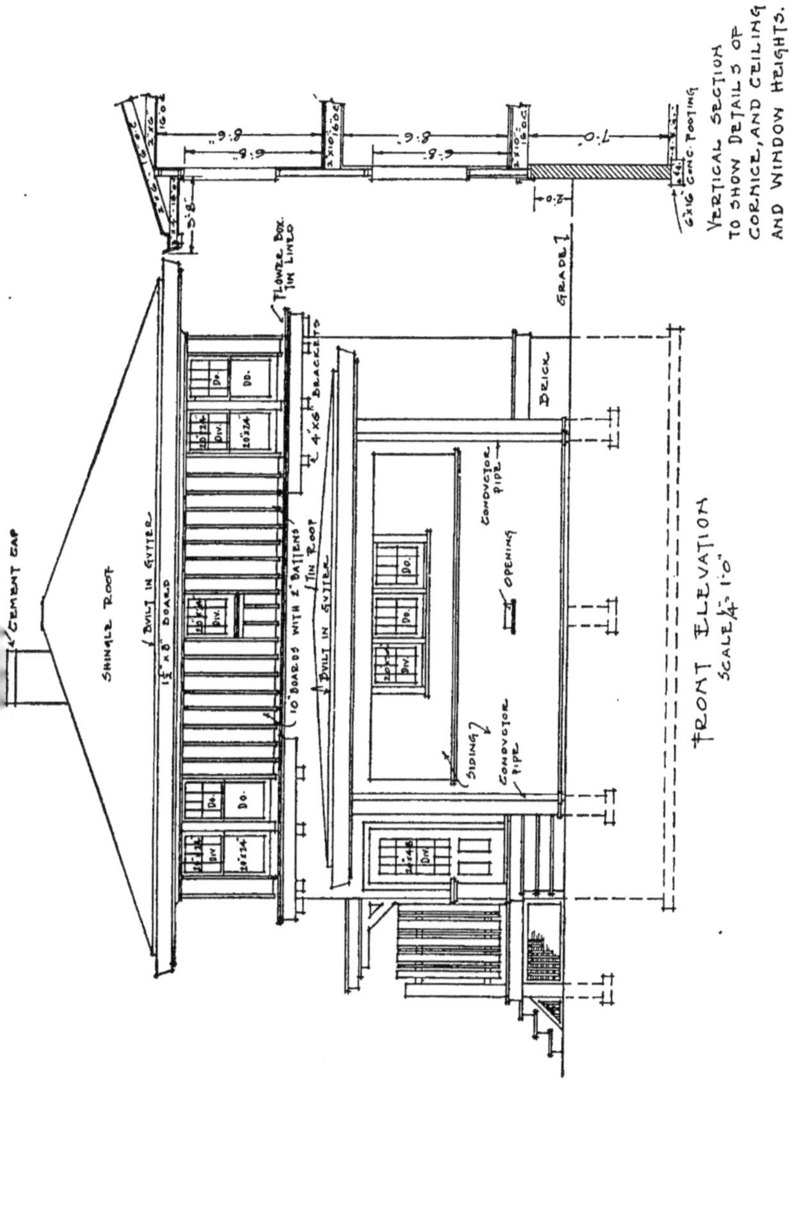

FIG. 334. FRONT ELEVATION OF HOUSE

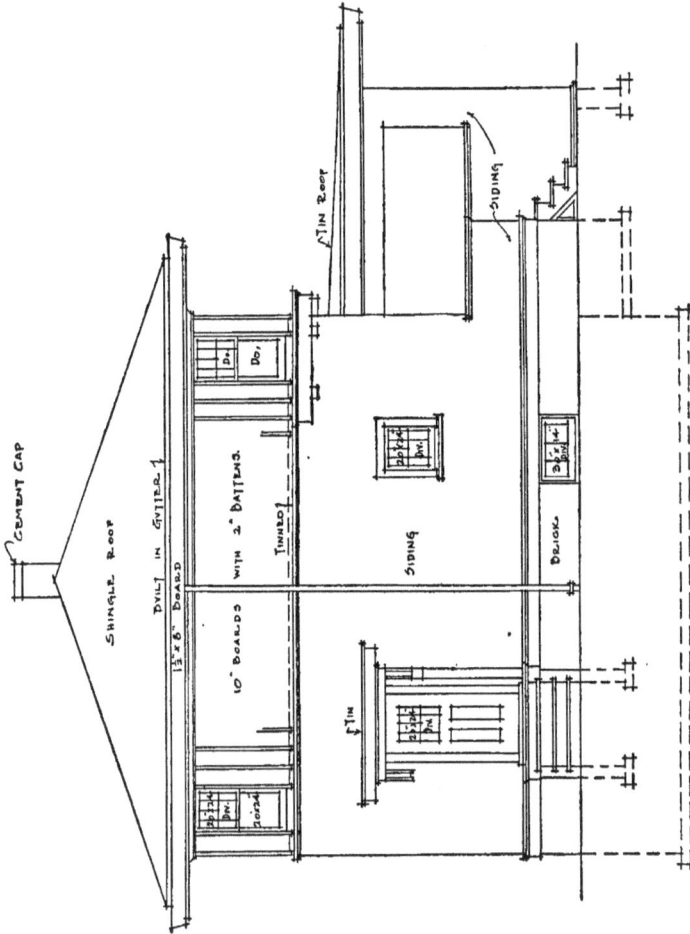

Fig. 335. Side Elevation of House

MEASUREMENTS 299

A bathroom should not be less than 5′ 6″ one way with not less than 49 sq. ft. of floor space. In planning bedrooms, there should be windows on two adjacent sides, if possible, to provide light and good ventilation. Bedrooms may be as small as 9′ 6″

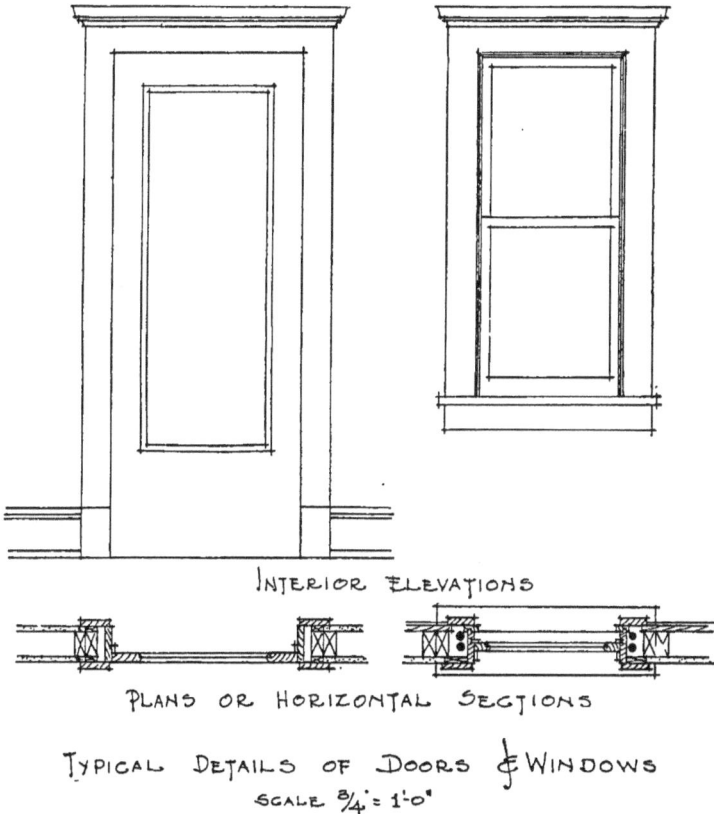

INTERIOR ELEVATIONS

PLANS OR HORIZONTAL SECTIONS

TYPICAL DETAILS OF DOORS & WINDOWS
SCALE ¾″ = 1′-0″

FIG. 336. DETAILS OF DOOR AND WINDOW

by 11′ 6″, if well planned. The spaces for beds, dressers, etc., should be considered with reference both to natural and artificial light and to necessary wall space. All upstairs rooms should be provided with ample closet room. Other considerations entering into the problem of planning the upstairs is the type of roof, position of chimneys, etc. Electric wiring, especially for outlets, and pipes for either hot air or for steam or hot water should be thought out as the plans for the two floors are made.

When the floor plans have been arranged, the elevation and possibly a perspective of the house should be drawn. If the appearance is not satisfactory, the plans for the floors may need altering. It is not uncommon for plans to be worked over a number of times to meet the needs of both owner and builder.

DATA FOR DRAWING PLATE 50

Given: The two floor plans, the side and front elevations of a house, with details, as shown in Figs. 332, 333, 334, 335, and 336.

Required: To rearrange the floor plans, if desired, and to design the other side elevation and the rear elevation, or any similar problems assigned by the instructor.

Note. The student may change the elevations shown in Figs. 334 and 335 if his second floor plan demands a different arrangement of windows or changes in the roof.

Instructions: It will be found very convenient to draw the first floor plan and then lay out the plan of the second floor on transparent paper stretched over the first floor drawing.

Many measurements are thus copied by tracing; an accurate register of all first floor data in comparison with that for the second floor is thus made, and mistakes are less apt to occur. In a similar manner the elevations may be drawn over the plans but of course in this case all the vertical measurements must be made with a scale. The particular advantage in this method of drawing elevation views is to secure correct horizontal dimensions, window and door positions, etc.

PREPARATORY INSTRUCTIONS FOR DRAWING PLATE 51

After the construction of a house has progressed until the walls are covered with plaster on the inside and with siding on the outside, and after the roof is in place, the framing construction is not apparent. But in order to build a house which will stand firm in wind and will not let in the rain, the carpenter

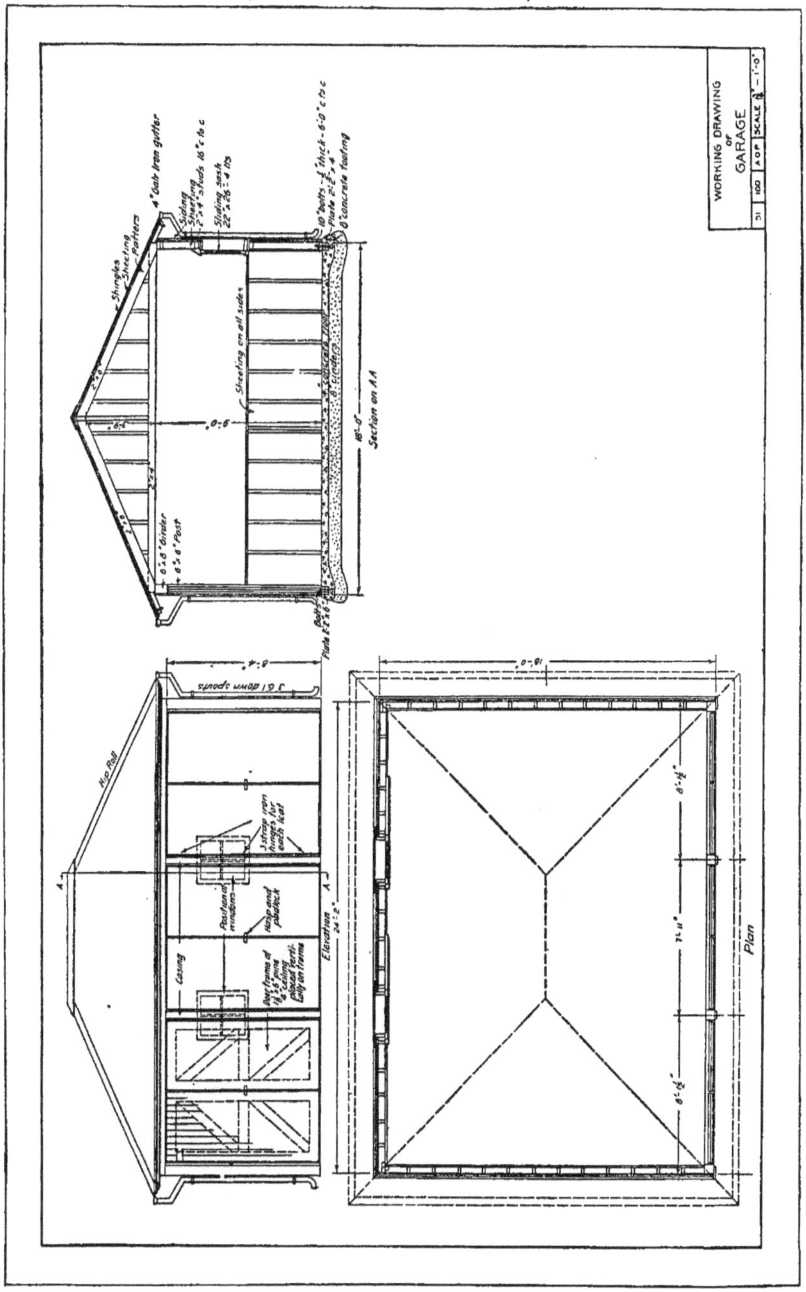

FIG. 337. TYPE PROBLEM. GARAGE

must consider carefully problems of framing construction. Figs. 338, 339, 340, and 341 give in detail typical construction for a small house.

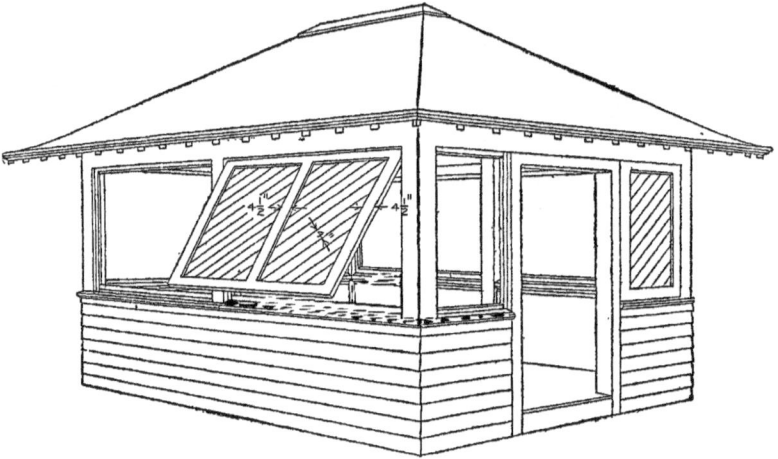

FIG. 338. PERSPECTIVE OF PARK HOUSE

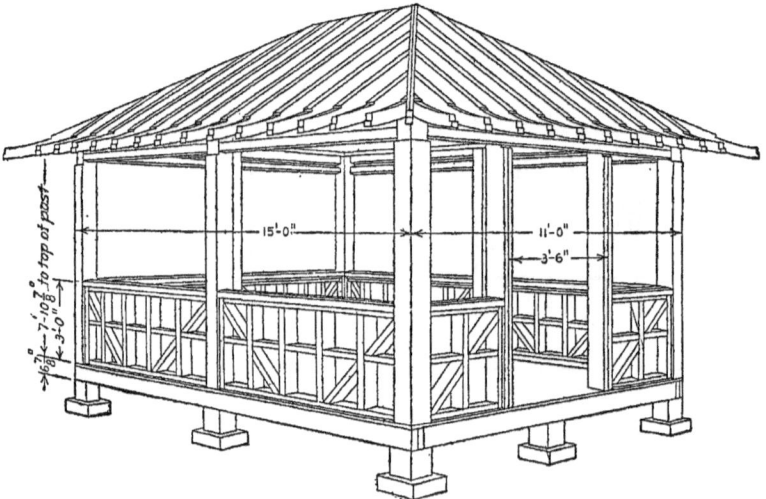

FIG. 339. PERSPECTIVE OF FRAMING CONSTRUCTION OF PARK HOUSE

DATA FOR DRAWING PLATE 51

Given: The perspective sketches, Figs. 338 and 339; the plan, Fig. 340; and the cornice detail, Fig. 341, for a small park house.

Required: To draw the orthographic views showing the framing and details of construction for the park house or any similar object as assigned by the instructor.

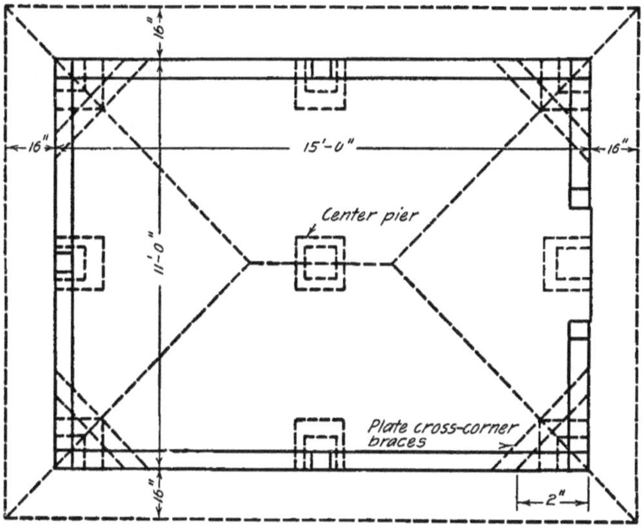

FIG. 340.   PLAN OF PARK HOUSE

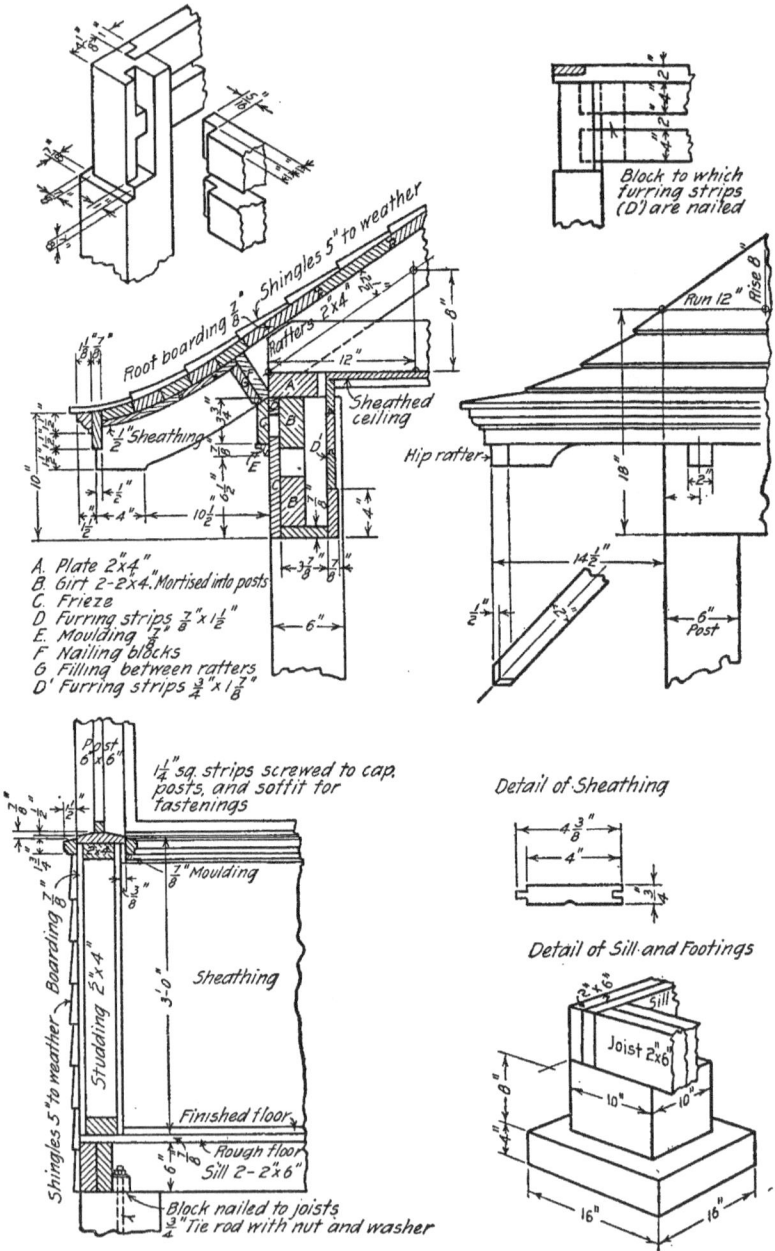

A. Plate 2"x4"
B. Girt 2-2"x4". Mortised into posts
C. Frieze
D. Furring strips $\frac{7}{8}$"x 1$\frac{1}{2}$"
E. Moulding $\frac{7}{8}$"
F. Nailing blocks
G. Filling between rafters
D' Furring strips $\frac{3}{4}$"x 1$\frac{7}{8}$"

Block to which
furring strips
(D) are nailed

Shingles 5" to weather
Root boarding $\frac{7}{8}$"
Rafters 2"x4"
Sheathed ceiling
$\frac{1}{2}$" Sheathing
Hip rafter
Run 12"
Rise 8
Post
6"

$1\frac{1}{4}$" sq. strips screwed to cap.
posts, and soffit for
fastenings

$\frac{7}{8}$" Moulding
Sheathing
Finished floor
Rough floor
Sill 2-2"x6"
Block nailed to joists
$\frac{3}{4}$" Tie rod with nut and washer

Post
6"x6"
Boarding $\frac{7}{8}$"
Studding 2"x4"
Shingles 5" to weather

Detail of Sheathing

Detail of Sill and Footings

Joist 2"x6"
Sill

FIG. 341.　CORNICE DETAILS

CHAPTER VI

ISOMETRIC AND CABINET DRAWING

PREPARATORY INSTRUCTIONS FOR DRAWING PLATE 52

Isometric Drawing is a mechanical method of representing objects pictorially. The object may appear somewhat distorted when drawn by this method, but to one who is not accustomed to reading orthographic drawings or for one who is unable to make a good freehand perspective drawing, it serves the purpose of making clear the general form of the object. By placing

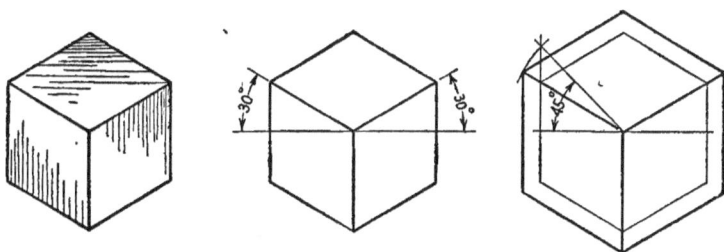

FIG. 342. THE CUBE IN ISOMETRIC

dimensions on it the isometric drawing may be used as a working drawing.

Fig. 342 shows the isometric of a cube. The cube appears as shown in this figure when it is viewed (as in orthographic drawing) with a diagonal of the cube coincident to, or parallel with, the line of sight. When in this position the three edges meeting in the near corner are represented by lines 120° apart.

The three lines 120° apart are the axes parallel to which all measurements are made in isometric drawing.

Non-Isometric Lines. A line which is not parallel to one of the three axes is a *non-isometric* line. A non-isometric line is drawn by referring points on the line to the axes by means of coördinates. Fig. 343 shows a rectangular solid on the top face of which is a non-isometric line having a curved and a straight portion. The position of the point D is determined in the isometric by transferring lengths A B and A C from the orthographic views with the dividers and drawing lines B D and C D parallel to the axes. In some cases where a figure containing non-isometric lines is to be drawn, it is convenient to enclose the figure in a rectangle. The hexagon in the side face of the rectangular solid, Fig. 343, and the circle enclosed in a square, Fig. 343, are illustrations of such cases.

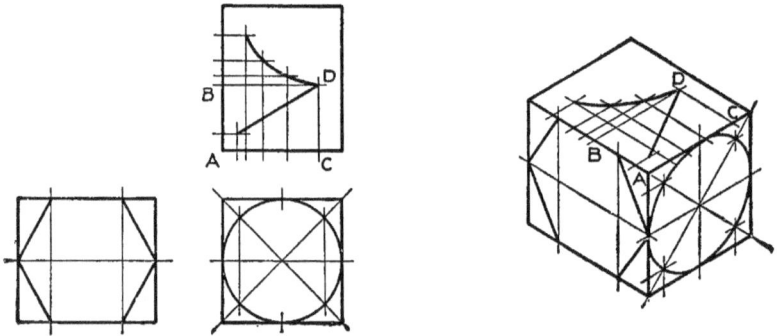

FIG. 343.   LOCATING POINTS ON NON-ISOMETRIC LINES USING TWO COORDINATES

When non-isometric lines do not lie in a face of a rectangular solid, three coördinates are necessary to locate points on each line. In drawing the isometric of the frustum of the hexagonal pyramid, Fig. 344, the base is first enclosed in a rectangle and the points on the top face are located by three coördinates as shown. The lengths A B, B D, and D E are taken from the orthographic views and laid off on the isometric in the directions parallel to each of the three axes, respectively.

Isometric Circles. Circles may be drawn by locating points as described above, or by the *four-center method* shown in Fig. 345. Point A, the center of the smaller arc, is located by laying off A B = B C. The center D for the larger arc is located by drawing A D through A perpendicular to B E. The other centers

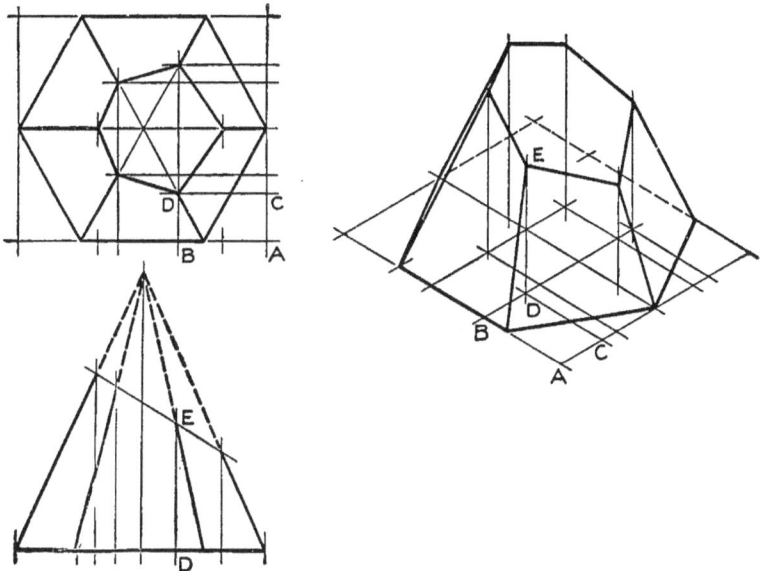

FIG. 344. LOCATING POINTS ON NON-ISOMETRIC LINES USING THREE
COORDINATES

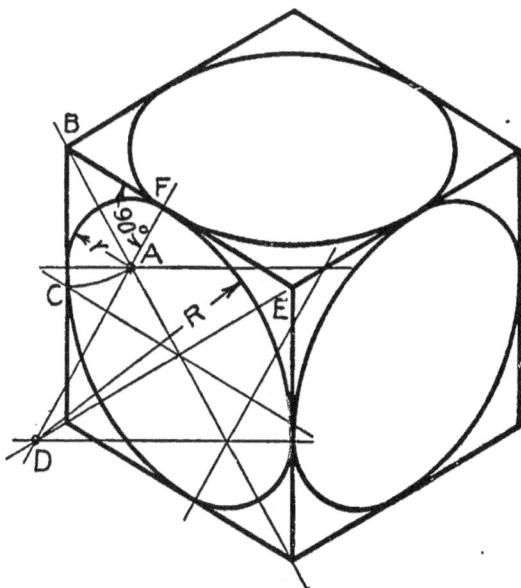

FIG. 345. A FOUR-CENTER METHOD FOR DRAWING CIRCLES IN ISOMETRIC

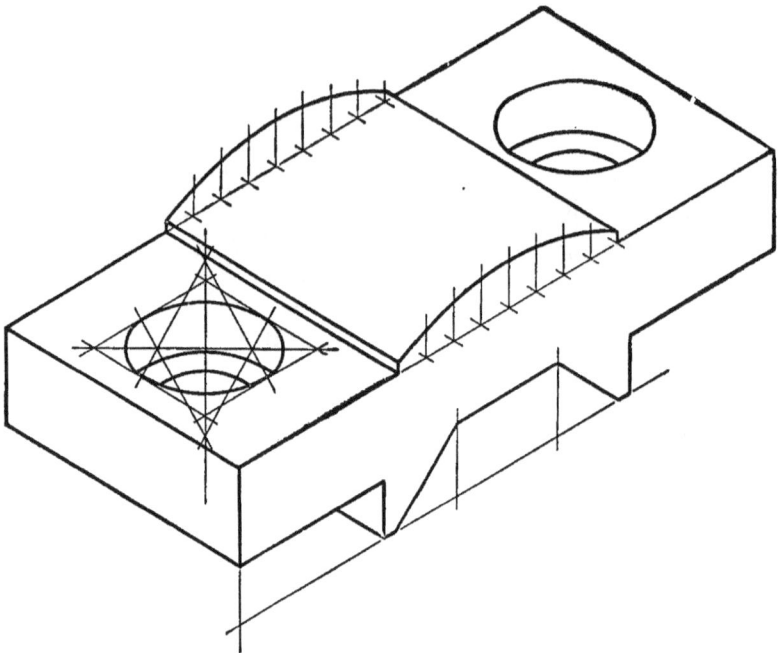

FIG. 346. METHODS OF DRAWING CIRCLES AND ARCS

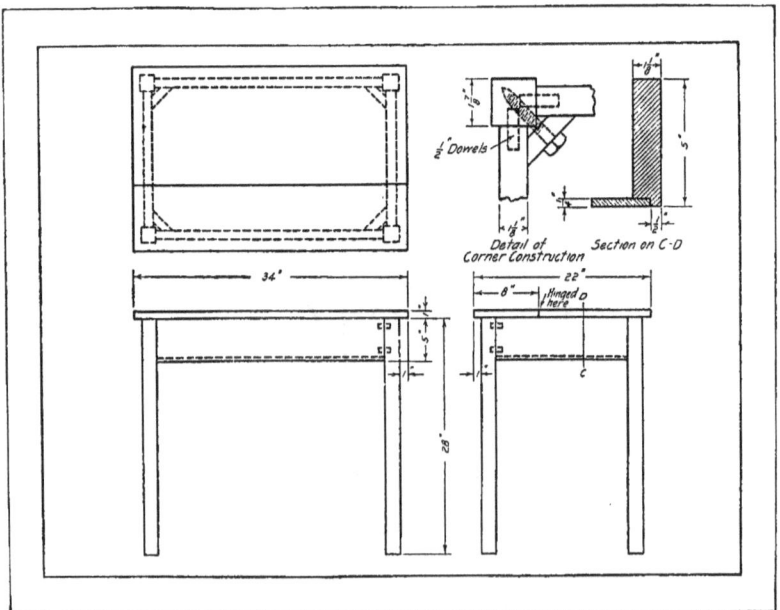

FIG. 347. TYPE PROBLEM. STUDY TABLE. ORTHOGRAPHIC VIEWS

(308)

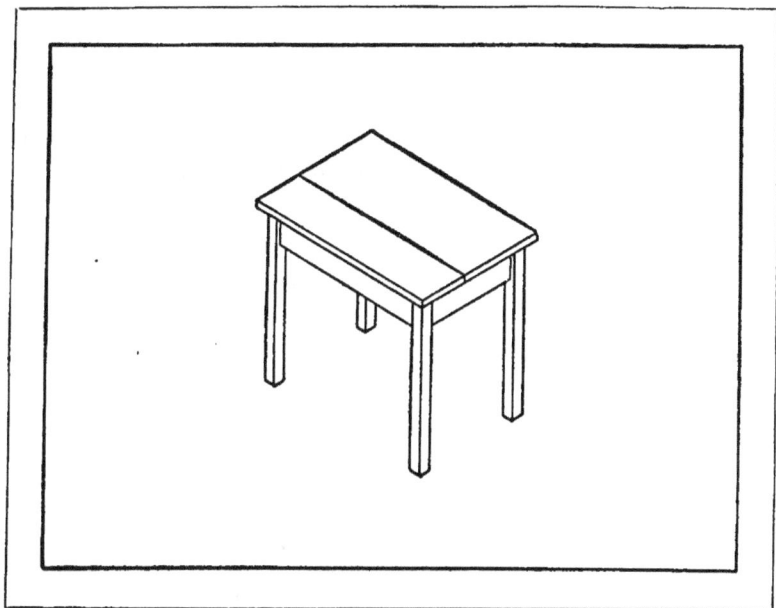

FIG. 348. TYPE PROBLEM. ISOMETRIC DRAWING OF STUDY TABLE

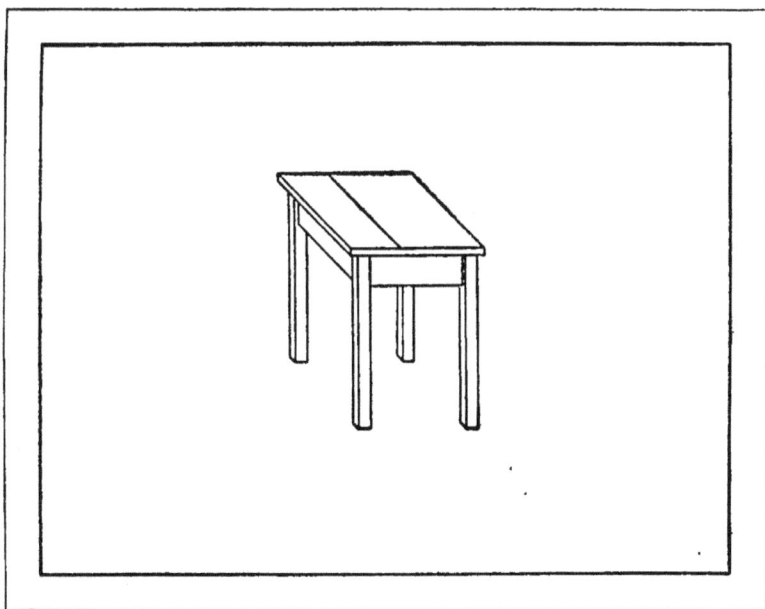

FIG. 349. TYPE PROBLEM. CABINET DRAWING OF STUDY TABLE

(309)

are located in a similar manner. The arcs are tangent at the point F. The four-center method is an approximation and is usually suitable only for full circles. When an arc is drawn which must pass through certain points the plotting method is preferable. Fig. 346.

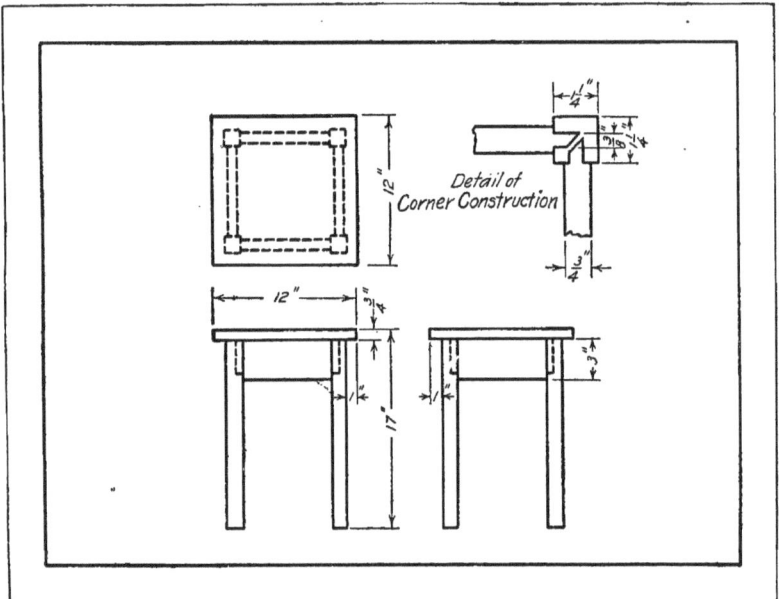

FIG. 350. TABORET

DATA FOR DRAWING PLATE 52

Given: The orthographic views of a taboret. Fig. 350.

Required: To make an isometric (or cabinet drawing) of the object shown in Fig. 350 or any similar object assigned by the instructor. For cabinet drawing instructions see page 314.

Instructions:

1. Make an orthographic drawing to scale.

2. Draw lines for the isometric drawing in the directions of the three axes: One vertical, and one each to the right and to the left, making 30° with the horizontal.

3. Decide which of the general dimensions of the object is to be measured in the direction of each axis.

4. Locate a certain point, usually an extreme corner of the object, at the intersection of the lines drawn as directed in 2.

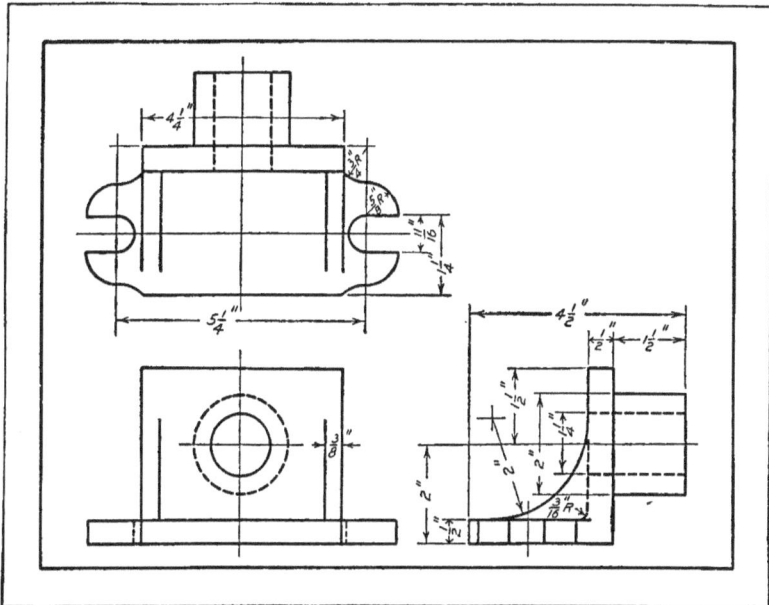

FIG. 351. TYPE PROBLEM. PLANING JIG FOR ROD BRASS. ORTHOGRAPHIC
VIEWS

5. Transfer actual lengths from the orthographic drawing with the dividers, taking care to lay them off in the direction of the proper axis.

6. Draw the necessary lines through the points located, remembering that lines of the object which are parallel to a general dimension of the object should be drawn parallel to the axis representing that dimension.

7. All other lines must be determined by locating points as described on page 306.

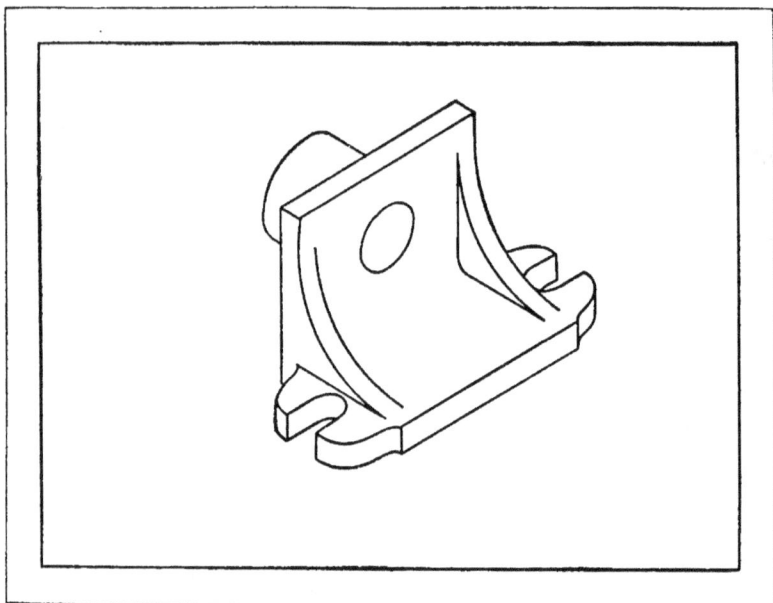

FIG. 352. TYPE PROBLEM. ISOMETRIC OF JIG FOR ROD BRASS

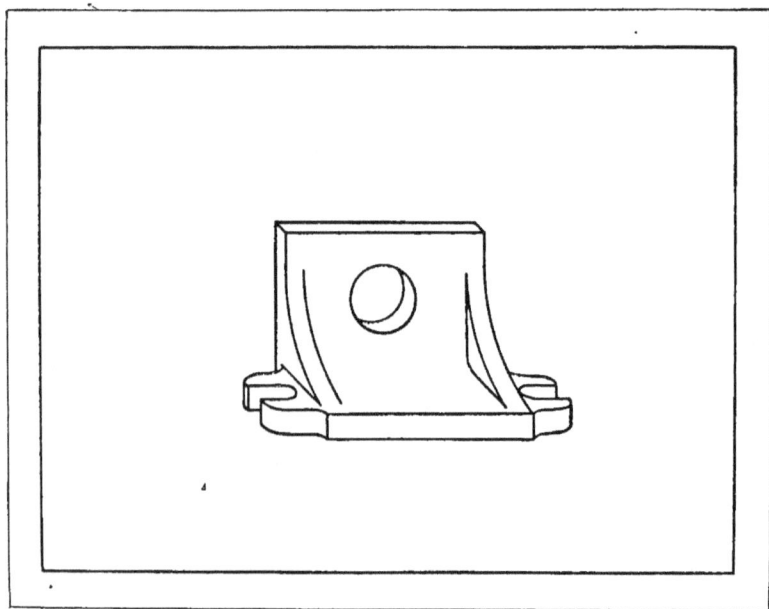

(312) FIG. 353. TYPE PROBLEM. CABINET OF JIG FOR ROD BRASS

DATA FOR DRAWING PLATE 53

Given: The orthographic views of a crosshead brass. Fig. 354.

Required: To make an isometric (or cabinet drawing) of the object shown in Fig. 354, or any similar problem assigned by the instructor. For cabinet drawing instructions see page 314.

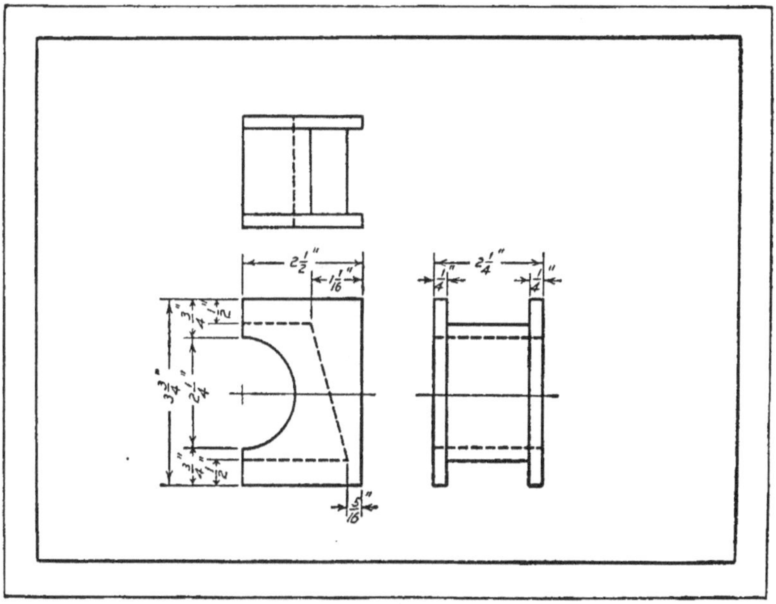

FIG. 354.   CROSSHEAD BRASS

Instructions:

1. Proceed in drawing the straight lines as for the preceding plate.

2. To draw the circles by the four center method, first determine the center by the method of coördinates.

3. Draw a figure representing a square which just circumscribes the circle. Care must be taken to draw this figure so that it will appear to lie in the same face of the object as the circle to be drawn. The direction of the sides of this figure will correspond to those representing one of the faces of the cube. Fig. 345.

4. Draw the curves by the method given under, "Isometric Circles."

CABINET DRAWING

PREPARATORY INSTRUCTIONS FOR DRAWING PLATE 54

Cabinet Drawing is similar to isometric in that measurements
are made parallel to three axes. One of the axes is horizontal,
the second vertical, and the third 45° to the horizontal. Fig. 355.
Actual lengths are measured parallel to the horizontal and vertical axes and one-half the actual lengths are measured parallel

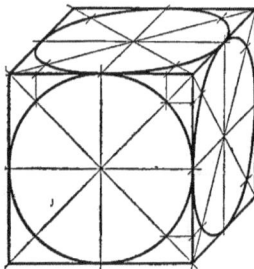

FIG. 355. CABINET DRAWING OF CUBE

FIG. 356. EXAMPLE OF AN OBJECT WITH CIRCLES AND ARCS PARALLEL TO
ONE PLANE

to the 45° axis. It should be evident that objects which involve
the drawing of irregular shaped figures which are located in or
parallel to the *front surface* of the object can be represented
more easily by cabinet than by isometric, inasmuch as such
figures will be drawn in their true form in a cabinet drawing.
For example, an object which has a number of circles parallel
to one plane is more easily represented in cabinet than in
isometric, since the circles can be drawn with the compass.
Fig. 356.

DATA FOR DRAWING PLATE 54

Given: The orthographic views of a taboret. Fig. 350.

Required: To make a cabinet drawing of the object shown in Fig. 350, or any similar object assigned by the instructor.

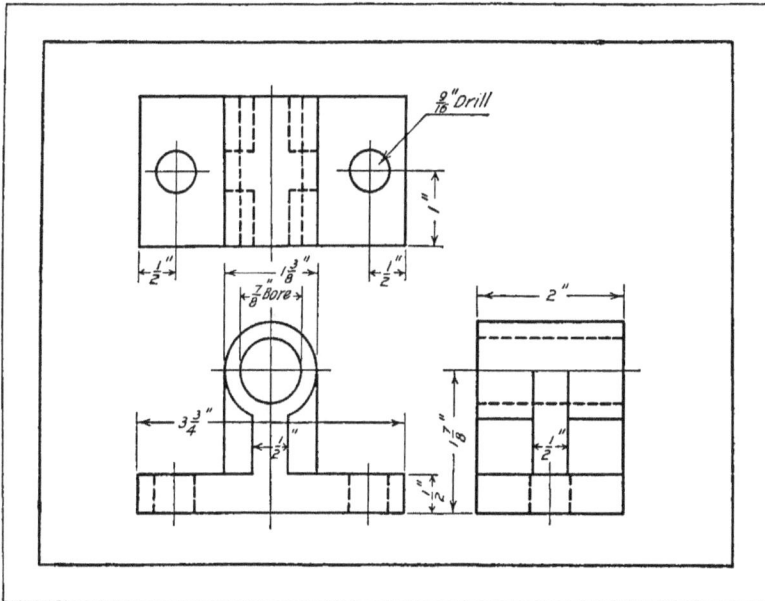

FIG. 357. AIR STARTER BEARING

Instructions:

1. Make an orthographic drawing to scale.

2. Draw lines for the cabinet drawing in the direction of the three axes; one vertical, one horizontal, and one at 45°.

3. Decide upon the general dimension of the object to be measured in the direction of each axis.

4. Locate a certain point, usually an extreme corner of the object, at the intersection of the lines drawn as directed in 2.

5. Transfer actual lengths from the orthographic drawing with the dividers for the measurements parallel to the horizontal and vertical axes, and one-half actual lengths for the measurements parallel to the 45° axis.

6. Draw the necessary lines through the points located, remembering that the lines of the object, which are parallel to a general dimension of the object, should be drawn parallel to the axis representing that dimension.

7. Lines not parallel to the axes must be located by the same method of plotting points used for the isometric drawing and described on page 306.

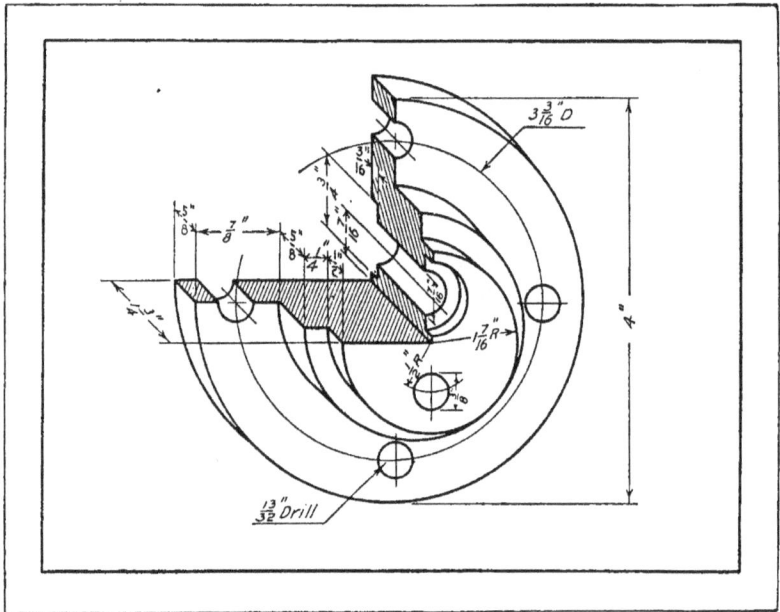

FIG. 358.   TYPE PROBLEM.   SPARKER BODY

DATA FOR DRAWING PLATE 55

Given: The orthographic views of a crosshead brass.   **Fig.** 354.

Required: To make a cabinet drawing of the object as shown in Fig. 354, or any similar problem assigned by the instructor.

Instructions:

1. Proceed in drawing the straight lines as for the preceding plate.

2. The circles which are in or parallel to the front face of the object may be drawn with the compass. All others must be determined by plotting points. The object should be placed, if possible, so that the circles can be drawn with the compass.

DATA FOR DRAWING PLATE 56

Given: The orthographic views of a sparker body. Fig. 358.

Required: To make a cabinet drawing of the object shown in Fig. 358, or any similar object assigned by the instructor.

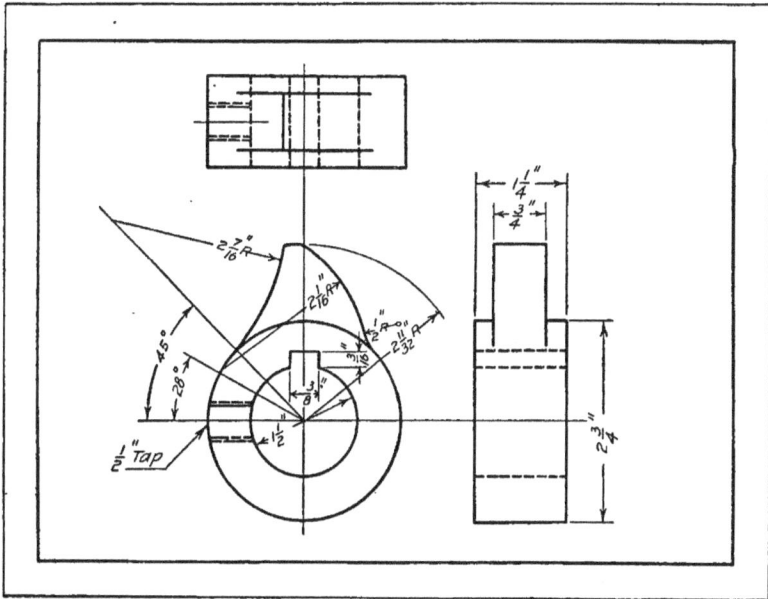

FIG. 359. BOSCH MAGNETO CAM

Instructions:

1. Make an orthographic drawing to scale. Draw lines for the cabinet drawing in the direction of the three axes; one vertical, one horizontal, and one to the right or the left, making 45° with the horizontal.

2. Transfer measurements by the same general method used in isometric, laying off only one-half the actual lengths in the direction of the 45° axis.

3. Circles or curves in the front face of the object or planes parallel to the front face should be drawn in their exact size and form. All other circles and curves must be plotted.

CHAPTER VII

GEOMETRICAL CONSTRUCTIONS

PROBLEM 1

Given: A straight line or arc A B.
Required: To bisect A B. Fig. 360.

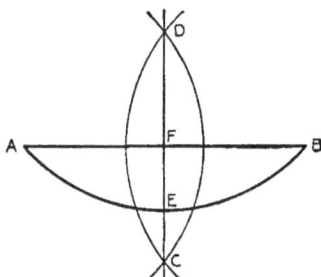

FIG. 360. TO BISECT A LINE OR ARC

Instructions: With A and B as centers describe arcs intersecting at C and D. The line C D bisects the straight line A B at F and the arc at E.

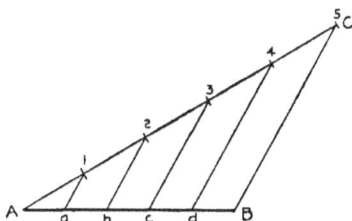

FIG. 361. TO DIVIDE A LINE INTO A NUMBER OF EQUAL PARTS

PROBLEM 2

Given: A straight line A B.
Required: To divide A B into any number of equal parts, as five. Fig. 361.

Instructions: Draw line A C at any angle with A B and lay off on it five equal spaces, using any convenient unit. Draw 5 B and parallels to it through 1, 2, 3, 4. a, b, c, d are the required divisions.

Note. A line such as A B in Problems 1 and 2 may be divided by means of the dividers, as described on page 129.

PROBLEM 3

Given: An angle A B C.
Required: To bisect angle A B C. Fig. 362.

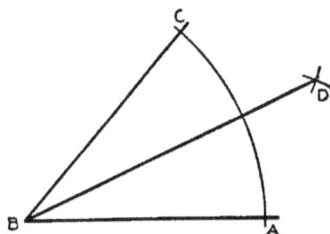

FIG. 362. TO BISECT AN ANGLE

Instructions: With B as a center, draw an arc A C of any radius. With A and C as centers describe arcs of equal radius intersecting in D. B D bisects the angle A B C.

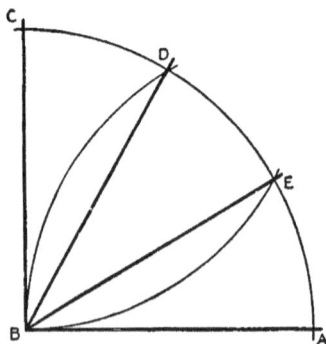

FIG. 363. TO TRISECT A RIGHT ANGLE

PROBLEM 4

Given: A *right* angle A B C.
Required: To trisect angle A B C. Fig. 363.

Instructions: First Method. With B as a center, draw an arc A C of any radius. With A and C as centers and the same radius, draw arcs B D and B E. The angles thus formed are 30°.

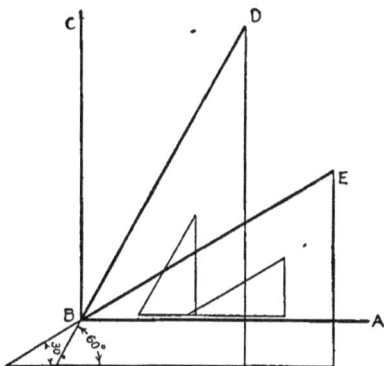

FIG. 364.   To TRISECT A RIGHT ANGLE WITH THE TRIANGLE

Second Method. An angle may be divided into any number of equal parts by drawing an arc such as A C in Problems 3 and 4 and stepping off equal distances on the arc, with the dividers.

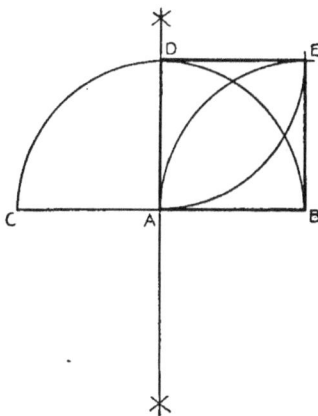

FIG. 365.   To CONSTRUCT A SQUARE

PROBLEM 5

Given: The length of the side of a square, A B.
Required: To construct the square. Fig. 365.

Instructions: First Method.   Draw arc B D C with A B as a radius.   Bisect arc B D C (Problem 1).   A D is now at right angles to A B.   With centers at D and B draw arcs of radius A B intersecting in the fourth corner of the square.

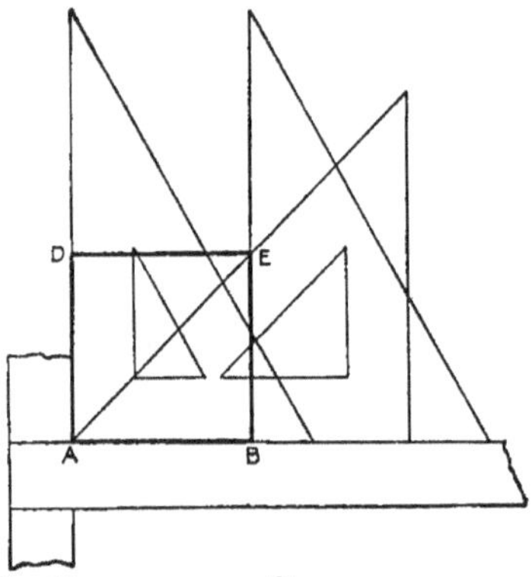

FIG. 366.   TO CONSTRUCT A SQUARE WITH THE TRIANGLE

Second Method.   The square may be constructed with the triangles and T-square.   Fig. 366.   B E and A D are drawn at right angles to A B and A E at 45° to A B.   D E is drawn through E parallel to A B.

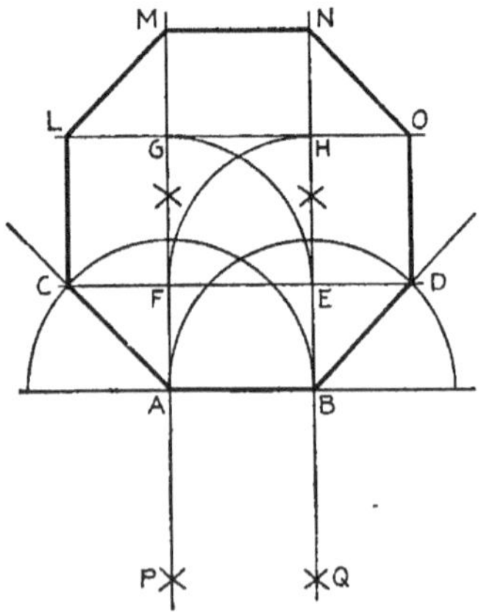

FIG. 367.   TO CONSTRUCT AN OCTAGON

PROBLEM 6

Given: The length of the side of a regular octagon.
Required: To construct the octagon.   Fig. 367.

Instructions: First Method. Draw arc B C and A D and bisect each as with the lines P M and Q N. Bisect the exterior right angles and draw A C and B D equal to A B. Connect C and D. Make F G and E H equal to F E and draw L O through G H. Make L G, G M, H N, and H O equal to C F or E D. Connect C, L, M, N, O, and D.

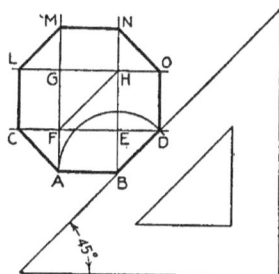

Fig. 368. To Construct an Octagon with the Triangle

Second Method. The octagon may be constructed with the T-square and 45° triangle, Fig. 368. B D is drawn at 45° to A B and equal in length to A B. A M and B N are drawn perpendicular to A B. C D is drawn parallel to A B. F H is at 45° to A B and L O parallel to A B. The figure may now be completed by drawing the following lines in the order given. A C, C L, D O, L M, O N, M N.

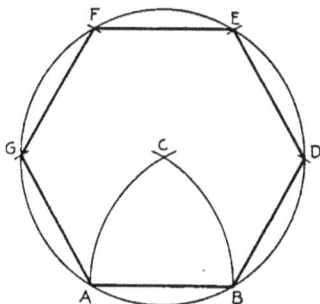

Fig. 369. To Construct a Hexagon

PROBLEM 7

Given: The length of the side of a regular hexagon.
Required: To construct the hexagon. Fig. 369.

Instructions: First Method. With a radius A B and centers at A and B, describe arcs meeting at C. With C as center and the same radius, draw a circle. With the same radius set off the arcs B D, D E, E F, F G, and G A. The side of the hexagon equals the radius of the circumscribed circle.

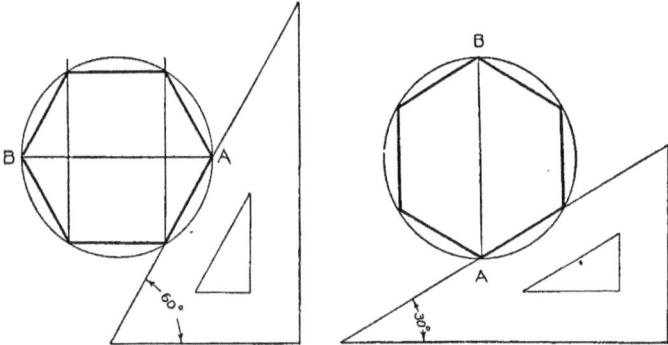

FIG. 370. To CONSTRUCT A HEXAGON WITH THE TRIANGLE

Second Method. Draw a circle with a radius equal to the side of the hexagon. Draw a horizontal or vertical diameter A B, Fig. 370, depending on the position of the hexagon. Draw lines with the 60° – 30° triangle through A and B, striking the circle. Complete the figure by horizontal or vertical lines, as the case may require.

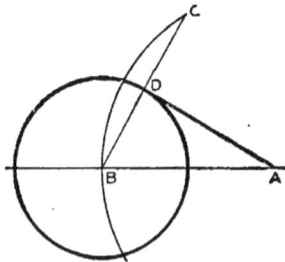

FIG. 371. To CONSTRUCT A TANGENT TO A CIRCLE THROUGH A POINT OUTSIDE
THE CIRCLE

PROBLEM 8

Given: A circle and a point outside the circle.

Required: To draw a tangent to the circle through the point.

GEOMETRICAL CONSTRUCTIONS 325

Instructions: First Method. With a radius A B and center at A, describe arc B C. With a radius equal to the diameter of the circle cut the arc at C. The chord B C strikes the circle in D, the point of tangency. B D is perpendicular to the tangent A D. Fig. 371.

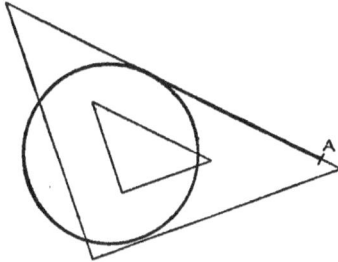

FIG. 372. To Construct a Tangent to a Circle Through a Point with the Triangle

Second Method. The tangent may be drawn with a straight edge, as shown in Fig. 372. The point of tangency may be found as shown in Fig. 195, page 184.

PROBLEM 9

Given: Two arcs.

Required: To draw a line tangent to both arcs.

Instructions: First Method. Make E F equal to A G, Fig. 373, and draw a tangent through A to the small circle C F, as in Problem 8. Extend B H and draw A G at right angles to A C. Join G K.

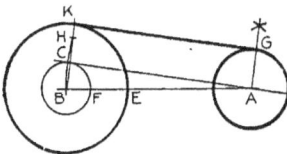

FIG. 373. To Draw a Line Tangent to Two Arcs

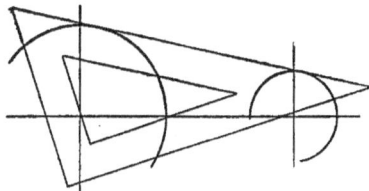

FIG. 374. To Draw a Line Tangent to Two Arcs with the Triangle

Second Method. The tangent may be drawn with a straight-edge, as shown in Fig. 374. The points of tangency may be found, as shown in Fig. 195, page 184.

PROBLEM 10

Given: Two straight lines intersecting at any angle.

Required: To draw an arc of given radius tangent to the two lines. Fig. 375.

Instructions: First Method. With radius equal to the given radius, draw arcs from two different points in each line. Draw tangents to each pair of arcs. The intersection of these lines, H, is the center of the tangent arc. Perpendiculars from this point to the tangent lines locate the points of tangency K, L.

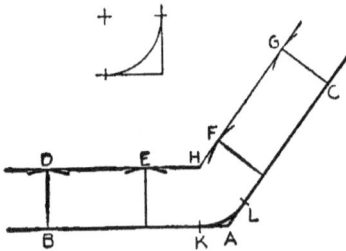

FIG. 375. To Draw an Arc Tangent to Two Intersecting Lines

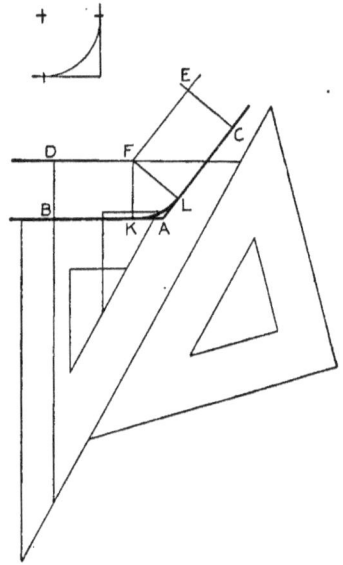

FIG. 376. To Draw an Arc Tangent to Two Intersecting Lines
with the Triangle

Second Method. Fig. 376. Draw a line at right angles to each of the intersecting lines with the triangles. as described on page 125. Lay off B D and E C equal to the radius of arc. Draw lines D F and E F parallel to A B and A C, respectively, as described on page 125. F is the center of the arc. The points of tangency K and L may be located as described on page 184.

PROBLEM 11

Given: Two circles of different diameters.

Required: To draw a circle of given radius tangent to both circles. Fig. 377.

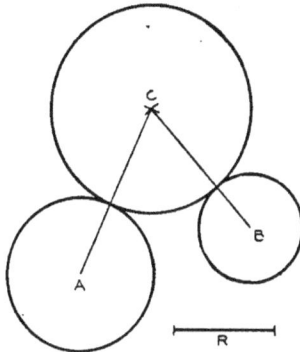

FIG. 377.   To Draw a Circle of Given Radius Tangent to Two Circles

Instructions: From the center of circle A with a radius equal to R plus the radius of A, and from the center of B with a radius equal to R plus the radius of B, draw two arcs intersecting at C, which is the center of the required circle. The points of tangency are found by joining the centers of the circles.

PROBLEM 12

Given: Two parallel straight lines A B and C D.

Required: To draw arcs of circles tangent to A B and C D and passing through E. Fig. 378.

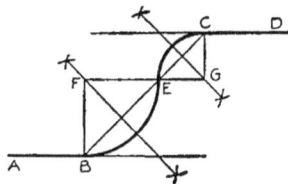

FIG. 378.   To Draw Arcs Tangent to Two Straight Lines Through a Point

Instructions: Bisect B E and C E and erect perpendiculars to A B and C D at B and C. F and G are the required centers. The arcs are tangent at E. This is called a reverse or an O. G. curve.

PROBLEM 13

Given: The length of the major and minor axes of the ellipse, O A and O B.

Required: To construct the ellipse.

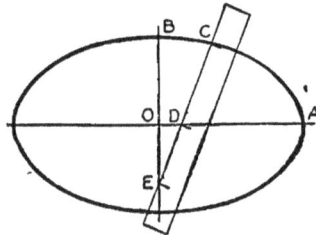

FIG. 379. TRAMMEL METHOD OF DRAWING AN ELLIPSE

Instructions: Trammel Method. Fig. 379. Mark off on a card, C D equal to O B and C E equal to O A. Keep the trammel with the point D always on the major axis and point E always on the minor axis. Move the trammel and mark points opposite C to form the curve.

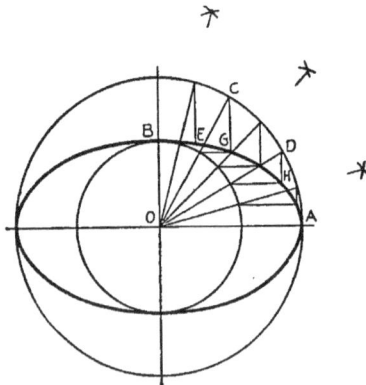

FIG. 380. CONSTRUCTION METHOD OF DRAWING AN ELLIPSE

Second Method. Fig. 380. Draw circles of radii equal to the major and minor axes. Draw any radii O C, O D, etc.; draw C G, D H, etc., perpendicular to O A, and E G, etc., parallel to O A. G, H, etc., are points on the curve.

INDEX

330 INDEX

www.ingramcontent.com/pod-product-compliance
Lightning Source LLC
Chambersburg PA
CBHW021500210326
41599CB00012B/1069